AF716969

Transposable Elements and Genome Evolution

To my unforgettable ~~cat~~ muse

Aurélie Hua-Van

SCIENCES

Ecosystems and Environment
Field Directors – Françoise Gaill and Dominique Joly

Evolution, Subject Head – Dominique Joly

Transposable Elements and Genome Evolution

Coordinated by
Aurélie Hua-Van
Pierre Capy

ISTE

WILEY

First published 2024 in Great Britain and the United States by ISTE Ltd and John Wiley & Sons, Inc.

Apart from any fair dealing for the purposes of research or private study, or criticism or review, as permitted under the Copyright, Designs and Patents Act 1988, this publication may only be reproduced, stored or transmitted, in any form or by any means, with the prior permission in writing of the publishers, or in the case of reprographic reproduction in accordance with the terms and licenses issued by the CLA. Enquiries concerning reproduction outside these terms should be sent to the publishers at the undermentioned address:

ISTE Ltd
27-37 St George's Road
London SW19 4EU
UK

www.iste.co.uk

John Wiley & Sons, Inc.
111 River Street
Hoboken, NJ 07030
USA

www.wiley.com

© ISTE Ltd 2024
The rights of Aurélie Hua-Van and Pierre Capy to be identified as the authors of this work have been asserted by them in accordance with the Copyright, Designs and Patents Act 1988.

Any opinions, findings, and conclusions or recommendations expressed in this material are those of the author(s), contributor(s) or editor(s) and do not necessarily reflect the views of ISTE Group.

Library of Congress Control Number: 2023949519

British Library Cataloguing-in-Publication Data
A CIP record for this book is available from the British Library
ISBN 978-1-78945-178-8

ERC code:
LS2 Genetics, 'Omics', Bioinformatics and Systems Biology
LS2_6 Genomics (e.g. comparative genomics, functional genomics)
LS8 Ecology, Evolution and Environmental Biology
LS8_5 Evolutionary genetics

Contents

Acknowledgments

We extend our heartfelt appreciation to all of the authors for their significant contributions to this book. Additionally, we would like to express our sincere gratitude to Wolfgang Miller for providing insightful comments on specific chapters, Arnaud Le Rouzic for his invaluable assistance during the production phase, and Dominique Joly for her meticulous reviews of the chapters and unwavering support throughout the entire process.

Aurélie HUA-VAN
Pierre CAPY

Introduction

Aurélie HUA-VAN and Pierre CAPY

Évolution, Génomes, Comportement et Écologie (EGCE), CNRS, IRD, Université Paris-Saclay, Gif-sur-Yvette, France

Any biologist interested in the genetic material which is occupying the cells of their preferred species has been or will be confronted one day to transposable elements (TEs). They can represent an important part of the genome, although their *raison d'être* is not to participate in the core genetic program necessary for the cell survival.

TE activities are the causative agent of numerous major scientific discoveries in genetics and evolution, and the basis of important technological advances in Life Science research. They are passively used by our species in its effort for diversifying the most important supplies it needs (like all heterotrophic species), that is to say food. More importantly, they are responsible for some spectacular evolutionary key innovations, but also for many diseases that affect our species. Some examples are shown in Figure I.1.

Discovery of the genetics laws: Gregor Mendel, the father of genetics, formalized in 1865 the basic laws that govern the transmission of phenotypic characters under the control of a single gene. Rediscovered simultaneously by several authors at the beginning of the 20th century, these laws still stand true. One of Mendel's models was the transmission of the wrinkled phenotype of pea seeds (Figure I.1). In 1999, it was found that the wrinkled phenotype is due the insertion of a TE in the promoter of a gene (Bhattacharyya et al. 1990).

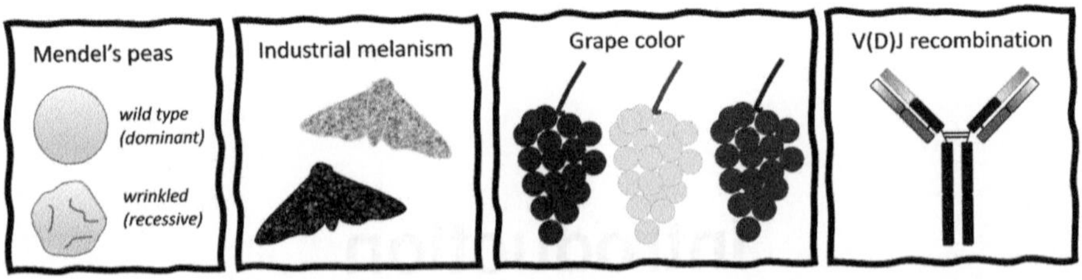

Figure I.1. *Famous examples of TE insertions that affect phenotypes or participate to evolution. By crossing smooth and wrinkled peas, Mendel discovered dominance/recessivity between alleles. Strong environmental pressure triggered rapid shift in morphotype in the pepper moth. Insertion of a TE in the promoter of a gene and subsequent rearrangement may change the color of grape. The programmed recombination process triggering diversity of immunoglobulin chains chains is mediated by an enzyme originating from a TE. For a color version of this figure, see www.iste.co.uk/huavan/transposable.zip*

Illustration of natural selection acting in real-time: industrial melanism is a textbook example of rapid evolution and adaptation by natural selection. During the industrial revolution in England (mid-1800s), the peppered moth *Biston betularia* was known for having quickly shifted from a predominant light colored wings morphotype to a blackened one (Figure I.1). This species rests on trees such as birches during the day, and its color serves as camouflage, preventing the moth from being preyed on by birds. The darkening of trunks due to the pollution may have favored the expansion of the dark morphotype, as supported by the famous capture-mark-recapture experiment of Kettlewell (1955). In this species, wings color is determined by a single gene. The mutated allele responsible for wing darkening has recently been characterized (Van't Hof et al. 2016). The mutation is due to an insertion originating from a TE. More examples concerning the involvement or TEs in adaptation to environmental stress are given in Chapter 5.

Input in biotechnology: as soon as TEs have been characterized, they have proven useful for researchers, as polymorphic markers, for typing strains, or for genome engineering, ranging from mutagenesis to gene therapy. Some examples of recent use of TEs in biotechnology and research are illustrated in Chapter 10.

Input in agronomy amelioration: artificial selection has been performed by human since the neolithic period and the start of domestication in order to improve quality and quantities of our supplies. Human-directed selection is still widely used nowadays, for examples, in agriculture and horticulture to create different varieties. Notably, the color diversity or variegation in color in cultivated flowers is often due to the activity of TEs, of which are also responsible for the diversity of the grape color (Figure I.1) or the varieties of blood oranges (Lisch 2013).

Example of evolutionary innovations: besides the genetic diversity of food and flowers enjoyed by humans, some TEs insertions have been naturally selected in the past to give rise to important genetic novelties contributing to species success, such as adaptive immunity by V(D)J recombination in jawed vertebrates (Figure I.1). Some famous examples of this phenomenon, called molecular domestication, will be illustrated in Chapter 6.

Input in human health: in eukaryotes, TEs are generally silenced by epigenetic mechanisms. The silencing is not perfect and low transposition activity may be observed, which can increase when epigenetic marks are altered. In humans, de novo insertions and recombination between the huge number of TE insertions are sometimes involved in diseases and cancers. Numerous examples are given in Chapter 3.

I.1. Almost 80 years of research on transposable elements

The existence of mobile genetic factors in eukaryotic genomes was suspected as early as 1944 when Barbara McClintock, a maize cytogeneticist, observed strange phenomena such as phenotypic variegation and phenotypic alterations to various degrees, sometimes associated with a precise timing or reversibility in plants having suffered some chromosome breakage-and-fusion cycles. At the chromosomal level, she observed rearrangements, modification of the locus size and transposition of the locus to another place, so she called it Ds (for dissociation). She realized that all of these were due to some controlling elements that could move (transpose) throughout the genome, modify the expression of nearby genes and cause chromosomal rearrangements. The system she described was bipartite, the effect of the Ds locus being only observed in the presence of the Ac (for Activator) locus, itself transposable (McClintock 1950).

Later on, at the beginning of the 1970s, puzzling observations such as male recombination (normally not occurring in *Drosophila*), sterility and high mutation rates were discovered in *Drosophila melanogaster* in some particular crosses between natural and laboratory strains, leading to the notion of hybrid dysgenesis. At about the same time, the first mobile insertion sequences (ISs), which are TEs in bacteria, were molecularly identified in *Escherichia coli*. Similar mobile sequences, the P element and the I factor, were finally found to be responsible for the hybrid dysgenesis phenomenon in Drosophila. It soon appeared that numerous TEs existed in *Drosophila* and that some of them transposed through a RNA intermediate.

TEs were then discovered in more and more species, after cloning of loci responsible for phenotypic changes. It became clear that most visible mutations in plants and animals were due to TE activities, in particular induced mutations (UV, X-rays, etc.), or selected ones. The noticeable effects are as follows (visible on multicellullar organisms): variegation (somatic transposition), reversibility (sometimes) and mild phenotypic effects. It appeared later on that the middle repetitive fraction of eukaryotic genomes was mainly composed of TEs (repeativeness of the genome could be measured by denaturation followed by slow renaturation, three compartments were detected, highly repeated sequences are the ones that renature very quickly, while it takes much more time for unique sequences to find the complementary strand). We now know that all eukaryotic genomes carry TEs, including unicellular eukaryotes. For most TEs, we have deciphered their transposition mechanisms and how they are regulated. We know their mutagenic consequences and their impact on genome evolution both at the functional and structural level. At the genomic era, a new challenge is to efficiently identify these TEs in new sequenced genomes (Chapter 11).

I.2. What is exactly a transposable element?

We can give a very minimal definition of transposable elements (TEs): a DNA sequence, able to move and replicate in the genome of a single cell. That is, TEs are mobile, repeated and dispersed in the genome. Although theoretically not false, this definition is too poor to encompass the extraordinary properties of such sequences.

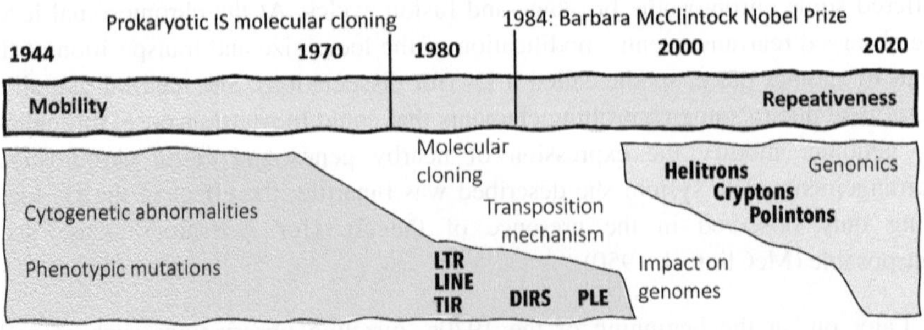

Figure I.2. *TEs identification: historical timeline. TEs have two properties: they are mobile and repeated. Their discoveries has first relied on the phenotypic changes associated with their mobility, mainly on model organisms. Since 20 years now, at the genomic era, new TEs are primarily identified in non-model organisms sequenced genomes because of their repeated nature. In between, the golden age of molecular biology has permitted a deep understanding of their biology and their impact on the genome. First identifications of main groups of eukaryotic TEs are indicated. For a color version of this figure, see www.iste.co.uk/huavan/transposable.zip*

Alternatively, you will often hear them described as junk DNA, parasitic elements, selfish DNA, jumping genes and mobile genetic elements. We will first try to define and distinguish all of these terms and determine how they fit into TEs.

Mobile genetic elements (MGEs): MGEs correspond to any nucleic acid sequence which is not permanently embedded into the genome of its host cell and able to replicate independently from the genome replication. TEs are part of the large group referred to as the MGEs that also includes viruses and plasmids. This group encompasses all nonliving genetic entities that spent some or all of their time outside the genomes of their host. Although some viruses are not integrative and will always been found outside the genome, TEs have to be within the genome at some point of their transposition process. Another difference is that viruses are infectious, they are able to exit a cell and enter another one from the same individual or from another one. Theoretically, TEs are restricted to within the cell. Yet TEs are not only able to escape the cell, but also to escape the individual and escape the species. The frontiers with viruses also fade when we examine the origin of these entities. Some viruses and some TEs have a common origin.

Jumping genes: this expression was initially used to describe short or longer DNA segments demonstrated to change position on the chromosome in bacteria, or to move from plasmid to plasmid. This led to the discovery of ISs (bacterial TEs similar to Class II TIR eukaryotic elements) and to composite Transposon (Tn) that often carry antibiotic resistance genes. These elements are able to form hairpin structure due to the complementary of the two ends of a single strand (terminal inverted repeats (TIRs)). The link was then made with eukaryotic DNA pieces able to move and to turn off expression of adjacent genes or to induce genomic rearrangements involved in rapid evolution. Molecular cloning of these eukaryotic jumping genes revealed that they share structural similarities (the TIRs) with IS.

Selfish DNA: selfish DNA has the intrinsic property to propagate with the host (cell, individual, population or species). The notion of selfish DNA refers to the selfish gene theory proposed by Dawkins (1976) in 1976. Yet there are differences between the two concepts. The selfish gene theory is a reformulation of Darwin's natural selection theory from the gene perspective. It merely states that a fitter allele will invade a population by progressively replacing alternative less fit alleles at the same locus. According to Dawkins, life began with the emergence of chemical entities able to replicate, which he called replicators. As a replicator, an allele can only survive on the long term if it replicates more than the others. Hence, genes (alleles) are selfish because non-selfish alleles have disappeared. To secure their survival, genes cooperate to construct their shared survival vehicle (an individual) capable of producing numerous next-generation vehicles that will propagate them. From this point of view, vehicles are selfish, species are selfish and well... life is selfish.

The fate of a TE is not different, for it TO BE, it has to propagate. Yet it uses quite different means. The way used by TEs to make copies of themselves (then to propagate) is to directly make copy of themselves (Orgel and Crick 1980). Hence, all autonomous TEs contain genes that encode proteins whose function is to copy the TE (their vehicle) or at least to move it to a better place. At the same time, TEs also use the genes vehicle to propagate in the population. Furthermore, since they are not forever attached to the rest of the genome (they have an extrachromosomal step), they sometimes profit of this to escape from the vehicle (the host) and reach a new one. This strategy is enough for them to propagate more efficiently than the genes. In brief, through these two features (creating a copy, or excising from the genome host), they behave more selfishly than selfish genes, and are therefore the most successful genetic entities on the planet, along with viruses.

Genomic parasite: TEs are called genomic parasites because they populate the genome and use (hijack) the cell machinery for their self-propagation, without bringing any apparent benefit to the host. Indeed, TEs encode genes that are structurally similar to any other protein-coding genes in the genome. They contain regulatory regions and promoter recognized by the transcription machinery. The resulting messenger RNA is translated by the cell ribosomes. The hijacking of host cell resources is likely a bit simpler compared to what we can find in some viruses, and thus less harmful. Indeed, TE activity is not responsible for the death or the collapsing of the cell that carries them. Yet, they may have some detrimental effects due to the fact that they are mobile or repeated.

Junk DNA: junk DNA was initially coined by Ohno (1972) as all of the noncoding DNA in the genome (or not transcribed regions). It included repeated fractions of the genomes, now known to mainly correspond to TEs or TE-derived sequences, as well as micro- or minisatellites sequences (sequences that are tandemly repeated, unlike TEs that are dispersed over the genome). TEs are not always noncoding since some of them contain genes. However, these genes do not perform any obvious useful functions for the cell (Doolittle and Sapienza 1980). A more accurate definition would be that junk DNA has no function for the cell, that is, it has no effect on the phenotype and then no potential advantage for the host. It is more accurate because it excludes regions that are not transcribed but are involved in gene regulation and then have an obvious advantage for the host.

I.3. How do transposable elements survive in the genome?

When considering that TEs are mildly harmful for the host, we wonder why they are still there, present in every genome. Natural selection should act and they should then disappear. Furthermore, if they have no function/no phenotype, they are not expected to be submitted to selection. Hence, they will freely accumulate mutations in their sequences and will ultimately become inactivated by mutation. It is tempting to bring forward the explanation that TEs are maintained because they can be beneficial, since they are major actors in evolution.

However, we have to consider several points: (i) the harmfulness to the cell because of insertion or ectopic rearrangement is associated with particular insertions and not the whole bunch of copies of a TE family; (ii) the TE family, as long as it is active, may continue to replicate, replacing copies that have been counterselected; (iii) even if over time a TE family accumulates mutations and becomes inactive, invasion by new TE families from the outside is still possible, and this seems to occur relatively frequently, through horizontal transfer, in bacteria of course but also in eukaryotes (see Chapter 7); (iv) the genome has developed sophisticated defense mechanisms to silence TEs (see Chapter 4), but not to actively get rid of the TE sequences. It is then expected that every genome, at any point of the evolution time, will contain at least some active families, and some (or many) inactive relics, even if TEs have no beneficial impact on the host. And this is indeed what we observe.

Yet, each trajectory of a TE family in a species is unique, and each species has a unique cocktail of TEs in term of diversity and abundance. Why is it so diverse? We still poorly understand the dynamics of TE families, and the interplay of the different factors involved. An important parameter in eukaryotes is sexual reproduction, which allows TEs to spread through non-Mendelian mechanisms (Chapter 8). Within the genome, TEs live together, cooperate and compete, and interact with genes and other genomic entities. This has lead to viewing the genome as an ecosystem (Chapter 9).

I.4. Structure, classification and origin

TEs have various structures that reflect their diverse origins and transposition mechanisms (see Chapters 1 and 2). They must contain genes encoding proteins, necessary for transposition and specific *cis* sequences for recognition of the TE *termini*. Biochemical activities are the basic activities involved in nucleic acid metabolism: polymerase, strand transferase (strand transfer), endonuclease and ligase. Polymerase is needed for all replicative TEs. Often this activity is carried by the element itself. Strand transferase activity is present in all TEs except the ones that directly replicates at the integration site. Endonuclease activity is present in all TEs, required for either for excision from the donor site or for preparing the target site. DNA cleavage can often be performed by the same catalytic core that also makes strand transfer (strand transferase). Ligase is usually provided by the cell and serves finishing the transposition process by restoring a continuous DNA molecule at the excision and donor site. Depending on the TE, some more enzymatic activity may be needed, such as helicase in helitrons, which helps to destroy hydrogen bonds between the two complementary DNA strands or RNAse H or Protease in Class I elements.

The origin of TEs is unknown. It is certainly not unique, considering the high diversity of structures. Both TEs and viruses can be viewed as purely parasitic, having an advantage over their host, that is to profit on it for expanding. Most likely these nucleic acid parasites emerge very early in the history of life. As early cheaters, they may have been present as soon as the first replicators emerged. Coupling an enzymatic activity involved in nucleic acid metabolism, with a DNA binding domain, will bring the needed specificity. Let us surround this gene by the ad hoc DNA sequences, and we will obtain a fair TE!

Among extant TEs, group II introns are probably the most interesting candidates to reflect how parasites may have evolved at the very beginning of life. They usually are neglected because they are not abundant in prokaryotes, are found in eukaryotes only in the organelle genomes (of some plants and fungi) and have a particular way of moving. Mobility occurs at the RNA level, in a way resembling spliceosomal introns. Hence, group II introns are suspected to be their progenitors. The excision (or splicing) is mediated by a ribozyme present in the sequence (ribozymes may have been present at the very beginning of life). The RNA is then reverse transcribed by a reverse transcriptase (RT) encoded in the sequence (and which is related to bacterial RT, such as RT in retrons) and reinserted in a sequence-specific manner in a empty homologous site. Interestingly, recently, a number of ribozymes have been identified in some TEs (Cervera and de la Peña 2020). Yet, tracing back the origin of TEs remains extremely challenging.

I.5. References

Bhattacharyya, M.K., Smith, A.M., Ellis, T.H., Hedley, C., Martin, C. (1990). The wrinkled-seed character of pea described by Mendel is caused by a transposon-like insertion in a gene encoding starch-branching enzyme. *Cell*, 60(1), 115–22.

Cervera, A. and de la Peña, M. (2020). Small circrnas with self-cleaving ribozymes are highly expressed in diverse metazoan transcriptomes. *Nucleic Acids Res.*, 48(9), 5054–5064.

Dawkins, R. (1976). *The Selfish Gene*. Oxford University Press, London.

Doolittle, W.F. and Sapienza, C. (1980), Selfish genes, the phenotype paradigm and genome evolution. *Nature*, 284(5757), 601–603.

Kettlewell, H.B.D. (1955). Selection experiments on industrial melanism in the lepidoptera. *Heredity*, 9(3), 323–342.

Lisch, D. (2013). How important are transposons for plant evolution? *Nature Reviews Genetics*, 14(1), 49–61.

McClintock, B. (1950). The origin and behavior of mutable loci in maize. *Proceedings of the National Academy of Sciences (PNAS) USA*, 36(6), 344–355.

Ohno, S. (1972). So much “junk” dna in our genome. *Brookhaven Symposium in Biology*, 23, 366–370.

Orgel, L.E. and Crick, F.H. (1980). Selfish DNA: The ultimate parasite. *Nature*, 284(5757), 604–607.

Van't Hof, A.E., Campagne, P., Rigden, D.J., Yung, C.J., Lingley, J., Quail, M.A., Hall, N., Darby, A.C., Saccheri, I.J. (2016). The industrial melanism mutation in British peppered moths is a transposable element. *Nature*, 534(7605), 102–105.

1

Transposable Elements in Eukaryotes

Aurélie HUA-VAN
Évolution, Génomes, Comportement et Écologie (EGCE), CNRS, IRD, Université Paris-Saclay, Gif-sur-Yvette, France

1.1. Introduction

While transposable elements (TEs) are found in all three domains of life, they have expanded the most over time in eukaryotes, such that they sometimes reach 80% of the genome, with copy number above millions. It is also in eukaryotes that their diversity is the highest. Different kinds of transposition strategies have been developed by these selfish DNAs. Each species can host thousands of different TE families, and even closely related species may have specific TE families. Their evolutionary history is quite complicated, as is their relationship with their respective host. At the mercy of their host's silencing strategy, they can manage to escape repression and subsequent extinction by invading new species, where they evolve and amplify freely until defenses are put in place by this new host. The old companionship they maintain with their host is still visible, unlike in prokaryotes, where the turnover is very rapid in order to keep their genomes small.

In this chapter, I will review the basics of structure and transposition mechanisms of eukaryotic TEs, their origins, what we know about their distribution and abundance in eukaryotes, and finally the impact of such selfish elements in genome size, structure, function and evolution.

Transposable Elements and Genome Evolution,
coordinated by Aurélie HUA-VAN and Pierre CAPY. © ISTE Ltd 2024.

1.2. Classification, structure and transposition mechanism

The first attempt at TE classification was by Finnegan (1989). At this time, three types of TEs were known, called LTR retroelements, non-LTR retroelements (LINEs) and DNA transposons. Two classes were defined, based on the nature of the transposition intermediate, with Class I (LTR and non-LTR retroelements) transposing through an RNA intermediate and Class II (DNA transposons) directly from DNA to DNA (Figure 1.1).

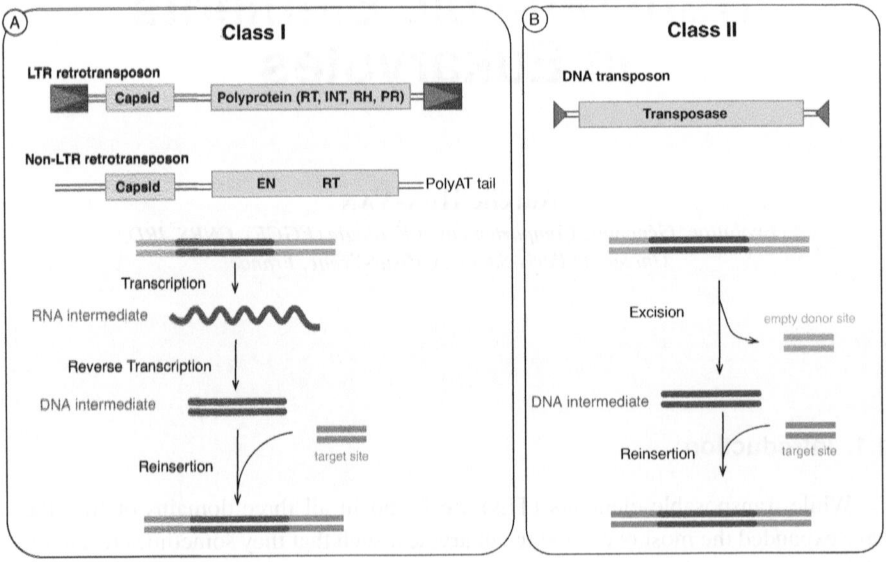

Figure 1.1. *The two classes of transposable elements as defined by Finnegan (1989). (A) Class I is represented by LTR retrotransposons and non-LTR retrotransposons, which transpose through a copy-and-paste mechanism. An inserted copy is transcribed, then reverse transcribed into DNA and integrated at a new genomic location. RT: reverse transcriptase; INT: integrase; RH: ribonuclease H; PR: protease; EN: endonuclease. (B) Class II transposons transpose through a cut-and-paste mechanism: The copy is excised from the genome and reinserted into the target site. For a color version of this figure, see www.iste.co.uk/huavan/transposable.zip*

For most non-specialist researchers, TEs classically belong to one of these three types. However, in the last few decades, a huge diversity of TEs has been uncovered in eukaryotes, which makes this classification too simplistic. Most of the new TE groups were identified through genome mining. These correspond to less active groups, and deciphering their transposition mechanism remains challenging. However, refinement of the former classification or new classification scheme was proposed.

In 2003, a classification based on the types of enzymes responsible for integration was proposed (Curcio and Derbyshire 2003). Interestingly, similar catalytic activities (represented by catalytic motifs) could be found independently in the two main classes. The classification integrated both eukaryotic and prokaryotic elements. However, this classification did not catch all of the diversity that can be found in TE transposition, since it focused mainly on the integration process.

A major revision was proposed by Wicker et al. (2007). While the two main classes remain valid, new hierarchical subdivisions such as subclasses, orders and superfamilies were added to take into account variation in the transposition mechanism (number of cut strands), the structure and finally the phylogeny/sequence homology. The classification system included the most recently discovered types of TEs, but remained focused on eukaryotic TEs; it did not integrate some particular elements, once considered as not belonging to the TEs, such as group II introns. It was then challenged (Piégu et al. 2015; Kapitonov and Jurka 2008), but still currently remains the most widely used.

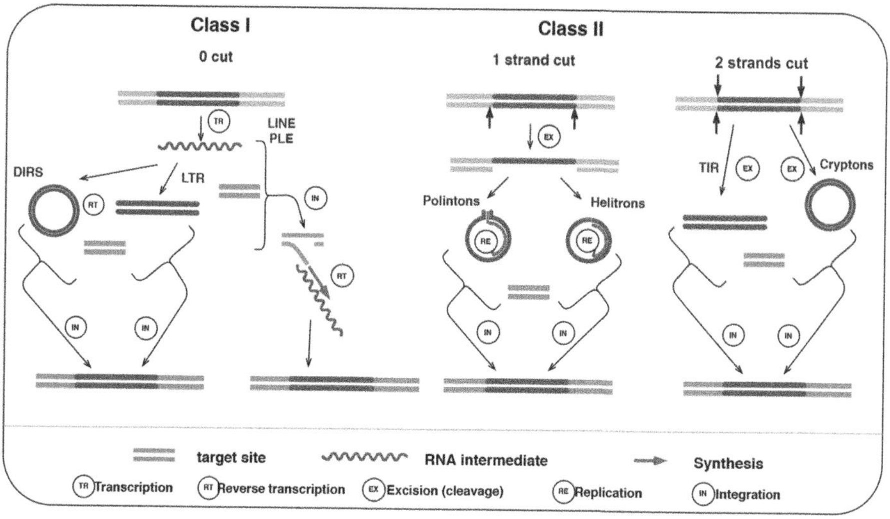

Figure 1.2. *The different transposition mechanisms of eukaryotic transposable elements. For a color version of this figure, see www.iste.co.uk/huavan/transposable.zip*

COMMENTS ON FIGURE 1.2.– *Class I TEs use an RNA intermediate, so no cleavage occurs at the donor site, unlike Class II TEs, for which one strand (Polintons, Helitrons) or both strands (TIR, Cryptons) are excised from the donor site. Whatever the number of cut, intermediates can be found as linear or circular. If DNA synthesis has to occur, it can be done on the extrachromosomal intermediate before integration, coupled with reintegration, or after integration. Linear intermediates usually reinsert after staggered cut of the target site, which creates small direct duplications at both ends of the insertion, called the TSD (target site duplication). Circular intermediates (DIRS, Helitrons, Cryptons) usually do not.*

Recently, Arkhipova (2017) proposed a revised classification that attempted to integrate all previous classification systems, and also to include prokaryotic elements, as well as other mobile genetic elements (see Introduction). Figure 1.2 illustrates the various transposition mechanisms that characterize the main groups of eukaryotic TEs.

1.2.1. *Class I*

This class is likely the most abundant in eukaryotes, due to the replicative nature of the transposition process. All TEs in this class transpose via a copy-and-paste mechanism, which involves the production of an RNA intermediate that is then reverse transcribed into a new DNA copy, ready to insert at another location in the genome. The initial copy at the donor site is always left intact. According to Wicker et al. (2007), this class is divided into several orders, which are characterized by different structures and transposition mechanisms. The two most famous are the LTR and the LINE orders, which correspond to the initial A and B groups defined by Finnegan. However, some other orders exist, showing a less typical structure, the diversity of which is less known.

1.2.1.1. *LTR retroelements*

LTR retroelements share their structure with LTR retroviruses. Their mechanism of transposition is well known. They are characterized by two long terminal repeats: LTRs that are in direct orientation and that frame a piece of DNA several kb long, which comprise of one or more open reading frames (ORFs) that encode proteins necessary for the transposition process. The first ORF usually encodes a structural protein that associates to form viral-like particles (VLP). A second gene encodes several proteins translated in a row and later cut by a protease. The reverse transcriptase (RT) will generate the cDNA complementary of the transcribed genomic RNA, synthetizing the new copies, while an RNAse H has the role of destroying the RNA in the hybrid DNA-RNA duplex. Reverse transcription usually takes place after formation of the VLP that encapsulates the genomic RNA and needed proteins, along with the two short RNAs needed for initiation of the cDNA. Finally, the integrase will perform the integration to the new site, through strand transfer, after target site cleavage. Several superfamilies have been recognized, based on gene order and on the phylogeny on the RT conserved domains: Copia, Gypsy/Ty3, ERV (Endogenous retrovirus) and BEL/Pao. These superfamilies have been included in the International

Virus Taxonomy, within the order of Ortervirales. For example, ERVs are classified in the class Retroviridae, which also includes retroviruses such as HIV. In fact, the main distinctions between LTR retroviruses and LTR retrotransposons is that the former are infectious, able to exit the cell/individual before entering a new one, while LTR retrotransposons realize all their cycles within the cell. Some of these contain a third ORF that encodes an envelope gene, theoretically allowing them to leave the cell. The *Gypsy* element in *D. melanogaster* was first considered as a retrotransposon but is actually able to change cells and can then be functionally considered as a virus. The distribution of Env genes in the phylogenetic tree (based on RT) is quite patchy. Different envelope genes have been acquired independently and then been subsequently lost. Those LTR retroelements are found in large numbers in all eukaryotic genomes. *Gypsy* and *Copia*-like elements are particularly frequent.

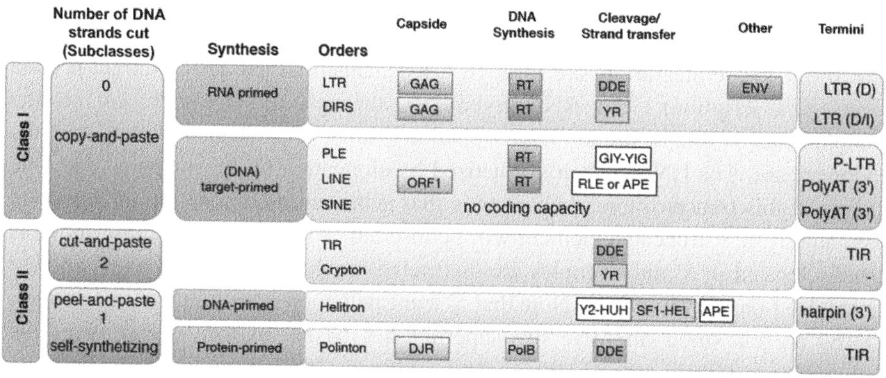

Figure 1.3. *Classification of eukaryotic transposable elements showing their main features and catalytic domains. GAG: capsid-like protein; HUH: Rep endonuclease domain; Hel: helicase domain; APE: apurinic/apyrimidinic endonuclease; RLE: restriction-like endonuclease; DDE: DDE integrase; YR: tyrosine recombinase; SF1-HEL: helicase (subfamily 1); T: reverse transcriptase; DJR: double jelly roll; PolB: DNA-dependant-DNA-polymerase B. D: direct; I: inverted. P-LTR: pseudo-LTR. For a color version of this figure, see www.iste.co.uk/huavan/transposable.zip*

1.2.1.2. *DIRS*

DIRS elements also possess LTRs, but present some specific features, and are clearly different from classical LTR retroelements. Their LTRs can be direct of inverted, depending on the DIRS family. An internal/subterminal region, homologous to the LTR sequence, is present, at least in the DIRS-like family; this appears to be crucial for the transposition mechanism. DIRS encode a reverse transcriptase, which seems to be phylogenetically close to that of the LTR, as well as an RNAse H and a GAG-like gene; however, they typically contain a gene coding for a tyrosine recombinase that replaces the DDE integrase. They also sometimes carry more specific genes (methylase, or hydrolase, of unknown function). The DIRS order remains a poorly characterized order among Class I. Four different superfamilies have been recognized (Malicki et al. 2020).

1.2.1.3. *Non-LTR retroelements*

Non-LTR retrotransposons lack LTRs and have few structural characteristics at their ends, including a TA rich stretch at the 3' end, which is necessary for transposition. Non-LTR retrotransposons include LINEs for long interspersed elements, which form a large order of Class I elements. They nevertheless share catalytic activities with LTR elements: a GAG-like protein (usually called ORF1), a reverse transcriptase and an endonuclease, which can be of various types, restriction-like endonuclease (RLE) or Apurinic-like endonuclease (APE), defining two subgroups. The transposition mechanism is quite different from the one from LTR elements. After transcription (and translation of proteins), transposition occurs directly at the reintegration site through a mechanism called TPRT (target-primed reverse transcription). The RNA hybridizes through the TA rich tail to the complementary single-strand DNA obtained after the staggered cut by the endonuclease. The DNA is then synthetized by elongation of the staggered cut. One feature of this transposition mechanism is that if the reverse transcription is aborted prematurely, a 5' truncated element will be inserted. Since signal for transcription is usually located in 5', those copies are themselves unable to transpose again and are called dead-on-arrival (DOA). Note that 5' truncated copies are frequent in genomes. LINEs are classified into five big superfamilies based on the specificity of the endonuclease domain and RT phylogeny (Arkhipova 2017; Kojima 2020). They are present in all genomes and can reach huge copy numbers (about 500,000 for L1 in the human genome).

Tightly associated with LINEs are the SINEs (short interspersed elements). These are highly abundant in some genomes (for example, in some mammals). They have no coding capacity and rely entirely on LINEs for their transposition. SINEs are derived from non-mobile short RNA sequences such as tRNA or 7SL transcripts. They do sometimes carry in 3' some short regions homologous to LINEs, suggesting de novo emergence by template switching during reverse transcription. In humans, the SINEs Alu have amplified up to 1 million copies.

1.2.1.4. *PLEs*

Penelope-like element (PLEs) are a particular subclass among Class I elements. They form an ancient group characterized by a reverse transcriptase, which is more related to the eukaryotic telomerase reverse transcriptase (TERTs). They contain atypical LTRs, called pseudo-LTRs, that can be direct or inverted. Among other strange features, we find sometimes the presence of introns and hammerhead-like ribozymes in the pseudo-LTRs. An endonuclease may sometimes be present. Absence of endonuclease correlates with a preference for insertion near telomeres, taking advantage of the short RNA sequences from which originate the telomeres repeats. The absence of integrase or recombinase suggests they may reinsert using a TPRT process. Recent studies uncovered a distribution larger than expected, with elements present in all major eukaryotic kingdoms (Craig et al. 2021).

1.2.2. *Class II*

1.2.2.1. *TIR*

TIRs elements are the most known TEs among Class II. While the first TIR elements had been identified after their insertion into a gene, triggering a phenotypic change, most recent elements have been exclusively identified through genome sequence analysis. About 20 different superfamilies are now recognized (Yuan and Wessler 2011; Kojima 2020) and are quite universal (present in almost every sequenced genome). All TIR TEs share characteristic features such as the presence of terminal inverted repeats (TIR), ranging from 11 to several hundreds of bp, that surround one or two genes encoding – at least for a DDE transposase. The DDE motif takes its name from three important residues in the catalytic core of the transposase. Initially identified in *Tc1-mariner* elements only, variant DDE motifs have finally been recognized in all known TIR superfamilies. DDE transposases are part of the large RNAse H fold superfamilies of protein. DDE-containing proteins are also found in Polintons, as well as in LTR retroelements (integrase). Interestingly, most of the bacterial TEs also contain DDE transposases, and several eukaryotic TIRs superfamilies share more homologies with their bacterial counterpart than with other eukaryotic TIRs superfamilies (Yuan and Wessler 2011). This is in favor of an ancient diversification, prior to the split between eubacteria and archaea/eukaryotes.

The mechanism of transposition is referred to as the "cut-and-paste" mechanism. It is obvious then that this mechanism per se is not responsible for the amplification of copies. However, TIRs elements are sometimes present in a large number of copies. In eukaryotes, two main processes have been proposed to explain this apparent contradiction. In prokaryotes, in which similar elements exist, transposition figures between two DNA molecules are usually resolved by replication. This is a common mechanism (see Chapter 3). In eukaryotes, the fate of a transposing copy depends partly on genome replication. Indeed, if a copy present in an already replicated site transposes into an unreplicated site, this copy will be duplicated in one of the daughter cells (Figure 1.4). Of course, if the contrary occurs (transposition from unreplicated to replicated site), one daughter cell will lose the copy. Hence, in order to be efficient, transposition of Class II elements should be synchronized with replication. In any case, there is another way for a TIR element to amplify. The excision from the donor site leaves a double strand DNA break (DSB) that must be repaired by the cellular machinery for the cell to survive. From this point of view,

Class II are quite harmful! Several ways exist in eukaryotic cells to repair a DSB (Figure 1.4). One is homology-dependant template repair, which uses an homologous DNA sequence to restore the intact site. Homologous sites are found in sister chromatids after replication, or in homologous chromosomes when the cell is at the diploid stage. If this homologous locus carries the TE copy, a full copy can be restored at the empty site. It often happens that the template-dependent repair is prematurely interrupted; the DSB is then repaired by an alternative mechanism known as non-homologous end-joining (NHEJ), which uses microhomology for direct ligation of the two broken ends. In such cases, an internally deleted copy can be generated, as shown for the P element in *Drosophila*. For other elements, NHEJ could be the main solution for repairing and this explains the excision footprint usually found at the empty donor site.

1.2.2.2. *Cryptons*

Cryptons were discovered in 2003 in several pathogenic fungi (Goodwin et al. 2003), and more recently in animals and some stramenopiles (Kojima and Jurka 2011). They have no repeats at the end and transpose using a tyrosine recombinase (YR), which supposedly excises them as a circular DNA intermediate and reinserts them in target sites with some homology. Indeed, short direct repeats are found at both termini. This group is poorly characterized, sharing relatively low similarity. They have been proposed to be at the origin of DIRS retroelements (Class I), and some domesticated versions have been identified in humans (Kojima and Jurka 2011).

1.2.2.3. *Helitrons*

Helitrons were discovered in silico in 2001, as a new group of Class II elements, characterized by the presence of a rolling-circle replication initiator protein (Rep) present in some prokaryotic TEs (such as IS91) and other mobile genetic elements that replicate through a rolling-circle mechanism (Kapitonov and Jurka 2001). The Rep domain corresponds to a HUH endonuclease and is associated with a C-terminal helicase domain. Helitrons have been found in almost all eukaryotic supergroups and are particularly prevalent in some species. Two groups are recognized that differ in their termini. Unlike TIR elements, Helitrons from group 1 do not have TIRs. They end with short sequences (LTS and RTS, for left and right terminal sequences, respectively) with several characteristics (5'TC, and 3'CTRR with a subterminal hairpin structure at the RTS which functions as a termination signal). In the second group, Helentrons contain an extra endonuclease domain similar to the one found in LINEs, and possess small subterminal inverted repeats in addition to the hairpin structures that can be present at both ends (Thomas and Pritham 2015).

Some aspects of the supposed transposition mechanism have been recently experimentally verified (Grabundzija et al. 2018; Kosek et al. 2021). The top strand of the donor DNA (coding strand) is cut at the LTS and displaced by the helicase activity, while synthesis of a new strand is initiated. The 3' cut at the RTS is elicited by a pause of the protein at the hairpin structure and allows the formation of a single-stranded circular intermediate that theoretically can undergo rolling circle replication. However, it seems probable that the reinsertion concerns single-stranded DNA (ssDNA), the double strand at the target site being reconstituted during genome replication. This mechanism is referred to as "peel-and-paste". More details are provided in Chapter 3 for prokaryotic TEs.

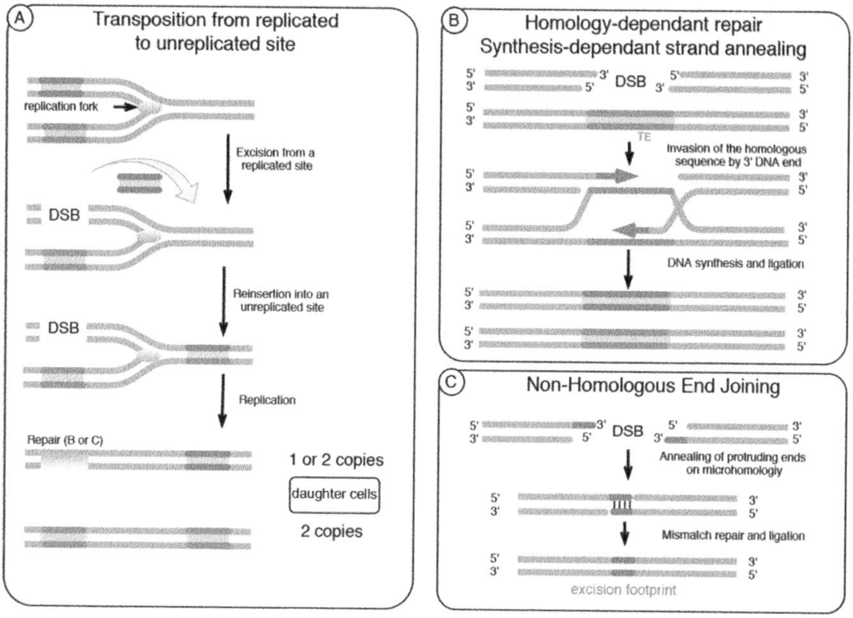

Figure 1.4. *Different ways to increase copy number and to repair the excision site. (A) Transposition can proceed from a replicated site to an unreplicated one. After replication is finished, at least one daughter cell will have a extra copy. (B) The excision leaves a double strand break (DSB) that can be repaired by using an homologous template. If this template contains the insertion, the TE is restored at the excision site. Processes A and B can combine. (C) Another way to repair the DSB is by direct ligation using a non-homologous end joining pathway. This relies on microhomology at both ends that allows annealing of the two protruding ends, followed by ligation. This repair mechanism can be used to repair aborted SDSA (Single-Dependant Strand Annealing), leading to internally deleted elements, and can result in an excision footprint, depending on the way the element is excised. For a color version of this figure, see www.iste.co.uk/huavan/transposable.zip*

The 3' termination of the strand displacement at the donor site is loosely defined by the short 3' hairpin, and adjacent sequences are frequently included in the transposition intermediate; therefore, these elements are able to capture genes, although this transduction-like mechanism may not account for all cases of gene capture observed. Several other models have been proposed but will not be discussed here (Kapitonov and Jurka 2007).

1.2.2.4. *Polintons*

Polintons (aka Mavericks) were discovered relatively recently (Kapitonov and Jurka 2006). They are long (around 20 kb) and possess TIRs as TIR elements. Between TIRs, several genes are present, including a DDE transposase gene and a DNA polymerase B gene. Hence, those elements are supposed to be replicative and have been called self-synthetizing. However, the precise mechanism of transpositions is not yet fully deciphered. Polintons may not be as active as TIRs, nor as universally spread, although they have been identified in several protists, fungi and animals, with the exception of plants (Haapa-Paananen et al. 2014). However, they can be abundant in some species (30% of the genome in *Trichomonas vaginalis*) (Pritham et al. 2007).

1.2.3. *Autonomous, non-autonomous and relics*

Most of these groups also contain shorter versions of TEs that contain *cis* sequences necessary for transposition, but lack coding capacity. These elements cannot transpose by themselves but can still move and amplify using the transposition machinery of the autonomous element. They are called non-autonomous and can be very short in size (between 200 bp and 1 kb). Non-autonomous elements derived from TIRs are very frequent and often outnumber the autonomous copies. They have been called MITEs (miniature inverted transposable elements); often, they have only kept the TIRs and the short central region appears unrelated or highly degenerated, compared to the autonomous. Every TIR superfamily has MITEs. Non-autonomous short versions also exist for Helitrons, for example. the DINE element of *Drosophila*, and for LTR retroelements (LARDs, TRIMs), LINEs and PLEs. In humans, the most numerous TEs are Alu elements that hijack the transposition machinery of L1 elements for transposing. Recently, a new group of non-autonomous elements have been identified in plants. Their peculiarities are the presence within LTRs of a hammerhead-like ribozyme. Such ribozymes are also found in non-LTR retroelements from animals. Those elements have been called retrozymes (Cervera and de la Peña 2020).

1.3. Abundance, diversity and distribution

Eukaryotic TEs are found everywhere but each species has a unique TE landscape. Abundance may vary from a few percent of the genome to more than

80%, and these characteristics are independent of the eukaryotic groups (Figure 1.5). Class I is usually more abundant, either because of LTR retroelements or of LINEs, but species with a majority of Class II elements can be found in any group. Hence, TE contents (proportion of the different types of TEs) are also quite variable (Wells and Feschotte 2020). Closely related species tend to have similar landscapes which can be explained by phylogenetic inertia (the TE landscape is ancestral). Shared TEs can also be more abundant in closely related species because of increased rates of horizontal transfer (Gilbert et al. 2021). However, very different patterns are sometimes observed between species belonging to a same phylogenetic clade. The huge increase in sequenced genomes should allow better deciphering of tendencies associated with phylogenetic clades. However, drawing an exhaustive picture is certainly impossible. The TE annotations can be performed using different tools, on assemblies of very variable quality, using a TE database more or less complete, which create lots of biases (see Chapter 11). Some recent analyses that combine homology-based de novo searches sometimes exhibit up to 50% of unclassified repetitive sequences. In brief, there are too many genomes and not enough comparative studies.

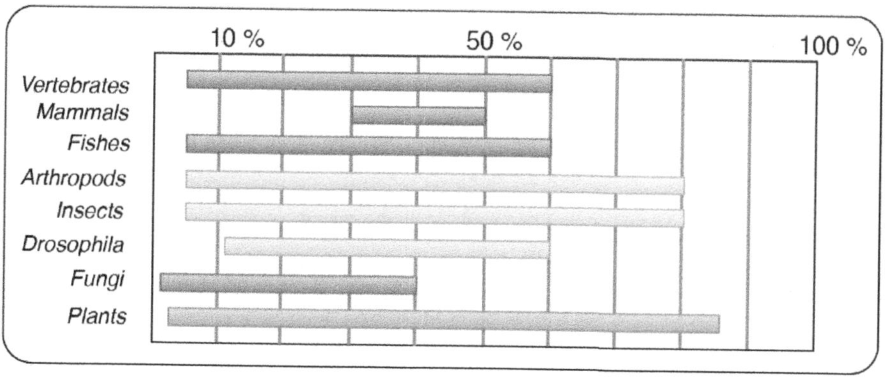

Figure 1.5. *Variation of TE contents (in percentage of the genome) in major Eukaryotic groups*

The diversity of TEs increases with the number of sequenced species. There is no exhaustive database but the most widespread superfamilies are divided into thousands of different TE families. For example, *Ty3/Gypsy* (Class I) and *Tc1-mariner* (Class II) are almost universal, and tens or hundreds are related; however, different families (lineages) often coexist in the same genome for each TE superfamily. Usually, only a few families are active and the remaining correspond to ancient, extinct (or little active) families. Why TEs have exploded in some genomes and not in others and why the success of any given superfamily is so variable are still unresolved questions.

Population size, lifestyle and demography are thought to impose some constraints, or at least have an influence. Genome ability to control TE transposition or to purge efficiently non-essential sequences or genome permissivity to horizontal transfer are likely important factors as well.

Within genomes, TEs are not randomly distributed. Some genomic regions, such as heterochromatin, are enriched in TEs. This may be due to a harsher selection on euchromatic arms rich in genes. Hence, TE will tend to accumulate in regions less important for functions or in specific regions (centromeres, telomeres, sexual chromosomes, dispensable chromosomes, low recombination regions). This also depends on the TE insertional preferences, which range from no preference (TEs insert almost randomly) to highly specific (like some LINE elements do by targeting a specific sequence in the rDNA genes – which are nevertheless repeated!).

1.4. Origins of transposable elements and evolutionary relationships with other genetic elements

TEs are very old components of genomes that might have emerged at the same time as the genes involved in nucleic acid metabolism and have since co-evolved with the genes. Basic catalytic activities involved in DNA metabolism include endonuclease, strand transferase and DNA/RNA-dependent DNA/RNA polymerase, all of which are found in TEs. We have to remember that TEs are parasites. Parasites cannot exist as parasitic entities before host appearance, and hosts obviously need parasitic sequences to evolve. As soon as these activities appeared, that is, very early, they must have been hijacked by sequences that could benefit, and became what we call parasites. In the long term, there must have been continuous back and forth exchanges between hosts and parasites, since the beginning of life. Hence, it is not surprising that links are detectable among TEs, and with all other mobile genetics entities. The previous descriptions draw the landscape of the major eukaryotic TEs based on Wicker et al.'s classification (2007). It does not consider all of Arkhipova's classification (the most recent to-date), which is more complicated, and noticeably has the merit to highlight links between TEs (eukaryotic and prokaryotic), as well as links with other genetic elements (mobile or not). Retracing the phylogenies of TEs is not possible because TEs are not a monophyletic group; however, it is still possible to infer the evolution at the protein domain level. The comparison of protein sequences shared between TEs can give insights on the evolution of TEs and their relationships. Such studies have mainly revealed that TEs evolve through recurrent module (protein domains) exchanges and the exact suite of events that have sequentially shaped the TE landscape cannot be determined.

While two classes of TEs coexist in eukaryotic genomes, prokaryotic TEs are mainly Class II elements. Reverse transcriptasea are nevertheless found in bacteria, for example, in group II introns (retroelements that can invade homologous empty loci, after self-splicing from the RNA by a ribozyme activity and reverse transcription by the self-encoded reverse transciptase). Reverse transcriptases are also at the heart of the mechanism described in retrons and retroplasmids. Reverse transcriptase could be one of the oldest activities, derived from an ancestral RNA-recognition motif (RRM) that may also be at the origin of polymerase B or rolling-circle replication endonuclease (Krupovic et al. 2019).

The DDE domain exists in the transposase of all TIR DNA elements (Yuan and Wessler 2011), Polintons (Class II), as well as in integrases of LTR retroelements. DDE domains are gathered into structurally distinct 3D structures that share some common motifs. A large number of prokaryotic TEs also rely on DDE transposases. Interestingly, several DDE domains are shared by eukaryotic and prokaryotic elements, suggesting an emergence prior to the last universal common ancestor. DDE domains belong to the RNAse H-like superfamilies of proteins that also comprise exonucleases (Majorek et al. 2014).

The recent discoveries of retrozymes elements, or ribozymes embedded in TEs, are additional arguments in favor of a very early emergence of TEs and other mobile genetic parasites that could be as early as during the RNA world.

In fact, protein module exchanges are observed between TEs and non-TEs entities, such as viruses. Based on the structure and the reverse transcriptase phylogeny, LTR retroelements and ERV are closely related to vertebrate retroviruses and caulimoviruses, and have integrated into the virus classification of the International Committee for Taxonomy of Viruses. Previous analysis showed that vertebrate LTR retroviruses are in fact a clade of LTR retroelements that have acquired an envelope gene before diversification. Other retroviruses have been related to various other TEs and have acquired env genes independently from several viral sources (Malik et al. 2000). However, things become complicated when analyzing the RNAse H domain. Vertebrate retroviruses have independently acquired a second RNAse H domain, and the initial gene has been subfunctionalized (to regulate the RNAse H activity). Several independent acquisitions may have occurred, from non-LTR retroelement, or from unknown origin. Furthermore, the same scenario seems to have occured for plant retrotransposons (Ustyantsev et al. 2015). In any case, it is usually thought that LTR retroelements (and then retroviruses) emerged relatively recently compared to DNA and non-LTR retroelements.

It has been hypothesized that LTR retroelements emerged from a fusion between a non-LTR and a DNA transposon. In particular, the Gypsy superfamily DDE integrase is phylogenetically closely related to the one from the Ginger superfamily (TIR) (Bao et al. 2010). However, a clear picture cannot be drawn because several events or gene replacements seem to have taken place. Other examples of shared genes include the tyrosine recombinases found in both Class I (DIRS) and Class II (Cryptons) elements; and it has been proposed that DIRS have emerged from a combination of a Crypton with the RH/RT of a Gypsy LTR retroelement (Poulter and Butler 2015). However, those elements are not as frequent as LTR retroelements, and alternative scenarios are also quite possible. In this case, the relationship with bacterial tyrosine recombinase is not very clear.

Finally, Polintons are remarkable for the links they share with several other mobile entities, which gives some clues about their origin. They are actually quite close to virophages, those viruses able to infect giant viruses, that themselves parasite unicellullar eukaryotes, such as amoeba. Virophages, Polintons and related Polintons-like viruses (PLVs) all share several common genes and a common origin. Since they also possess capsid genes, Polintons may well have a dual life style (virus and TE), although Polinton virions have never been detected. Polintons are found in eukaryotes, while PLV, virophages and shorter versions of Polintons (called transpovirons, considered as linear integrative plasmids) are specific to eukaryotic giant viruses. On the one hand, giant viruses are supposed to have emerged before LECA (last common eukaryotic ancestor). On the other hand, Polintons are suspected to be at the origin of giants viruses, adenoviruses by virophages and linear cytoplasmic plasmids. This incredible network illustrates the extraordinary entangled relationships between all mobile genetic elements (Koonin and Krupovic 2017).

When speaking of shared domains, numerous exchanges have also occurred between TEs and host genes. In particular, DNA binding domains of Class II elements are recurrently identified within genes (Feschotte 2008). We can wonder about the direction of the transfer. It is always difficult to decide between the two alternatives: which is the one derived from the other one? It is usually thought that it is from the TEs toward the genes. However, recent TE families may have emerged after acquiring gene or viral domains (de Mendoza et al. 2018).

1.5. Genomic impact

Eukaryotic TEs are diversified, abundant and have a huge impact in the genome that hosts them. They greatly participate to its evolution, whether its structure or its function. The activity of TEs make genomes very dynamic. Hence, insertion polymorphism of recent copies increases a lot of the genetic diversity observed among populations, and hence the adaptive capacity of the species.

1.5.1. *Genome size*

Of course, TE copies are finally not that big (usually) but they may be numerous; this impacts the genome size. In eukaryotes, genome size is highly variable ranging for a few Mb for organisms such as obligate parasites, that relies on their host genes for many functions, to several hundreds of gigabases. Genome size is not correlated to the complexity of the organisms, and some unicellular organisms exhibit genomes much bigger than most multicellular complex organisms. Among multicellular organisms, plants are the ones that contain more genes; we also know that gene number is not either correlated to complexity, and more surprisingly, gene number is not correlated to genome size. This has led to what is called the C-value paradox, the C-value being the amount of (haploid) DNA in an organism's cell. What has been found is that genome size correlates extremely well with the genomic TE content. Hence, in eukaryotes, the bigger the genome, the higher the TE contents! Why do some species allow such accumulation of TEs in their genomes? It is still debated: a larger genome will take longer to replicate, and the nucleus, and then the cell, will be larger. These characteristics are likely to influence the organism's lifecycle.

TE accumulation could be a consequence of populational parameters. It has been suggested that the effective population size of a species (Ne), which determines the force of the natural selection relative to the genetic drift, could be a key factor for putting the limit to the genome size (Lynch and Conery 2003).

1.5.2. *Genome structure*

Genome was once seen as a static, unmovable suite of genes that has to be transmitted intact to the next generation. In fact, the genome is quite fluid, and allows for many changes, even if related species usually display a good synteny (gene colinearity) between their genomes. However, major rearrangements (chromosomal inversion, translocation) are observed between closely related species. Within genome, genes also evolve, by substitution or small indels, but also by more important changes such as exon shuffling, or changes in the promoter region. Mutations are the initial evolutionary force that create the diversity necessary for natural selection to act. Mere substitutions are finally quite rare, especially in species with large genomes, and a good part of mutations come from small or big indels and rearrangements. By intrinsic nature (mobile, and repeated), TEs are a major source of mutation: by transposing, they trigger insertions, deletions and recombination between similar copies settled at different loci (a process known as ectopic recombination); they trigger large rearrangements (Figure 1.6). Hence, TEs are major actors in genome structure evolution.

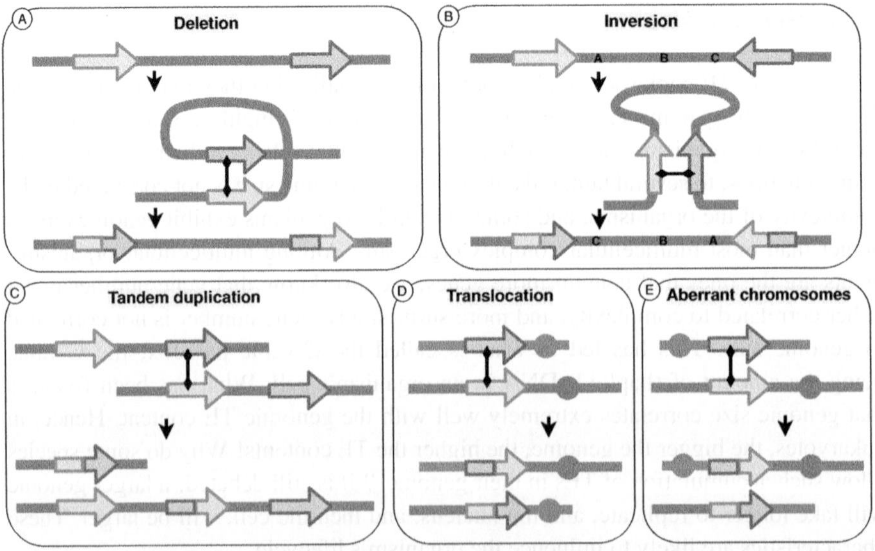

Figure 1.6. *Chromosomal rearrangements triggered by ectopic recombination between TEs. TEs copies can be on the same DNA molecules, on sister chromatids, homologous chromosomes or different chromosomes. Depending on the orientation between the two repeats, or relative to the centromere, different outputs are expected: deletion (A), inversion (B), tandem duplication (C), translocation (D) and aberrant chromosomes (E). For a color version of this figure, see www.iste.co.uk/huavan/transposable.zip*

1.5.3. *Genome function and evolution*

Inevitably, TEs interact with the part of the genome that insures cell and organism survival, which is the genes. More details are provided in other chapters, but TEs have been involved in gene evolution (exon shuffling, exonization) and gene regulation (rewiring of regulation networks by exaptation of TEs short sequences). Numerous processes have emerged in response to TEs (silencing mechanisms via small interfering RNA, or deposition of epigenetics marks). Finally, TE protein domains (such as the DNA binding domains) from numerous elements are found in genes, coding for transcription factors, or proteins that interact with DNA. Finally, TEs are at the origin of several evolutionary innovations such as mass excision of short DNA fragments at the origin of the macronuclei of paramecia, V(D)J recombination in jawed vertebrates, or placenta emergence in mammals, which we call molecular domestication (see Chapter 6).

1.6. References

Arkhipova, I.R. (2017). Using bioinformatic and phylogenetic approaches to classify transposable elements and understand their complex evolutionary histories. *Mob. DNA*, 8, 19.

Bao, W., Kapitonov, V.V., Jurka, J. (2010). *Ginger* DNA transposons in eukaryotes and their evolutionary relationships with long terminal repeat retrotransposons. *Mob. DNA*, 1(1), 3.

Cervera, A. and de la Peña, M. (2020). Small circRNAs with self-cleaving ribozymes are highly expressed in diverse metazoan transcriptomes. *Nucleic Acids Res.*, 48(9), 5054–5064.

Craig, R.J., Yushenova, I.A., Rodriguez, F., Arkhipova, I.R. (2021). An ancient clade of *Penelope*-like retroelements with permuted domains is present in the green lineage and protists, and dominates many invertebrate genomes. *Mol. Biol. Evol.*, 38(11), 5005–5020.

Curcio, M.J. and Derbyshire, K.M. (2003). The outs and ins of transposition: From mu to kangaroo. *Nat. Rev. Mol. Cell Biol.*, 4(11), 865–877.

Feschotte, C. (2008). Transposable elements and the evolution of regulatory networks. *Nat. Rev. Genet.*, 9(5), 397–405.

Finnegan, D.J. (1989). Eukaryotic transposable elements and genome evolution. *Trends Genet.*, 5(4), 103–107.

Gilbert, C., Peccoud, J., Cordaux, R. (2021). Transposable elements and the evolution of insects. *Annu. Rev. Entomol.*, 66, 355–372.

Goodwin, T.J.D., Butler, M.I., Poulter, R.T.M. (2003). Cryptons: A group of tyrosine-recombinase-encoding dna transposons from pathogenic fungi. *Microbiology (Reading)*, 149(Pt 11), 3099–3109.

Grabundzija, I., Hickman, A.B., Dyda, F. (2018). Helraiser intermediates provide insight into the mechanism of eukaryotic replicative transposition. *Nat. Commun.*, 9(1), 1278.

Haapa-Paananen, S., Wahlberg, N., Savilahti, H. (2014). Phylogenetic analysis of Maverick/Polinton giant transposons across organisms. *Mol. Phylogenet. Evol.*, 78, 271–274.

Kapitonov, V.V. and Jurka, J. (2001). Rolling-circle transposons in eukaryotes. *Proc. Natl. Acad. Sci. USA*, 98(15), 8714–8719.

Kapitonov, V.V. and Jurka, J. (2006). Self-synthesizing dna transposons in eukaryotes. *Proc. Natl. Acad. Sci. USA*, 103(12), 4540–4545.

Kapitonov, V.V. and Jurka, J. (2007). Helitrons on a roll: Eukaryotic rolling-circle transposons. *Trends Genet.*, 23(10), 521–529.

Kapitonov, V.V. and Jurka, J. (2008). A universal classification of eukaryotic transposable elements implemented in repbase. *Nat. Rev. Genet.*, 9(5), 411–412.

Kojima, K.K. (2020). Structural and sequence diversity of eukaryotic transposable elements. *Genes Genet. Syst.*, 94(6), 233–252.

Kojima, K.K. and Jurka, J. (2011). Crypton transposons: Identification of new diverse families and ancient domestication events. *Mob. DNA*, 2(1), 12.

Koonin, E.V. and Krupovic, M. (2017). Polintons, virophages and transpovirons: A tangled web linking viruses, transposons and immunity. *Curr. Opin. Virol.*, 25, 7–15.

Kosek, D., Grabundzija, I., Lei, H., Bilic, I., Wang, H., Jin, Y., Peaslee, G.F., Hickman, A.B., Dyda, F. (2021). The large bat helitron DNA transposase forms a compact monomeric assembly that buries and protects its covalently bound 5'-transposon end. *Mol. Cell*, 81(20), 4271–4286.

Krupovic, M., Dolja, V.V., Koonin, E.V. (2019). Origin of viruses: Primordial replicators recruiting capsids from hosts. *Nat. Rev. Microbiol.*, 17(7), 449–458.

Lynch, M. and Conery, J.S. (2003). The origins of genome complexity. *Science*, 302(5649), 1401–1404.

Majorek, K.A., Dunin-Horkawicz, S., Steczkiewicz, K., Muszewska, A., Nowotny, M., Ginalski, K., Bujnicki, J.M. (2014). The RNase H-like superfamily: New members, comparative structural analysis and evolutionary classification. *Nucleic Acids Res.*, 42(7), 4160–4179.

Malicki, M., Spaller, T., Winckler, T., Hammann, C. (2020). DIRS retrotransposons amplify via linear, single-stranded cDNA intermediates. *Nucleic Acids Res.*, 48(8), 4230–4243.

Malik, H.S., Henikoff, S., Eickbush, T.H. (2000). Poised for contagion: Evolutionary origins of the infectious abilities of invertebrate retroviruses. *Genome Res.*, 10(9), 1307–1318.

de Mendoza, A., Bonnet, A., Vargas-Landin, D.B., Ji, N., Li, H., Yang, F., Li, L., Hori, K., Pflueger, J., Buckberry, S. et al. (2018). Recurrent acquisition of cytosine methyltransferases into eukaryotic retrotransposons. *Nat. Commun.*, 9(1), 1341.

Piégu, B., Bire, S., Arensburger, P., Bigot, Y. (2015). A survey of transposable element classification systems—A call for a fundamental update to meet the challenge of their diversity and complexity. *Mol. Phylogenet. Evol.*, 86, 90–109.

Poulter, R.T.M. and Butler, M.I. (2015). Tyrosine recombinase retrotransposons and transposons. *Microbiol. Spectr.*, 3(2), MDNA3–0036–2014.

Pritham, E.J., Putliwala, T., Feschotte, C. (2007). Mavericks, a novel class of giant transposable elements widespread in eukaryotes and related to DNA viruses. *Gene*, 390(1-2), 3–17.

Thomas, J. and Pritham, E.J. (2015). Helitrons, the eukaryotic rolling-circle transposable elements. *Microbiol Spectr*, 3(4), 1–32.

Ustyantsev, K., Novikova, O., Blinov, A., Smyshlyaev, G. (2015). Convergent evolution of ribonuclease H in LTR retrotransposons and retroviruses. *Mol. Biol. Evol.*, 32(5), 1197–1207.

Wells, J.N. and Feschotte, C. (2020). A field guide to eukaryotic transposable elements. *Annu. Rev. Genet.*, 54, 539–561.

Wicker, T., Sabot, F., Hua-Van, A., Bennetzen, J.L., Capy, P., Chalhoub, B., Flavell, A., Leroy, P., Morgante, M., Panaud, O. et al. (2007). A unified classification system for eukaryotic transposable elements. *Nat. Rev. Genet.*, 8(12), 973–982.

Yuan, Y.-W. and Wessler, S.R. (2011). The catalytic domain of all eukaryotic cut-and-paste transposase superfamilies. *Proc. Natl. Acad. Sci. USA*, 108(19), 7884–7889.

2

Prokaryotic Transposable Elements

Alessandro M. VARANI[1], Karen E. ROSS[2] and Mick CHANDLER[3]

[1] *Department of Agricultural and Environmental Biotechnology, School of Agricultural and Veterinary Sciences, UNESP - São Paulo State University, Jaboticabal, Brazil*

[2] *Protein Information Resource, Department of Biochemistry and Molecular & Cellular Biology, Georgetown University Medical Center, Washington, DC, USA*

[3] *Department of Biochemistry and Molecular & Cellular Biology, Georgetown University Medical Center, Washington, DC, USA*

2.1. Introduction

Prokaryotes have provided an abundance of transposable elements (TEs), which have greatly contributed to our present understanding of the importance of TE in remodeling their host genomes and regulating gene expression, and have been key to our present understanding of transposition mechanism(s) and its regulation.

2.1.1. *Historical*

It has now been over 40 years since the first bacterial insertion sequences (ISs), arguably the simplest autonomous prokaryotic TE, were described. They were identified as short DNA segments found repeatedly associated with mutations in the *gal* operon of *Escherichia coli* and in bacteriophage lambda (Fiandt et al. 1972; Hirsch et al. 1972; Malamy et al. 1972). Shortly after, it was established that ISs were normal residents of the *E. coli* genome (Saedler and Heiss 1973), sometimes

For a color version of all figures in this chapter, see www.iste.co.uk/huavan/transposable.zip.

Transposable Elements and Genome Evolution,
coordinated by Aurélie HUA-VAN and Pierre CAPY. © ISTE Ltd 2024.

present in multiple copies. They were shown to be involved in generating deletions (Reif and Saedler 1974) and activating gene expression (Jordan et al. 1968; Saedler et al. 1974). They were also identified as frequent constituents of bacterial plasmids (Hu et al. 1975a, 1975b, 1975c).

2.1.2. *Relationship between IS and transposons*

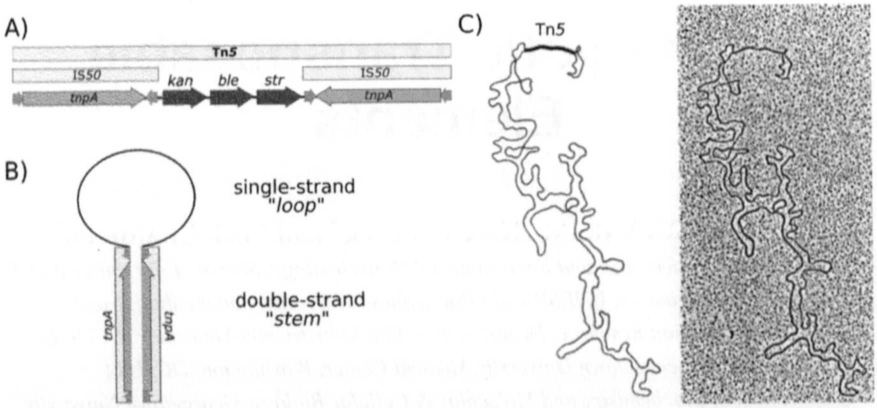

Figure 2.1. *Homoduplex electron microscopy for transposon visualization. The compound transposon Tn5 is used as an example. The left-hand panel shows the structure of Tn5 and the way in which the flanking inverted IS50 insertion sequences would form a double-strand stem on denaturation and renaturation. The right-hand panel shows the structure of plasmid pOX38 with a Tn5 insertion, as revealed by electron microscopy. Plasmid DNA was denatured and renatured to form a snap-back structure that locates the Tn5 insertion (electron microscopy by Edouard Boy de la Tour and Lucien Caro)*

At about the same time, it was observed that antibiotic resistance genes could also be transferred or "transposed" from one plasmid to another (Hedges and Jacob 1974; Hedges et al. 1974; Barth et al. 1976) (Figure 2.1). It was recognized that IS and the movable resistance gene machinery, "transposons" (Tn) (a term originally coined by Hedges and Jacob (1974)), better describing their activity than "controlling elements" in Maize (McClintock 1950, 1956), were both members of a group of genetic entities: TEs or mobile genetic elements (MGEs). Traditionally, transposons are defined as TEs, which carry passenger genes. Transposons and IS were identified by electron microscopy of heteroduplexes between plasmids or bacteriophage with or without the insertion, or by homoduplexes. The relationship between IS and transposons was reinforced by the observation that different genes could be translocated by two flanking IS in the form of compound transposons (Gottesman and Rosner 1975; Arber et al. 1979; So et al. 1979) (e.g. Figure 2.1).

2.1.3. *The prokaryotic TE landscape*

Over the last 50 years, a large variety of prokaryotic TEs have been identified. These include 28 families of IS (nearly 6,000 different IS in the ISfinder database https://www-is.biotoul.fr; see TnPedia https://tncentral.ncc.unesp.br/TnPedia/ for a detailed description), many compound transposons with a number of different flanking IS, so-called unit transposons which include at least four major families (Tn*3*, Tn*7*, Tn*402*, Tn*554*) and casposons. In a broader sense, integrons, integrative conjugative elements (ICE), p*dif* and genomic islands also constitute part of the prokaryotic mobilome. The majority of IS and Tn transpose using a double strand DNA intermediate and a transposase (the element-specific enzyme which catalyzes the DNA strand cleavages and transfers necessary for TE movement from a donor DNA molecule to a target) (Figures 2.2(A) and 2.2(B)).

However, a small number use rolling circle mechanisms (for review, see del Pilar Garcillán-Barcia et al. (2001) and Chandler et al. (2013)) (Figure 2.2(C)) or circular single-strand DNA intermediates (e.g. IS*200*-IS*605*) (for review, see He et al. (2015) and Figure 2.2(D)), while others, such as integrons, ICE, p*dif* and genomic islands, use site-specific recombination in their movement.

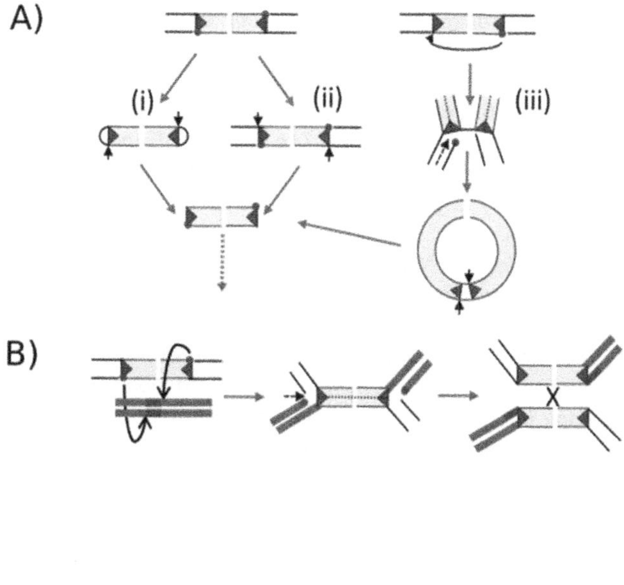

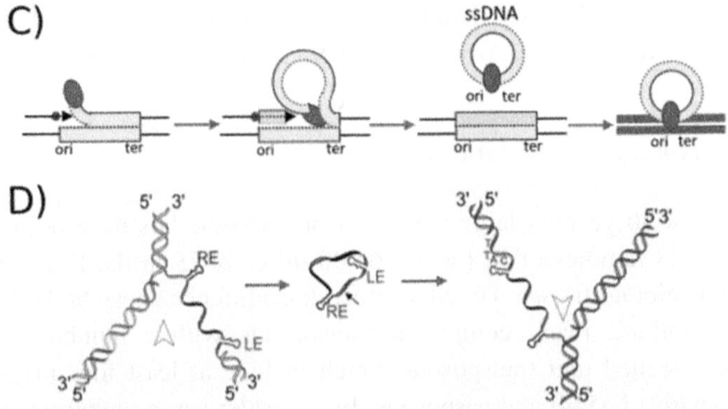

Figure 2.2. *(A and B). The major DDE mechanisms involved in transposition. (A) DDE cut out-paste in (i and ii) and copy out-paste in (iii). In cut out-paste in transposition, cleavage occurs at both ends to generate a 3'OH (small circle). These then either attack the opposite (complementary strand) (i) to separate the transposon from flanking donor DNA and generate a highly constrained hairpin structure, which then undergoes a second cleavage to form the final transposition intermediate or second strand cleavage occurs using a second enzyme (ii) to generate the intermediate. In copy out-paste in, transposition cleavage occurs at one end and the 3'OH then attacks the second end to generate a bridge and a free 3'OH in the donor. This then directs replication of the IS to form a double strand circular form, which then undergoes cleavage to generate the final transposition intermediate. This then continues (dotted arrow) to integrate into a suitable target. (B) DDE copy in transposition. Here, the transposase cleaves both ends and the free 3'OH then attacks the target, and in-so-doing (left), fuses both the donor and target molecules, generating a replication fork (middle). Replication creates two transposon copies (right), one at each donor-target junction, and these then undergo recombination (X) to a separate donor and target. (C and D) The major DDE mechanisms involved in transposition. (C) Rolling circle transposition. The HUH transposase cleaves at one IS end (the origin "ori" end) to generate a Tpase-5' DNA covalent complex and a 3' end on the donor DNA (left). This primes replication to complete the second transposon strand (termination "ter" end) and "roll" out a single (or double) strand circlular copy which acts as a transposition intermediate to form a covalent complex with the target and undergo "roll" in replication. The precise steps in this pathway are not known. (D) IS608 peel-and-paste transposition. Left (LE) and right (RE) ends with subterminal transposase recognition hairpins (HP) transposition excision. (Left) A single-strand IS copy located on the lagging strand is uncovered at the replication fork during replication. Cleavage at each end and strand transfer yields a single-strand circle (middle), which subsequently attacks a target sequence in the lagging strand of the target DNA replication fork (right). Note that insertion is directional with LE located distally to the replication fork*

2.2. Transposases: the enzymes driving transposition

There are a limited number of transposase types (Tpases) based on the chemical reactions they catalyze during strand cleavage, and they are often named after key amino acids in their catalytic sites. These include DDE (for the triad aspartate, aspartate, glutamate), HUH (for the triad histidine, bulky hydrophobic, histidine) and enzymes resembling the site-specific tyrosine and serine recombinases (Smith and Thorpe 2002; Rice 2015; Smyshlyaev et al. 2021).

2.2.1. *DDE enzymes*

A large fraction of IS encodes a Tpase, which includes a DDE catalytic motif originally observed to be a common feature of retroviral integrases and IS*3* family members (Fayet et al. 1990; Khan et al. 1991; Rezsöhazy et al. 1993). The catalytic mechanism of these enzymes has been well documented both in prokaryotic and eukaryotic transposition (for review, see Hickman and Dyda (2015a)). However, in spite of identical chemistries, DDE enzymes catalyze the cleavage (hydrolysis which uses H_2O as a nucleophile) of a single DNA strand, known as the transferred strand, to generate a 3′OH and a 5′ phosphate. There is a wide diversity in the overall transposition mechanism. This depends on the order of cleavage and strand transfer, and the intervention (or not) of a replicative step (Figures 2.2(A) and (B)). Moreover, the spacing between the second D and the E residue can be very variable. This is often due to so-called insertion domains (in both prokaryotes and eukaryotes) whose function is to modulate the reaction mechanism (Dyda et al. 2012; Kosek et al. 2021). This has been reviewed in detail a number of times (Mizuuchi and Baker 2002; Hickman and Dyda 2015a).

Transposition involves two transposon ends (four DNA strands) and the target site. Note that no covalent TE-transposase intermediate is generated during the transposition process.

2.2.1.1. *DDE enzymes: conservative cut out-paste in*

The most straightforward transposition pathway is known as conservative or cut out-paste in. It requires cleavage of both DNA strands at each transposon end. This can be accomplished in two ways:

1) Transposase catalyzes cleavage of the transferred strand (i.e. the DNA strand carrying the 3′OH at each TE end, which is transferred into the target DNA) at each IS end to generate a 3′OH. This then attacks the opposite, complementary (non-transferred) strand, liberating the IS from flanking donor DNA and creating a hairpin bridge between the two strands at each end (Figure 2.2(Ai)). This structure

then undergoes a second Tpase-catalyzed cleavage to regenerate the 3′OH and a linear transposition intermediate, which then undergoes insertion into the target DNA. Members of the IS*4* family are known to transpose in this way (Kennedy et al. 1998; Bhasin et al. 1999).

2) An alternative pathway adopted by members of the Tn*7* transposon family (see below) is to use two proteins: a DDE transposase to catalyze cleavage of the transferred strand and a protein resembling a restriction endonuclease to catalyze cleavage of the complementary strand (see Peters (2014)) (Figure 2.2(Aii)).

2.2.1.2. *DDE enzymes: replicative copy out-paste in*

Another major pathway, copy out-paste in (Chandler et al. 2015), involves Tpase-catalyzed cleavage at one end and attack of the other IS end by the resulting 3′OH to generate a single strand bridge between the two ends. Replication of the IS (copy out) using the 3′OH generated at the bridge generates a double strand IS circle as a transposition intermediate, which can then integrate into a target using the two abutted IS ends (paste in). This mechanism has been adopted by a large number of IS families (Figure 2.2(Aiii)).

2.2.1.3. *DDE enzymes: replicative (classical copy-in)*

The fourth principal DDE-catalyzed pathway, copy in, involves cleavage at both ends and simultaneous transfer of these to the target (Figure 2.2(B)). This creates two potential replication forks, one at each TE end. Replication then generates a replicon fusion in which donor and recipient are linked by two directly repeated copies of the TE, one at each junction. This structure is then "resolved" by recombination between the two TE copies using either an efficient element-encoded sequence-specific recombination system, or the host homologous recombination system (Shapiro 1979). One IS family, IS*6* (Trieu-Cuot and Courvalin 1985; Varani et al. 2021), and one transposon family, Tn*3* (see Nicolas et al. 2015) have adopted this mechanism, although it is probable that a second Tn family, Tn*402*, may also follow this pathway. In intermolecular transposition, two separate DNA molecules generate a small direct target repeat sequence (DR) flanking the TE because each TE strand attacks the target DNA in a staggered fashion. However, this is not the case with intramolecular transposition within the same DNA molecule, which can occur in one of two orientations into the target site: in one orientation, it results in inversion to the DNA region between the donor site and the target site, while in the other, it simply results in deletion of this intervening DNA. In both cases, the DR does not appear as a flank to one Tn, but is distributed between the two Tn copies (e.g. Varani et al. 2021).

2.2.2. *HUH enzymes*

A second group of TE encode transposases known as HUH transposases carry an active-site tyrosine and an amino acid triad of two histidine residues and a bulky hydrophobic amino acid (U). These are related to relaxases, which are involved in the transfer of conjugative plasmids and Rep proteins, which are involved in the replication of some plasmids, and both bacterial and eukaryotic viruses (see Chandler et al. 2013). The tyrosine acts as the nucleophile in cleaving the TE end and forms a transitory covalent intermediate. Two types of IS employ this type of enzyme: members of the IS*200*/IS*605* group and of the IS*91* family. IS of the former use a mechanism termed peel-and-paste, while members of the latter use a rolling circle mode (Figures 2.2(C) and (D)).

2.2.2.1. *HUH enzymes (rolling circle)*

The second major pathway used by IS with HUH enzymes is rolling circle transposition, where one end of the IS (*ori*) is cleaved and transferred to the target, and the entire IS is replicated into the insertion site and stops at a site (*ter*) at the second IS end. This type of pathway has been adopted by the IS*91* family (Figures 2.2(C) and 2.3). Few detailed biochemical details have been published, although single-strand circles have been identified, but these may be by-products rather than intermediates in transposition. In vivo, transposition events suggest that recognition of *ter* is relatively inefficient and elements of this family are often found to transpose genes neighboring *ter* in the donor in transposition events known as single end transposition. A related group of IS and ISCR is often located upstream of groups of antibiotic-resistant genes in complex multi-resistance genomic regions, leading to the idea that they may be important vehicles in antibiotic resistance gene spread (Toleman et al. 2006), although there is no firm evidence for their transposition.

Figure 2.3. *IS and their derivatives. An IS91 family IS showing the secondary structures (arrows) on the left (ori) and right (ter) ends, together with the HUH transposase*

2.2.2.2. *HUH enzymes (peel-and-paste)*

The transposition mechanism of IS*200*/IS*605* family members has been well characterized (see He et al. (2015) and https://tncentral.ncc.unesp.br/TnPedia/index.php/IS_Families/IS200-IS605_family). The transposase cleaves a specific IS end and joins it to the opposite end to form a single-strand circular intermediate. It is thought that the single strand, which is removed from the donor, is replaced by replication. The insertion of the transposition intermediate occurs at the host chromosome replication fork in a strand- and orientation-specific manner (Figure 2.2(D)). This constraint is evident when analyzing the distribution of these IS in entire bacterial genomes where chromosome replication is bidirectional from a common origin: the majority of insertions in each replication arm of bidirectionally replicating chromosomes are oriented in the direction of replication.

2.3. Insertion sequences

2.3.1. *Overview*

ISs (Figure 2.4) are among the smallest autonomous TE. They range in length from about 700 bp to 2.5 kb (see Siguier et al. 2015) and are genetically compact. Most are characterized by short, flanking terminal imperfect inverted repeats (IRs), generate short direct target repeats (DR) on insertion and include a transposase gene (*tnp*), which occupies almost the entire length of the IS (Figure 2.4(A)).

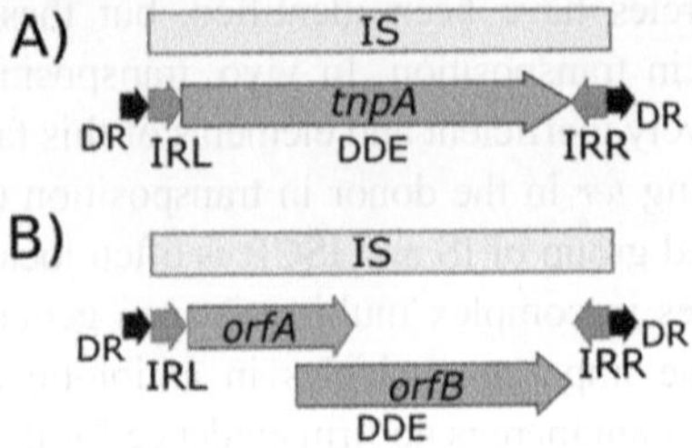

Figure 2.4. *IS. Direct target repeats (DR), terminal inverted repeats (IR), transposases (DDE) and passenger genes are indicated. (A) A typical IS with a single transposase open reading frame. (B) An IS with two open reading frames in which the transposase is formed by programmed -1 translational frameshifting (Figure 2.5(A))*

A number carry two transposition-related genes (Figure 2.4(B)): a short upstream open-reading frame that includes a sequence-specific binding function and is expressed as a unit, while a longer downstream frame located in the -1 frame with respect to the upstream frame and which includes the catalytic DDE motif is expressed by a programmed -1 frameshift as a fusion protein with the upstream frame (Figure 2.5(A)). This allows economical coding of different functions in one translation unit. This is relatively common in prokaryotic TE compared to other genes and provides effective transposition regulation at the translational level (Chandler and Fayet 1993; Sharma et al. 2011; Antonov et al. 2013). Another alternative translational regulatory mechanism is translational coupling, where the termination codon of an upstream reading frame overlaps the initiation codon of the downstream *orf* (see Huber et al. (2019)) (Figure 2.5(B)).

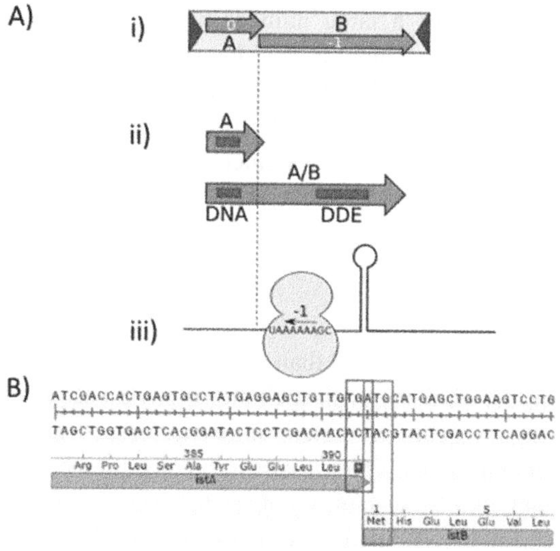

Figure 2.5. *Common translational regulation in IS. (A) Programmed -1 translational frameshifting. (i) The IS carries 2 orfs upstream (A, generally a DNA binding protein) in translational phase 0 and the overlapping downstream (B, generally with the catalytic site) frame in the -1 phase. (ii) Upstream and downstream translation products together with a fusion protein composed of the upstream and downstream products created by translational fusion. (iii) Messenger RNA structures which lead to frameshifting. The ribosome is shown to move backwards by a single base when it encounters a run of "so-called" slippery codons aided by a downstream secondary structure. (B) Translational coupling showing overlapping stop and start codon allowing the ribosome to reinitiate translation of the downstream gene by -1 slippage without a ribosome binding site*

2.3.2. *Impact*

The initial identification of IS was effectively due to the type of genetic impact they have on their host genomes: to create mutations by insertion, deletion and other DNA rearrangements. However, the impact of specific IS groups on their host genome depends largely on the level of their transposition activities and on their exact transposition mechanism (see below). A general observation is that this activity appears to be tightly regulated. Activation of transposition by impinging transcription is prevented by a variety of mechanisms including sequestration of translation initiation signals in secondary structures facilitated by external transcription and differential expression of a "repressor" (Nagy and Chandler 2004).

One important property of a large number of ISs is their ability to activate neighboring gene expression. This can be the result of the presence of a -35 promoter hexamer element directed outwards from one IS end which, on insertion of the IS, can combine with a suitably placed -10 hexamer in the neighboring target DNA to generate a hybrid promoter (see Siguier et al. (2014) and Vandecraen et al. (2017)), or simply by transcription which "escapes" from internal TE promoters.

2.3.3. *IS diversity*

ISs have been classified into more than 28 different families according to their genetic organization, the similarities in their transposase sequences, their terminal IRs and the direct target repeats generated on insertion (see Mahillon and Chandler (1998) and Siguier et al. (2015)). This classification has often been reinforced by the observation that members of one family appear to share common features in their transposition mechanism. A detailed description of each family can be found at: https://tncentral.ncc.unesp.br/TnPedia/index.php/IS_Families.

2.3.4. *IS-related elements: tIS, MITEs and MICs*

A number of TEs have been identified, which are very closely related to known IS, but carry passenger genes and are simpler than known Tns. They have been called transporter ISs (tISs) (Siguier et al. 2009) (Figure 2.6). They can include passenger genes upstream or downstream of the transposase and can carry a significant amount of DNA with no clear coding capacity. The mechanisms involved in the acquisition of additional genes are at present unclear, although studies on IS*101* from the pSC101 plasmid indicated that transposition can occur using one established IS end and a second naturally occurring surrogate end located at some distance from the IS (Machida et al. 1982). Moreover, isolated individual IS ends are often observed in sequenced genomes and could provide such a source of surrogate ends. A number of potential TEs of this type, containing a single IS, a passenger gene and a short flanking sequence resembling an additional IS end, have been observed in nature (Poirel et al. 2003; Liu et al. 2016; Snesrud et al. 2018).

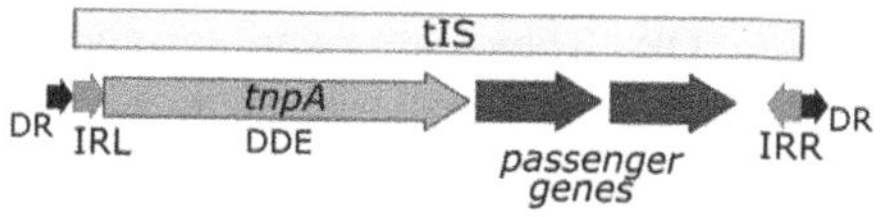

Figure 2.6. *IS and their derivatives: transporter IS (tIS). Direct target repeats (DRs), terminal inverted repeats (IRs), transposases (DDE) and passenger genes are indicated*

In some cases, such as IS*1380* family members (Figure 2.7), the transposase is quite promiscuous and will recognize rather distantly related sequences, which can be used as a second surrogate end. This has proved important in the sequestration and transmission of antibiotic resistance, particularly resistance to beta-lactams (Karim et al. 2001; Poirel et al. 2003; https://tncentral.ncc.unesp.br/TnPedia/index.php/IS_Families/IS1380_family).

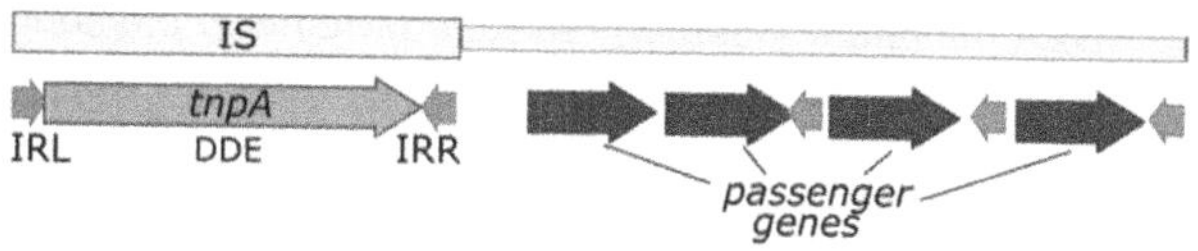

Figure 2.7. *IS and their derivatives: multiple IR (IS1380). Direct target repeats (DRs), terminal inverted repeats (IRs), transposases (DDE) and passenger genes are indicated. Alternative copies of IRR (small triangles) and additional passenger genes are indicated*

Other, non-autonomous IS derivatives called miniature inverted repeat transposable elements (MITEs) (Figure 2.8(A)) were first identified in *Neisseria* (Correia et al. 1988) and are found in many other prokaryotes and eukaryotes (e.g. Bureau and Wessler 1992). They are generally less than 300 bp long, include appropriately oriented left and right terminal IR but no Tpase, and generally generate flanking DR. MITEs are probably derived from IS by internal deletion. Some carry short non-coding sequences between these IRs, which may or may not be IS-derived. MITEs are considered to be non-autonomous TEs mobilizable in trans by Tpases of full-length parental genomic copies. Similar structures related to Tn*3* family transposons, TIMEs and *IS200*/IS*605* family ISs, called PATEs (palindrome-associated transposable elements), have also been identified (Dyall-Smith et al. 2011; Szuplewska et al. 2014). Another group of IS derivatives related to MITEs are called MICs (mobile insertion cassettes) (De Palmenaer et al. 2004) (Figure 2.8(B)). These, like MITEs, are flanked by IR, do not include a Tpase gene and generate flanking DR. They carry various coding sequences and are present in relatively low copy number. They appear to be particularly prevalent in Xanthomonads, where they include TALEs (transcriptional activator like effectors) genes involved in Xanthomonad pathogenicity (Ferreira et al. 2015).

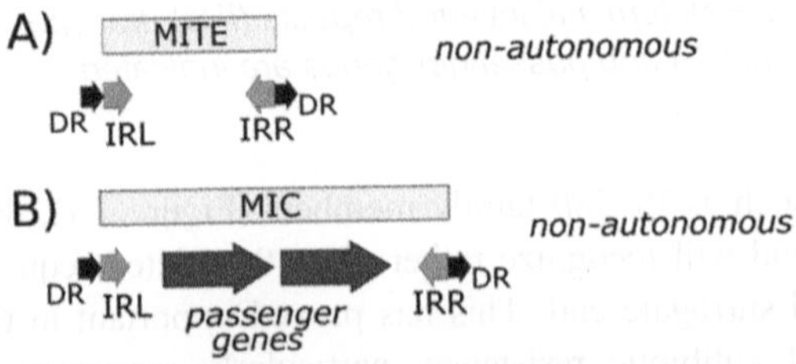

Figure 2.8. *IS and their derivatives. Direct target repeats (DRs), terminal inverted repeats (IRs), transposases (DDE) and passenger genes are indicated. (A) Miniature inverted repeat transposable elements (MITEs) composed of two IRs with no intervening gene. (B) A miniature insertion cassette (MIC) with passenger genes, but without transposition enzymes requiring transposase from an autonomous IS for transposition*

2.4. Transposons (Tn)

Apart from IS-derived compound transposons, there are only a limited number of transposon families based on their distinct sets of transposition-related genes and DNA, features such as the end sequences and recombination sites. Transposons play a pivotal role in the sequestration and transmission of antibiotic resistance genes, resistance to heavy metals, metabolic enzymes and various virulence factors.

2.4.1. *Compound transposons*

Compound transposons were among the first to be identified (e.g. Tn*9*, Tn*10*, Tn*5*) (Berg et al. 1975; Gottesman and Rosner 1975; Kleckner et al. 1978; Figures 2.1 and 2.9) and are relatively simple structures in which two similar IS flanking any DNA segment, either in direct repeat or IR, are capable of mobilizing the entire assembly. If the component IS generates direct target repeats on insertion, transposition of the compound transposon is characterized by a direct repeat flanking the entire transposon. Although each flanking IS can often transpose independently of the compound structure, in a number of these transposon types, one of the IS pairs is inactivated or has reduced activity (Rothstein and Reznikoff 1981;

Halling et al. 1982). This tends to increase the coherence and capacity of the entire structure to transpose.

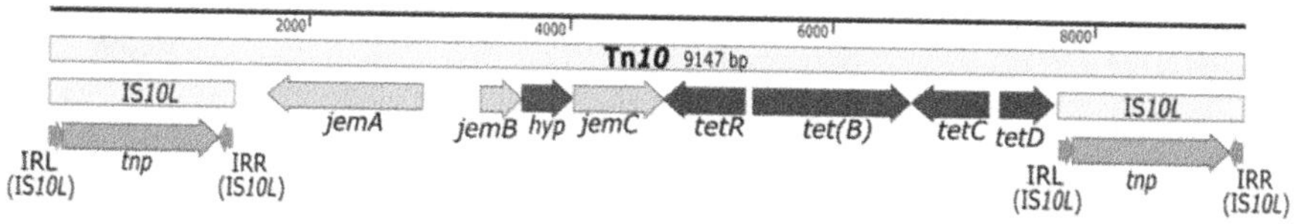

Figure 2.9. *The major transposon families: the compound transposon Tn10. The horizontal black bar shows the length. Transposition and passenger genes are represented by arrows with the direction of expression shown by the arrow heads*

2.4.2. *Pseudocompound transposons, targeted insertion and translocatable units (TU)*

IS*6* also forms part of structures resembling compound transposons. However, early studies on IS*6* family members revealed that they transposed using a cointegrate formation pathway (see Varani et al. (2021), generating replicon fusions with the donor and target replicons, separated at each junction by a directly repeated IS copy, although more recent results, principally with IS*26*, have suggested that they may also be able to transpose using alternative pathways (Harmer et al. 2014, 2020).

This IS family is able to form transposons that resemble compound transposons with the flanking IS in direct repeat, but because of the particular IS6 family transposition mechanism, these were called pseudo-compound transposons (Galas and Chandler 1989; Harmer et al. 2020). This is a direct consequence of the (homologous) recombination event required to resolve the cointegrate structure (Brown et al. 1984; Trieu-Cuot and Courvalin 1985). Transposition is initiated by one of the flanking IS to generate a cointegrate structure with three directly repeated IS copies (Figure 2.10). “Resolution” resulting in transfer of the transposon passenger gene requires recombination between the “new” IS copy and the copy that was not involved in generating the cointegrate. The implications of this model are that the transposon passenger gene(s) are simply transferred from donor to target molecules in the “resolution” event and are therefore lost from the donor “transposon”. Transposons of this type include Tn*610* (flanked by IS*6100*; Martin et al. 1990), Tn*4003* (flanked by IS*257*; Partridge et al. 2018), Tn*2680* (Mollet et al. 1985) and Tn*6023* (flanked by IS*26*; Cain and Hall 2011).

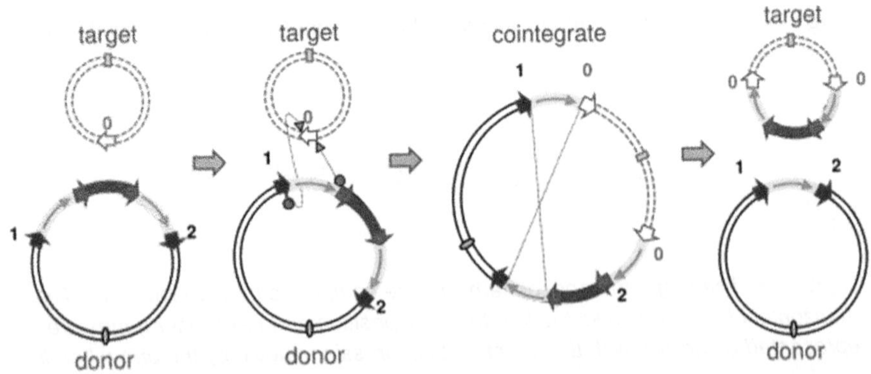

Figure 2.10. *Intermolecular transposition models. Replicative cointegration by an IS6-family pseudo-transposon. The transposon is composed of two directly repeated copies of the IS flanking a DNA segment carrying the passenger gene, with the internal flanks represented by arrows. Transposase-mediated replicon fusion of the donor and target occurs using one of the two flanking IS copies and generates a third IS copy in the same orientation as the original pair. Homologous recombination, using the recA system, between any two copies will either regenerate the donor plasmid, leaving a single IS copy in the target, delete the transposon, or transfer the transposon to the target (as shown), leaving a single copy of the IS in the donor molecule*

However, IS*6* family members appear to possess an alternative type of transposition, which has yet to be understood at the molecular and biochemical levels. This has been called conservative targeted transposition, in which an IS copy located on the donor molecule inserts directly, and in a transposase-dependent manner, next to a resident IS in the target to produce a cointegrate, again with directly repeated IS copies. Moreover, it has been suggested that this is a pathway which facilitates movement of resistance markers in a process that generates double-strand circles composed of a single IS and a resistance gene. These structures have been called translocatable units or TU (Harmer et al. 2014, 2020). Formal biochemical proof for this mechanism has yet to be provided.

2.4.3. *Unit transposons with DDE transposases*

2.4.3.1. *The Tn3 family*

2.4.3.1.1. Overview

Members of the Tn*3* family were among the earliest transposons to be identified (Figures 2.11(A–D)) (Hedges et al. 1974; Hedges and Jacob 1974; Barth et al. 1976).

They form a tightly knit group and transpose using the two-step cointegrate formation pathway involving replicative fusion. The basic Tn*3* family transposition module is composed of transposase (*tnpA*) and resolvase genes and two ends with well conserved IRs of 38-40bp, or sometimes even longer (Ferreira et al. 2015). There is a large (~1,000 aa) DDE transposase, TnpA, which is significantly longer than the DDE transposases normally associated with ISs (see Mahillon and Chandler (1998)).

A second feature of members of this transposon family is that they carry short (~100–150 bp) DNA segments, *res* (for *res*olution) (Figures 2.11(A) and (B)) or *rst* (Yano et al. 2013) (Figure 2.11(C)) and *irs*, (Figure 2.11(D)) at which site-specific recombination between each of the two Tn copies occurs to "resolve" the cointegrate into individual copies of the transposon donor and the target molecules each containing a single transposon copy (see Nicolas et al. (2015)). This highly efficient recombination system is assured by a transposon specified sequence-specific recombinase enzyme: the resolvase.

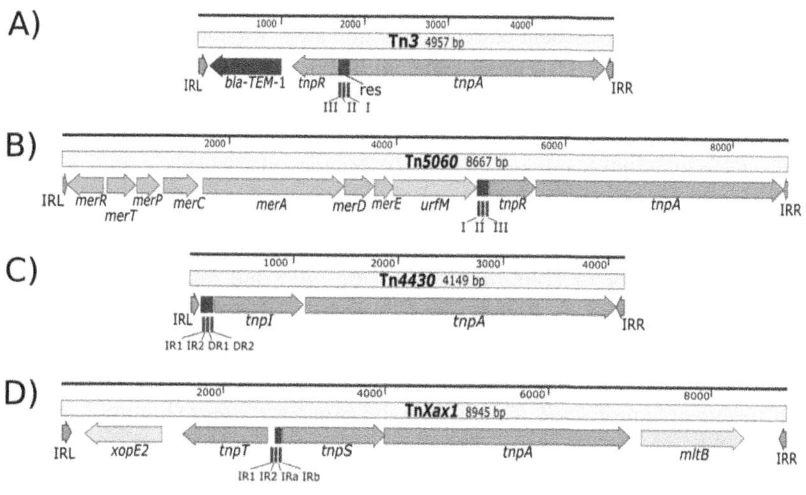

Figure 2.11. *The major transposon families: Tn3. The transposons are not to scale. The horizontal black bars above each show the relative lengths. Transposition and passenger genes are represented by arrows with the direction of expression shown by the arrow heads. (A) A Tn3 family transposon of the Tn3 group, Tn3 with its transposase, tnpA and resolvase, tnpR, genes and the three sites I, II and III comprising the res site. (B) A Tn3 family transposon of the Tn21 group, Tn5060. (C) A Tn3 family transposon with an irs/tnpI, Tn4330, with its tnpA and resolvase tnpI with the four sites IR1, IR2, DR1 and Dr2 comprising the irs site. (D) A Tn3 family transposon with an rst/tnpT/S, TnXax1, with its tnpA and resolvase tnpT/S, and the four sites IR1, IR2, IRa and IRb comprising the rst site*

2.4.3.1.2. Resolvases and res sites

There are three known major resolvase types: TnpR, TnpI and TnpS + TnpT, distinguished by the catalytic nucleophile involved in DNA phosphate bond cleavage and rejoining during recombination: TnpR, a classic serine (S)-site-specific recombinase (e.g. Stark et al. 1989a, 1989b); TnpI, a tyrosine (Y) recombinase similar to phage integrases (Mahillon and Lereclus 1988; Nicolas et al. 2015); and a heteromeric resolvase combining a tyrosine recombinase, TnpS, and a divergently expressed helper protein, TnpT, with no apparent homology to other proteins (Genka et al. 2002; Yano et al. 2013). Each *res* includes a number of short DNA sub-sequences that are recognized and bound by the cognate resolvases. These are different for different resolvase systems (Figures 2.11(A–D)). For *tnpR*-associated *res* sites, three conserved sub-sites (I, II and III) have been identified as TnpR binding sites. The intertwined synaptic complex between the two sites in the cointegrate intermediate and the pathway leading to productive recombination have been determined in detail at the molecular level (see Nicolas et al. (2015))..

The *tnpI*- and *tnpS/T*-associated recombination sites, *irs* and *rst*, respectively, are more complex (Figures 2.11(C) and (D)), with four repeated elements each (*tnpI*: IR1, IR2, DR1, DR2; Baum 1995); (*tnpS/T*: IR1, IR2, IRa, IRB; Genka et al. 2002; Yano et al. 2013). These systems are less well characterized than *res/tnpR* (Baum 1995; Genka et al. 2002; Vanhooff et al. 2010; Yano et al. 2013).

2.4.3.1.3. Overall organization

There are a number of configurations of the *tnpA* and resolvase genes in this family. In those encoding *tnpR*, the gene is generally located upstream of *tnpA* and can be either divergent (Figure 2.11(A)) or colinear (Figures 2.11(B) and (C)). In the former case, the *res* site lies between the divergent *tnpR* and *tnpA* genes and in the latter, it lies upstream of *tnpR*. This is also the case for *irs/tnpI* (Figure 2.11(C)). In both cases, the *res* site abuts the 5′ end of *tnpR*. Where analyzed, *res* sites also include promoters that drive both transposase and resolvase expression. Indeed, TnpR from Tn*3* was originally named for its ability to repress transposase expression by binding to these sites (Gill et al. 1978; Heffron et al. 1979).

For relatives encoding TnpS and TnpT, the corresponding genes are divergent and the *res* (*rst*) site lies between *tnpS* and *tnpT* (Figure 2.11(D)). The *tnpS* and *tnpT* genes can be located either upstream (Figure 2.11(D)) or downstream of *tnpA*.

2.4.3.1.4. Tn3 family groups based on transposase TnpA sequence

A phylogenetic tree defines seven deeply branching clades (Nicolas et al. 2015; Lima-Mendez et al. 2020) named after a representative Tn from each clade: Tn*3*,

Tn*21*, Tn*163*, Tn*1071*, Tn*3000*, Tn*4330* and Tn*4651*. The vast majority encode a *tnpR*/*res* resolution system. However, a significant sub-group of the Tn*4651* clade encodes the *tnpS*/*tnpT*/*rst* system, while the *tnpI*/res is represented in only three cases (in the Tn*3* and Tn*4330* clades).

2.4.3.1.5. Passenger genes

The complexity of these Tn resides in the diversity of other mobile elements incorporated into their structures (such as ISs and integrons), as well as other Tn*3* family members – see Nicolas et al. (2015) – and other passenger genes. The most notorious of these genes are those for antibiotic and heavy metal resistance (Figure 2.11(B)), although other genes involved in organic catabolite degradation and virulence function for both animals and plants (Figure 2.11(D)); they also form part of the Tn*3* family arsenal of passenger genes. One major family clade, Tn*21*, contains the majority of members with genes for mercury resistance, although it is unclear at present how these were acquired.

2.4.3.1.6. Antibiotic resistance

However, the particularly successful Tn*21* clade also carries class 1 integrons and is therefore an important player in the acquisition and transmission of antibiotic resistance. Class 1 integron acquisition appears to have occurred at least twice, both in ancestors carrying the mercury resistance operon. In both cases, the acquired integron platform was transmitted by a relative of the Tn*402* transposon family (see section 2.4.3.3). For the lineage Tn*1696*, insertion occurred into the *res* site, a known preferred IS*402* target, between site I and site II. This would presumably reduce the activity of *res* and have a negative impact on the resolution step, reducing the "fitness" of the host transposon for further transposition. However, Tn*402* insertion can occur elsewhere and for the Tn*21* lineage, insertion occurred into a gene of unknown function located at the 3′ end of the mercury operon (Figure 2.11(A)), splitting the gene. This insertion would not be expected to affect resolution and therefore not incapacitate the host Tn.

While some of these inserted integron platforms have retained a full complement of Tn*402* transposition-related genes (as shown in the hypothetical case, Figures 2.12(B) and 2.17), over time, these have gradually decayed by deletion and insertion of other TE and examples can now be observed with a range of Tn*402* transposition functions (e.g. Figure 2.11(C)).

This means most class 1 integrons are locked into the vector Tn*3* derivative unless they are complemented *in trans* by functional Tn*402* transposition genes (Petrovski and Stanisich 2010), although they can acquire (and lose) integron gene cassettes and have been very successful in the accumulation of multiple antibiotic resistances.

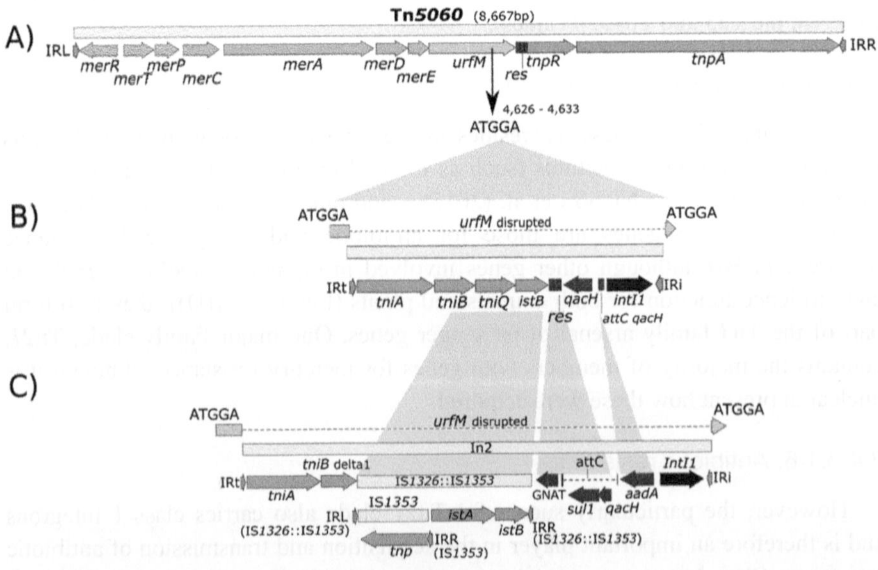

Figure 2.12. *Evolution of the integron-containing Tn21 clade. (A) Ancestor Tn5060 with a mercury resistance operon, the target gene ufrM and the 5bp target site in which (B) a (hypothetical) ancestral Tn402 transposon has been inserted carrying a class 1 integron and generating a 5bp target DR. (C) Accumulation of integron cassettes and decay of the Tn402 transposition functions to generate the present day Tn21. The regions in which renominations occurred are indicated*

2.4.3.1.7. Toxin antitoxin genes and transposition

Interestingly, a number of family members also carry type II toxin-antitoxin gene pairs of different types (Lima-Mendez et al. 2020). Often associated with plasmids, these encode a relatively stable toxin and a labile antitoxin. Cells, which lose the plasmid, no longer produce antitoxin to counteract the stable toxin and die. This sacrifice of plasmid-less cells assures that the plasmid will be maintained in the bacterial population (Hayes and Van Melderen 2011). Unlike other passenger genes in this transposon family, they are located at potentially functionally important positions within the transposition module, invariably close to the *res* site. Although they may have their own promoter, their expression could also be regulated by expression signals involved in TnpA/TnpR expression, ensuring that newly transposed Tn copies are stabilized on insertion and transfer into a naïve host (Lima-Mendez et al. 2020).

2.4.3.1.8. Recombination between res sites: inter-transposon passenger gene exchange

Inter-transposon recombination at *res* sites has been identified in which the associated passenger genes have been exchanged. In view of the recombination properties of these site-specific recombinase systems, it is perhaps not surprising if plasmids carry two related transposon copies in the correct orientation (Kholodii 2001). This was first observed in engineered hybrid plasmids containing copies of Tn*3* and Tn*1000* (Reed 1981), but has been observed in a number of other Tn with TnpR/*res* (Lima-Mendez et al. 2020) and with *rst*/TnpS/T (Yano et al. 2013) systems.

2.4.3.1.9. Immunity

The Tn*3* family exhibits a property called immunity, where a replicon with a resident Tn*3* is "immune" to insertion of a second copy. The mechanism for this in the case of Tn*3* is not understood. It has been reported that a single IR is sufficient to confer transposition immunity and that mutant IR copies, which are non-functional in transposition, are also unable to confer transposition immunity (Huang et al. 1986).

2.4.3.2. *The Tn7 family*

Like Tn*3*, Tn*7* (Figure 2.5(F)) is another example of an early-identified transposon able to transpose trimethoprim and streptomycin resistances (Barth et al. 1976). In contrast to Tn*3*, however, its transposition mechanism and regulation have been worked out in detail at the molecular level (see Peters (2014)).

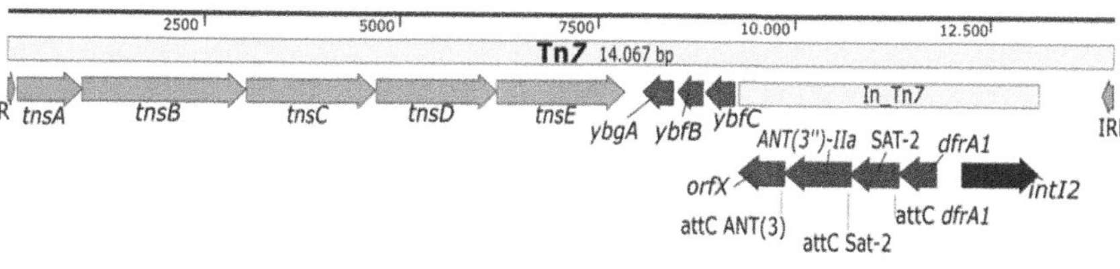

Figure 2.13. *The major transposon families: Tn7. The horizontal black bar above shows the length. Transposition and passenger genes are represented by arrows, with the direction of expression shown by the arrow heads. Tn7 is shown with its five transposition-related genes, tnsA, B, C, D and E, and its class 2 integron In_Tn7, att recombination sites, attC and integrase Int2*

2.4.3.2.1. *The Tn7 transposase and a switch to replicative transposition*

Tn*7* possesses a complex set of transposition proteins (Figures 2.13 and 2.15). The transposase is composed of two proteins: TnsA, which resembles a restriction endonuclease and cleaves the 5′ ("non-transferred") ends, and TnsB, a DDE transposase, which cleaves the 3′ ends. Their concerted action allows Tn*7* to shed its flanking donor DNA (Figure 2.2(Aii)).

It is noteworthy that TnsA active site mutations allow continued cleavage and strand transfer activity of TnpB, but convert Tn*7* transposition from conservative cut-and-paste to replicative (May and Craig 1996).

2.4.3.2.2. Tn7 ends

Tn*7* has 30 bp ends which, like those of Tn*402*, carry multiple transposase (TnpB) binding sites: four closely spaced *on* the left, with the Tpase-proximal ends all oriented in the same inward orientation, and three on the right end, more spaced and also orientated inward (Figure 2.14(A)). These different arrangements imply that TnpB binding creates a different architecture at each end and this must permit a distinction between the ends during transposition, allowing oriented insertion.

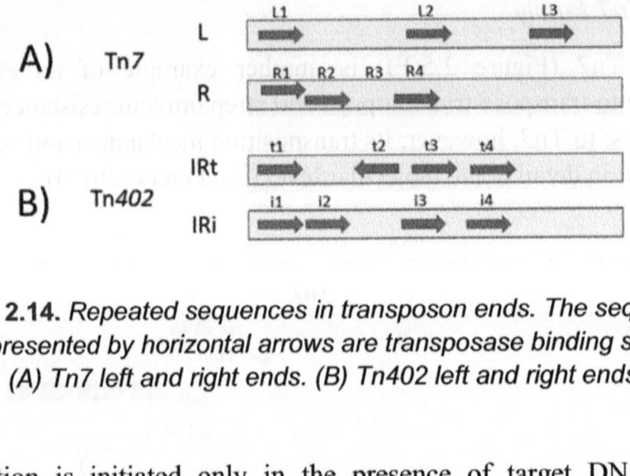

Figure 2.14. *Repeated sequences in transposon ends. The sequences, represented by horizontal arrows are transposase binding sites. (A) Tn7 left and right ends. (B) Tn402 left and right ends*

Transposition is initiated only in the presence of target DNA when TnsB catalyzes target cleavage with a 5 bp spacing on each strand, giving rise to the 5 bp direct target repeats on completion of transposition.

2.4.3.2.3. Tn7 targeting and transposon insertion

In addition to the heteromultimeric transposase, Tn*7* encodes three supplementary components of the transposition apparatus (Figures 2.13 and 2.15): TnpC, a DNA-dependent ATPase, acting as an immunity protein and as an adaptor for interaction between TnsAB, and two alternative targeting proteins, TnsD and TnsE (Figures 2.15 and 2.16).

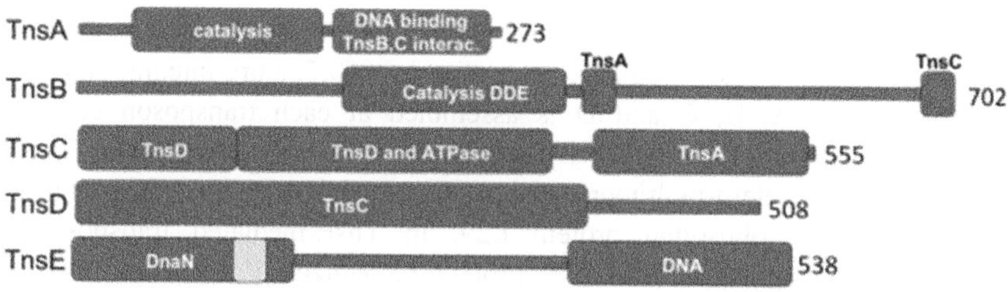

Figure 2.15. *Organization of Tn7 transposition proteins. The relative positions of regions involved in protein–protein interactions, DNA binding and catalytic domains are indicated*

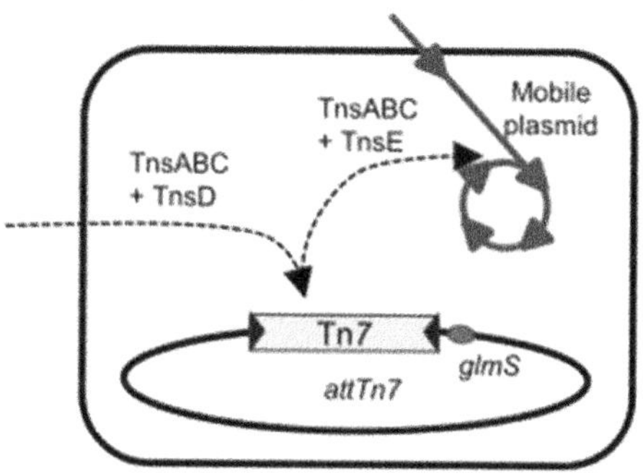

Figure 2.16. *Tn7 transposition pathways. Tn7 infecting a naïve cell using TnsA, TnsB, TnsC and TnsD. TnsD binds to a region in the glmS gene and TnsAB direct cleavage at the attTn7 site outside the gene. Once established, Tn7 then uses TnsA, TnsB, TnsC and TnsE for insertion into replication forks or other structures with 3' recessed DNA ends. It also interacts with the sliding clamp, DnaN, protein*

TnsD recognizes and binds a sequence within the highly conserved *glmS* gene carried by a wide range of bacterial species and directs TnsAB to target a downstream sequence, *attTn7*, thus leaving *glmS* intact (Figures 2.15 and 2.16). A large number of Tn*7* derivatives have been reported. TnsD, a member of a large protein family, the TnsD/TniQ family, has evolved to accommodate a variety of sequence-specific insertion sites (Peters 2019). TnsE is an alternative targeting protein promoting insertion into forked DNA structures such as those in conjugating plasmids. TnsD is involved in Tn*7* acquisition. On entering a cell, it would be directed to a safe haven 25 bp downstream from the *glmS* gene, whereas TnsE would be active in promoting Tn*7* dispersion via conjugation (Figure 2.8(C)).

Complex formation has been probed using alternative divalent metal ions that block assembly. For TnsD-mediated integration at *attTn7*, a pre-integration complex composed of TnsA, B, C and D is assembled at each transposon end. TnsA stimulates its formation and stability. For integration, TnsD recruits about 25 copies of a $TnsA_2TnsC_2$ hetero-multimer, the host factors ACP (acyl carrier protein) and the large subunit ribosomal protein L29. In TnsE-mediated transposition to replication fork lagging strands, the protein recognizes 3′ recessed DNA ends and the host sliding clamp processivity factor, DnaN. It is stimulated by DNA double-strand breaks. The complex interactions of this set of proteins have been probed by mutagenesis and crystallography (Figure 2.15). The catalytic domain of TnsA is located toward the N-terminal end, while the DNA binding and TnsB and TnsC interaction domain is located at the C-terminal end. The TnsB catalytic center (DDE) is located toward the protein centre, and small patches that interact with TnsA and TnsC are located just downstream from the DDE and at the C-terminus, respectively. The ATpase domain of TnsC is also located toward the middle and overlaps with the N-terminal TnsD interacting domain and a domain that interacts with TnsA located at the C-terminal end. TnsD includes a large domain that interacts with TnsC. The organization of TnsE is not well characterized (Figure 2.16). Clearly, Tn*7* transpososomes are complex and the overall dynamics are not fully understood.

2.4.3.2.4. Immunity

Like the Tn*3* family, Tn*7* shows target immunity. This involves TnsA, B and C, and it has been proposed that the relatively high concentration of TnsB bound to the multiple sites in Tn*7* ends prevents the formation of a productive target capture complex including either TnsD or TnsE. TnsB in this role can redistribute TnsC to a plasmid target without Tn*7* ends in vitro. This involves the TnsC ATpase activity since immunity is not observed if a non-hydrolyzable ATP derivative is used.

2.4.3.2.5. Tn7 and CRISPRs

A bioinformatic analysis revealed a very large inventory of Tn*7* derivatives, a number of which carry CRISPR/*cas* systems and these seem to have been recruited more than once (see Peters 2019). Different Tn*7* clades carry one of the two most frequently observed systems, I-F (Klompe et al. 2019) and I-B (Peters et al. 2017). Although these systems are generally involved in an adaptive immune response which recognizes and degrades invading bacteriophages or other mobile elements, those associated with the Tn*7* derivatives appear to be lacking the Cas3 nuclease that destroys the invading DNA after its recognition by the guide RNA. This suggests that they play an alternative role in this context. The observation that they appear in derivatives lacking the targeting *tnsE* gene suggested that they may have replaced this function. This proved to be the case. Recent evidence has shown that the two most common CRISPR/cas systems are used in targeting transposon insertion.

2.4.3.2.6. Passenger genes

The Tn*7* group of transposons carry a large variety of passenger genes including a class 2 integron platform, which is capable of acquiring integron cassettes (Figure 2.13).

2.4.3.3. *The Tn402 family and class 1 integrons*

2.4.3.3.1. Overview

Tn*402* was first identified (Figure 2.17) (Shapiro and Sporn 1977) as an insertion of trimethoprim resistance from plasmid R751 into a bacteriophage lambda derivative phage. This Tn family is intimately linked to class 1 integrons and, indeed, Tn*402*-like transposons have proved instrumental in integron dispersal. This first became apparent when studying integron structures, which include an integron integrase gene, *intI*, and various antibiotic resistance gene cassettes. These structures were found in different sequence environments, suggesting that they are mobile. Since they did not contain typical transposition-related genes or obvious terminal IRs, it was suggested that integrons were unusual mobile elements distinct from typical transposons. However, identification of an integron structure embedded in the Tn*3* family transposon, Tn*21*, and flanked by 25bp IR, in turn flanked by 5bp DR, suggested that the integron was part of a precursor transposon (Brown et al. 1986) whose transposition genes may have decayed following insertion into Tn*21*. Subsequent studies have identified a large number of these derivatives which present an extraordinary range and number of antibiotic resistance genes as cassettes.

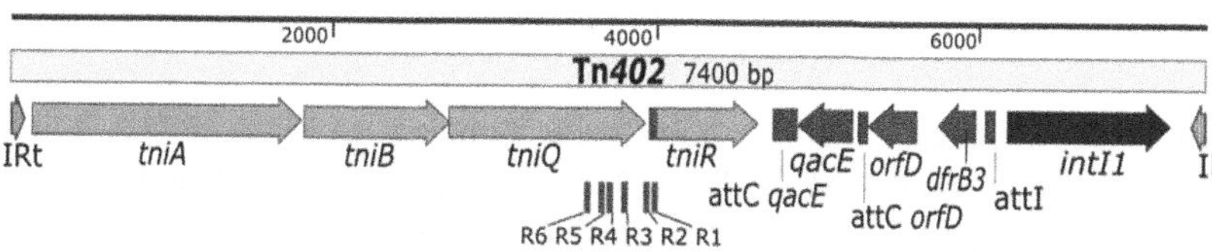

Figure 2.17. *The major transposon families: Tn402. The horizontal black bar above shows the length. Transposition and passenger genes are represented by arrows, with the direction of expression shown by the arrow heads. Tn402 is shown, including five transposition-related genes, tniA, B and Q, its res site, resolvase, att recombination sites, attC and integrase Int1*

There are a number of similarities between the Tn*402* and Tn*7* families (see below).

2.4.3.3.2. Organization

Further analysis showed that in addition to the conserved 25 bp IR, each end also included four repeated sub-sequences, spaced differently at each end, which presumably act as transposase binding sites in a similar way to those of Tn*7* (Figure 2.17). The end proximal to the transposition genes is called IRt and the end located proximal to the integron integrase, IRi. Tn*402* includes four transposition-related genes *tniA*, *tniB*, *tniQ* and *tniR*, together with a *res* site distinct from the Tn*3* family sites, and it is associated with six short sub repeats. Like those of Tn*3*, it abuts *tniR*, a presumed resolvase containing a typical serine recombinase motif.

2.4.3.3.3. Relationship with Tn7 and reaction mechanism

Tn*402* shows many similarities with Tn*7*. TniA and TniB show a low level of identity with Tn*7* TnsB and TnsC proteins, but they do not encode a TnsA equivalent, which, for Tn*7*, catalyzes cleavage of the non-transferred strand, liberating the transposon from its flanking donor sequences (see Peters (2014)). In addition, the *tniQ* gene product also shows some identity with the Tn*7* attTn*7*-targeting protein TnsD and forms part of a large TnsD/TniQ family of proteins (Peters 2019). Another important similarity between the two families is the presence of sub-terminal short repeats at both ends (Figure 2.14B). Both generate 5 bp target repeats on insertion.

TnsA is important for Tn*7* transposition. In its absence, Tn*7* adopts a replicative cointegrate generating pathway (May and Craig 1996). It is therefore significant that, in the absence of a TnsA equivalent to complete excision of the transposon, Tn*402* has acquired a *res* site and site-specific TniR to resolve the cointegrate structures, which are generated in the absence of a dedicated enzyme to catalyze cleavage of the non-transferred strand. Indeed, in the absence of TnpR, Tn*402* derivatives have been shown to generate cointegrate structures (Kholodii et al. 1995).

Tn*402* also encodes TniB that shows some similarity to TnsC of Tn*7*, a protein involved in both interaction of the TnsA/B transposase and the TnsE and D targeting proteins and in the phenomenon of transposition immunity. It is not known whether Tn*402* exhibits immunity.

2.4.3.3.4. Insertion specificity

This transposon group exhibits a very specific target choice. Early studies showed that they insert site-specifically into the partition locus (*par*) involved in plasmid stabilization following replication (see Kamali-Moghaddam and Sundström (2000)). This region is organized similarly to Tn*3* family *res* sites with three binding sites (*resI*, *resII* and *resIII*) for the ParA site-specific recombinase. Insertion requires ParA. Tn*402* family members show a high preference for insertion between *resI* and *resII* in Tn*3* family elements, and decayed Tn*402* derivatives such as the integron platforms In0 and In2 also insert preferentially into this region when supplied with appropriate Tn*402* family transposition proteins *in trans* (Petrovski and Stanisich 2010).

All Tn*402* family members tested appear to target plasmid Tn*3* family *res* sites and require the cognate resolvase protein, but not its catalytic activity (Kamali-Moghaddam and Sundström 2000). The role of TnpR could therefore be structural, providing an architecturally precise nucleoprotein complex for integration and/or furnishing an interface for direct protein–protein interaction with a TniABQ component.

2.4.3.3.5. Tn402 sub-groups

Tn*402* derivatives can be divided into three sub-groups based on the type of passenger genes they carry. The most prominent is that of Tn*402* itself, which carries a class 1 integron platform. A second group, Tn*Pfu1*, with only three examples, carries a class 3 integron platform, while a third group, Tn*5053*, carries a mercury resistance operon similar to those carried by Tn*3* family members. Moreover, the Tn*5053* group includes a sequence that closely resembles a Tn*3* family IR located between the beginning of the *mer* operon and the Tn*5053* IRi sequence, suggesting that the mercury resistance genes may have been acquired from a Tn*3*-like ancestor.

2.4.4. *Transposon using CRISPR components*

2.4.4.1. *Casposons*

In addition to the case of certain Tn*7* family members which use CRISPR components lacking their endonucleolytic functions for targeting, members of a newly recognized TE group, casposons (Krupovic et al. 2014; Hickman and Dyda 2015b; Béguin et al. 2016) (Figure 2.18), include a *cas1* (CRISPR-ASsociated protein 1) derivative encoding the Cas1 endonuclease, a key component of CRISPR elements. Casposons were initially recognized in a bioinformatics study of *cas* genes not associated with CRISPR loci and have been identified in both archaea and bacteria. In a CRISPR context, Cas1 catalyzes integration of new DNA fragments (protospacers) into specific sites between 25 and 50 bp repeats. In a casposon context, Cas1 is thought to catalyze insertion of specific DNA sequences (the casposon ends) into "random" DNA sites, the casposon target sites. This has led to the proposal that CRISPR systems have in part evolved from casposons. Like most classical double-strand DNA transposons, casposons are bordered by IR and flanked by 14 bp TSD, resembling the DR generated by many double-strand DNA transposons. They appear to insert at tRNA genes.

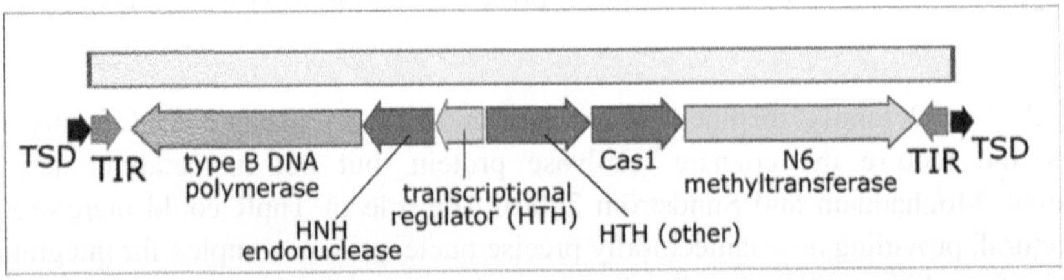

Figure 2.18. *Casposons. Target Site Duplication (TSD) terminal inverted repeats (TIRH). A casposon showing the Cas1 transposase and other genes*

Purified casposon-encoded Cas1 from the archaeon *Aciduliprofundum boonei* catalyzed concerted insertion (or strand transfer) of two precleaved casposon ends into a target plasmid generating 14 bp TSD. Cas1 includes a C-terminal region resembling an HTH domain not present in CRISPR-associated *cas1* genes. Casposon insertion also occurs preferentially into the tRNA-Pro gene, generating a 15 bp DR. This 15 bp and an adjacent 18 bp segment comprising the 3′ end of the tRNA-Pro gene is sufficient for insertion targeting. Satisfyingly, this sequence bears some similarity to the target for protospacer integration by the CRISPR-Cas system.

2.4.5. *Mobile elements which move using site-specific recombinases*

Generally, elements which use site-specific recombinases to move are not considered to be transposons due to their strictly sequence specificity. However, a number presents a less stringent sequence requirement, while others, by their interaction with transposons, are clearly an important part of the "mobilome".

2.4.5.1. *Class 1, 2 and 3 integron platforms*

All integron platforms (including the class 1 carried spread by Tn*402* derivatives) encode tyrosine site-specific integrase. The major difference between class 1, 2 and 3 integrons is the extent of divergence of these integrases.

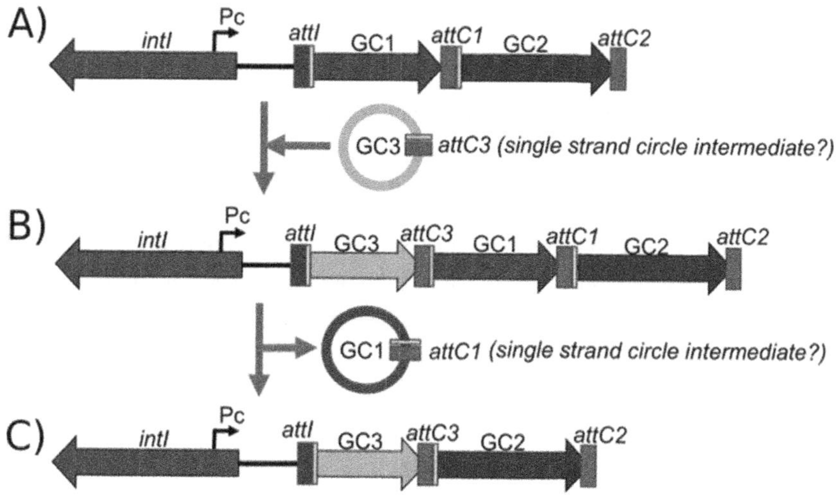

Figure 2.19. *Capture and excision of integron gene cassettes. The integrase gene (Int1) catalyzes integration of integron cassettes (GC) (B) as circular copies into the attI site proximal to a promoter which drives their expression (middle). It also catalyzes cassette excision (C)*

In these structures, generally associated with transposons, the integrase gene is preceded by a site, *attI*, which can acquire a variety of integron cassettes, often carrying single promoter-less genes for antibiotic resistance. The cassettes are acquired by integrase-catalyzed recombination between the integrase *attI* site (Figures 2.12, 2.17 and 2.19) and circular cassette copies carrying a short cassette recombination site, *attC*. The integration reaction involves single-strand regions of the *att* sites, which assume an intricate secondary structure (Escudero et al. 2015). This has been compared with integration of the single strand *Vibrio* phage, CTX (Val et al. 2005). Successive integration at *attI* of a variety of cassettes can generate extensive cassette arrays (Figure 2.19). The cassettes can also undergo excision, presumably in the same way. It is not yet clear whether the recombination intermediates are entirely single or double stranded. A strong promoter within the 5′ end of the Int gene directed in the opposite orientation drives expression of the promoter-less cassettes genes. Expression is more pronounced for the promoter-proximal gene(s) and decreases the further away these become by sequential cassette integration at the *attI* site. The promoter/*att* site region also carries a sequence for binding of the LexA repressor, and this renders expression sensitive to control by the host S.O.S. system.

2.4.5.2. *Unit transposons: the Tn554 family*

2.4.5.2.1. Overview

Tn*554* (Figure 2.20) was originally identified in Staphylococcus aureus as a determinant specifying spectinomycin resistance and inducible erythromycin (macrolide–lincosamide–streptogramin B (MLS); Phillips and Novick 1979). A limited number of Tn*554* family members have now been identified and all are restricted to the Firmicutes (in particular Staphylococci, Enterococci, Listeria and Bacillus), but are widely present in many different strains. They are all organized in a very similar way, with the transposition genes located to the left and a variety of AMR genes located to the right.

Figure 2.20. *The major transposon families. The horizontal black bar above shows the length. Transposition and passenger genes are represented by arrows, with the direction of expression shown by the arrow heads. Tn554 is shown with its three transposition-related genes, tnpA, B and C, and its att integration site, att*

2.4.5.2.2. Organization and transposition

Tn*554* transposons encode three transposition-related genes, *tnpA*, *B* and *C*, directly implicated in transposition (Bastos and Murphy 1988). None are DDE-type transposases. TnpA and TnpB are both tyrosine recombinase/integrase family members, whereas TnpC assists in integration (Bastos and Murphy 1988). Integration therefore presumably occurs using a mechanism related to site-specific recombination. Indeed, integration is sequence specific and generally occurs in a unique orientation at the 3′ side of a site, 5′ GATGTA 3′, in the *radC* gene (Murphy and Löfdahl 1984). In *tnpC* mutants, transposition is reduced to 2%, but although insertion remains sequence specific, it occurs in either Tn orientation. The *radC* gene is common to most Firmicutes and various family members occur at the same position in various firmicute species. In most studies, insertions include flanking copies of the target sequence, which resemble DR, and one of these is often shown as part of the transposon, with the observation that Tn*554* family members do not have IR. However, alignment of transposon ends from which this hexanucleotide is removed indicates a conserved imperfect repeat common to all members of the family:

5' **TTA** ACCAT **AGTA** -*Tn*- **TACT** CTTGA **TAA** 3'

Moreover, free circular copies of various family members have been observed, in which the two ends are abutted but separated by a copy of the 6 bp "att" sequence (e.g. Kehrenberg and Schwarz 2005).

–**TACT** CTTGA **TAA** -XXXXXX- **TTA** ACCAT **AGTA**–

This suggests that the circles are transposition intermediates that integrate in a second step in a similar fashion to certain bacteriophage.

Further details can be found at: https://tncentral.ncc.unesp.br/TnPedia/index.php/Transposons_families/Tn554_family.

2.4.5.3. *Integrative conjugative elements (ICE)*

Early examples of ICEs were initially thought to be resistance plasmids and were assigned an incompatibility group, incJ. These are maintained as an integrated copy in the host chromosome, but can nevertheless give rise to circular copies. ICEs were first clearly identified in Gram-positive bacteria as elements capable of transmitting tetracycline resistance. They use a site-specific recombination reaction for integration and excision (Figures 2.21(A) and (B)). The enzymes are generally tyrosine recombinases (Figure 2.21(A)), similar to those used by phage or the integrons. Although phage and integrons show pronounced sequence specificity for integration, ICEs are less exacting in their choice of integration sites. In addition to the recombinase and antibiotic resistance genes, they encode functions for conjugative transfer. Tn*916* and Tn*1545* are the best-characterized Tns of this type.

Several examples of ICEs encoding serine-transposases are also known (Figure 2.21(B)), while others encode with typical DDE transposases (see Siguier et al. (2014)) related to known IS and have IS-associated IR (Figure 2.21(C)).

Naturally occurring transfer-deficient ICE derivatives have also been identified and transfer can be complemented in trans by active ICE.

ICEs undergo excision from the donor chromosome as a double-strand DNA circle. This is processed at a transfer origin, transferred by rolling circle replication into a recipient cell, as are conjugative plasmids, where they are recircularized and integrated. Indeed, to ensure sufficient stability of the transposition intermediates to enable their subsequent integration, some ICE encode plasmid-related replication genes.

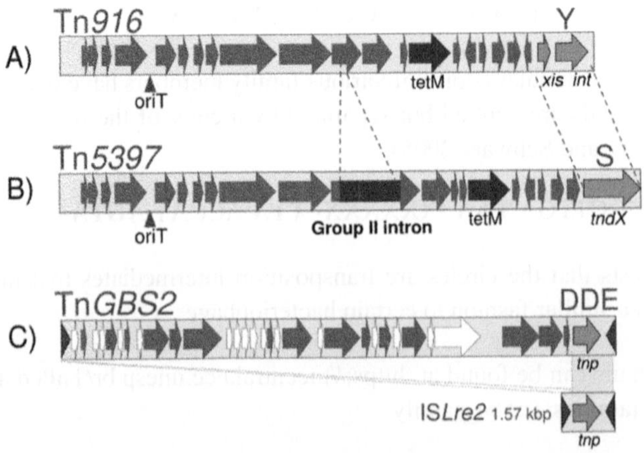

Figure 2.21. *Integrative conjugative elements (ICE). (A) Tn916 encodes a tyrosine recombinase, Y (Int) and an excisase (xis) together with the tetM resistance gene, various genes involved in conjugative transfer and a site which acts as an origin of transfer, oriT. (B) Tn5397 encodes a serine, tndX recombinase, S (Int) and a group II intron. It shares significant homology with Tn916. (C) TnGBS2 encodes a DDE enzyme similar to that of the insertion sequence ISLre2 with similar IR*

This is yet another example of the increasingly indistinct frontiers between phages, plasmids and transposons.

2.4.5.4. *Potential Xer mobility units: pdif*

Sequence analysis of a number of plasmids from clinically important multiantibiotic resistant *Acinetobacter baumannii* revealed a number of genes flanked by *xer*-like (plasmid-*dif*, or p*dif*) recombination sites (D'Andrea et al. 2009). Since the interstitial gene in each plasmid varied, it was suggested that they could be exchanged by site-specific recombination at these sites (D'Andrea et al. 2009; Balalovski and Grainge 2020). Dif sites were initially observed in bacterial chromosomes and plasmids as sites which undergo efficient recombination using a pair of tyrosine site-specific recombinases, XerC and XerD. These were implicated in the resolution of chromosome and plasmid dimers generated during replication and the recombination reactions have been well characterized. Dif sites were also implicated in the integration of multiple antibiotic resistance plasmids into the *Escherichia coli* chromosome (Clerget 1991) and subsequently shown to promote integration of the cholera toxin-carrying CTX bacteriophage of *Vibrio cholerae* (Val et al. 2005).

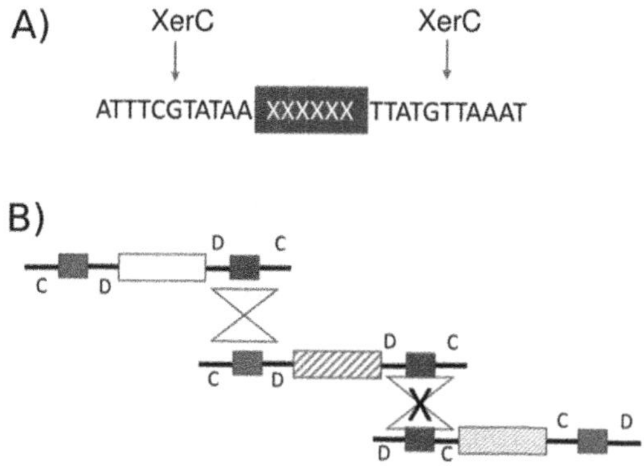

Figure 2.22. pdif. *The figure shows (A) a typical recombination site composed of flanking sequences, which are bound by the two recombinases XerC and D, and a hexanucleotide spacer region of more variable composition. (B) Alignment and recombination reaction allowing exchange and accumulation of pdif cassettes*

In the basic p*dif* module, the flanking p*dif* sites are inverted with respect to each other. The 28bp *dif* sites are composed of two 11bp flanks separated by a 6 bp spacer, which is generally different in the flanking p*dif* sequences (Figure 2.22(A)) (Balalovski and Grainge 2020). Recombination between these sites can result in acquisition (or loss) of p*dif* units (Figure 2.22(B)) and rearrangements between them. The associated genes are not limited to antibiotic resistance. Other "passenger" genes include those for chromate and tellurite resistance, as well as toxin-antitoxin (TA) systems, organic hydroperoxide resistance, a sulfate permease and a potassium uptake transporter. They have been found in a number of *Acenitobacter* species isolated from permafrost in Russia and can be present in multiple copies (Mindlin et al. 2019). Little is known at present about the mobility of these p*dif* elements.

2.4.5.5. *Genomic islands*

The first genomic islands identified were called "pathogenicity islands" since their DNA composition was different from the surrounding host chromosome and they encoded a number of virulence factors (Hacker et al. 1997) (Figure 2.23). This definition has now been expanded to regions carrying diverse functions (metabolic, resistance islands, virulence, symbiosis islands, etc.), depending on their gene content, and are now grouped under the general name: genomic islands (GI). They are rather long (from 10 up to 200 kb) and show an enormous range of diversity in their gene content, but have one common feature, the presence of a recombination module which consists of a tyrosine site-specific integrase and flanking recombination (*att* sites), *attL* and *attR*. Moreover, they invariably integrate in a site-specific manner into tRNA genes, in the same way as some bacteriophages. The integrase appears to be phylogenetically distinct from other integrases, including those associated with phages, ICE and integrons (Boyd et al. 2009). Excised circular copies have been identified (e.g. Murphy and Boyd 2008) and it is thought that they insert using a Campbell-type mechanism, in the same way as the lambdoid phages. They are often identified by their atypical GC content compared to the surrounding host DNA.

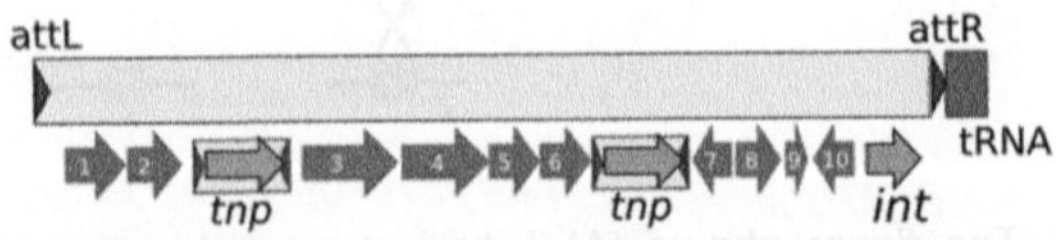

Figure 2.23. *Genomic islands. Passenger genes are shown labeled 1-10, together with the integrase gene and the flanking target tRNA gene. Insertion sequences are often present within these complex islands*

2.5. Conclusion

We have provided an overview of the diverse types of TE that have been identified in prokaryotes. These all move using a DNA intermediate. A large class of these TEs, those encoding a DDE transposase, use identical chemistry, whether they are present in prokaryotes or eukaryotes, but there are a number of other processing enzymes that catalyze DNA transactions, resulting in movement of DNA segments. Perhaps of these, the most unexpected are the CRISPR-associated proteins acquired by both Tn*7* and Casposon elements, used for targeting in the former case and for insertion and targeting in the second.

The dividing line between what is classically considered a TE and the more site-specific phage-related type of insertion is relatively fuzzy. There is a wide spectrum of TE behavior. Some TEs with DDE transposases show highly sequence specific insertion thanks to dedicated targeting proteins (e.g. Tn*7* and Tn*402*), while some elements which use phage-related site-specific recombinases are less stringent in their target choice (e.g. ICE). We have tried to underline how these TE have interacted extensively and become intertwined with different elements inserted into others.

2.6. References

Antonov, I., Baranov, P., Borodovsky, M. (2013). GeneTack database: Genes with frameshifts in prokaryotic genomes and eukaryotic mRNA sequences. *Nucleic Acids Research*, 41(Database issue), D152–D156.

Arber, W., Iida, S., Jütte, H., Caspers, P., Meyer, J., Hänni, C. (1979). Rearrangements of genetic material in *Escherichia coli* as observed on the bacteriophage P1 plasmid. *Cold Spring Harbor Symposia on Quantitative Biology*, 43(Pt 2), 1197–1208.

Balalovski, P. and Grainge, I. (2020). Mobilization of pdif modules in Acinetobacter: A novel mechanism for antibiotic resistance gene shuffling? *Molecular Microbiology*, 114(5), 699–709.

Barth, P.T., Datta, N., Hedges, R.W., Grinter, N.J. (1976). Transposition of a deoxyribonucleic acid sequence encoding trimethoprim and streptomycin resistances from R483 to other replicons. *Journal of Bacteriology*, 125(3), 800–810.

Bastos, M.C. and Murphy, E. (1988). Transposon Tn554 encodes three products required for transposition. *The EMBO Journal*, 7(9), 2935–2941.

Baum, J.A. (1995). TnpI recombinase: Identification of sites within Tn5401 required for TnpI binding and site-specific recombination. *J Bacteriol.*, 177, 4036–4042.

Béguin, P., Charpin, N., Koonin, E.V., Forterre, P., Krupovic, M. (2016). Casposon integration shows strong target site preference and recapitulates protospacer integration by CRISPR-Cas systems. *Nucleic Acids Research*, 44(21), 10367–10376.

Berg, D.E., Davies, J., Allet, B., Rochaix, J.D. (1975). Transposition of R factor genes to bacteriophage lambda. *Proceedings of the National Academy of Sciences of the United States of America*, 72(9), 3628–3632.

Bhasin, A., Goryshin, I.Y., Reznikoff, W.S. (1999). Hairpin formation in Tn5 transposition. *The Journal of Biological Chemistry*, 274(52), 37021–37029.

Boyd, E.F., Almagro-Moreno, S., Parent, M.A. (2009). Genomic islands are dynamic, ancient integrative elements in bacterial evolution. *Trends in Microbiology*, 17(2), 47–53.

Brown, A.M., Coupland, G.M., Willetts, N.S. (1984). Characterization of IS46, an insertion sequence found on two IncN plasmids. *Journal of Bacteriology*, 159(2), 472–481.

Brown, N.L., Misra, T.K., Winnie, J.N., Schmidt, A., Seiff, M., Silver, S. (1986). The nucleotide sequence of the mercuric resistance operons of plasmid R100 and transposon Tn501: Further evidence for mer genes which enhance the activity of the mercuric ion detoxification system. *Molecular & General Genetics: MGG*, 202, 143–151.

Bureau, T. and Wessler, S. (1992). Tourist: A large family of small inverted repeat. *The Plant Cell*, 4, 1283–1294.

Cain, A.K. and Hall, R.M. (2011). Transposon Tn5393e carrying the aphA1-containing transposon Tn6023 upstream of strAB does not confer resistance to streptomycin. *Microbial Drug Resistance*, 17(3), 389–394.

Chandler, M. and Fayet, O. (1993). Translational frameshifting in the control of transposition in bacteria. *Molecular Microbiology*, 7(4), 497–503.

Chandler, M., de la Cruz, F., Dyda, F., Hickman, A.B., Moncalian, G., Ton-Hoang, B. (2013). Breaking and joining single-stranded DNA: The HUH endonuclease superfamily. *Nature Reviews Microbiology*, 11(8), 525–538.

Chandler, M., Fayet, O., Rousseau, P., Ton Hoang, B., Duval-Valentin, G. (2015). Copy-out-paste-in transposition of IS911: A major transposition pathway. *Microbiology Spectrum*, 3(4).

Clerget, M. (1991). Site-specific recombination promoted by a short DNA segment of plasmid R1 and by a homologous segment in the terminus region of the *Escherichia coli* chromosome. *The New Biologist*, 3(8), 780–788.

Correia, F.F., Inouye, S., Inouye, M. (1988). A family of small repeated elements with some transposon-like properties in the genome of *Neisseria gonorrhoeae*. *The Journal of Biological Chemistry*, 263(25), 12194–12198.

D'Andrea, M.M., Giani, T., D'Arezzo, S., Capone, A., Petrosillo, N., Visca, P., Luzzaro, F., Rossolini, G.M. (2009). Characterization of pABVA01, a plasmid encoding the OXA-24 carbapenemase from Italian isolates of *Acinetobacter baumannii*. *Antimicrobial Agents and Chemotherapy*, 53(8), 3528–3533.

De Palmenaer, D., Vermeiren, C., Mahillon, J. (2004). IS231-MIC231 elements from *Bacillus cereus sensu lato* are modular. *Molecular Microbiology*, 53(2), 457–467.

Del Pilar Garcillán-Barcia, M., Bernales, I., Mendiola, M.V., de la Cruz, F. (2001). Single-stranded DNA intermediates in IS91 rolling-circle transposition. *Molecular Microbiology*, 39(2), 494–501.

Dyall-Smith, M.L., Pfeiffer, F., Klee, K., Palm, P., Gross, K., Schuster, S.C., Rampp, M., Oesterhelt, D. (2011). *Haloquadratum walsbyi*: Limited diversity in a global pond. *PLoS One*, 6(6), e20968.

Dyda, F., Chandler, M., Hickman, A.B. (2012). The emerging diversity of transpososome architectures. *Quarterly Reviews of Biophysics*, 45(4), 493–521.

Escudero, J.A., Loot, C., Nivina, A., Mazel, D. (2015). The integron: Adaptation on demand. *Microbiology Spectrum*, 3(2), MDNA3–0019.

Fayet, O., Ramond, P., Polard, P., Prère, M.F., Chandler, M. (1990). Functional similarities between retroviruses and the IS3 family of bacterial insertion sequences? *Molecular Microbiology*, 4(10), 1771–1777.

Ferreira, R.M., de Oliveira, A.C.P., Moreira, L.M., Belasque, J., Gourbeyre, E., Siguier, P., Ferro, M.I.T., Ferro, J.A., Chandler, M., Varani, A.M. (2015). A TALE of transposition: Tn3-like transposons play a major role in the spread of pathogenicity determinants of *Xanthomonas citri* and other xanthomonads. *mBio*, 6(1), e02505–e02514.

Fiandt, M., Szybalski, W., Malamy, M.H. (1972). Polar mutations in lac, gal and phage lambda consist of a few IS-DNA sequences inserted with either orientation. *Mol. Gen. Genet.*, 119, 223–231.

Galas, D.J. and Chandler, M. (1989). Bacterial insertion sequences. In *Mobile DNA*, Berg, D.E. and Howe, M.M. (eds). American Society for Microbiology Publications, Washington, DC, 109–162.

Genka, H., Nagata, Y., Tsuda, M. (2002). Site-specific recombination system encoded by toluene catabolic transposon Tn4651. *Journal of Bacteriology*, 184(17), 4757–4766.

Gill, R., Heffron, F., Dougan, G., Falkow, S. (1978). Analysis of sequences transposed by complementation of two classes of transposition-deficient mutants of Tn3. *Journal of Bacteriology*, 136(2), 742–756.

Gottesman, M.M. and Rosner, J.L. (1975). Acquisition of a determinant for chloramphenicol resistance by coliphage lambda. *Proceedings of the National Academy of Sciences of the United States of America*, 72(12), 5041–5045.

Hacker, J., Blum-Oehler, G., Mühldorfer, I., Tschäpe, H. (1997). Pathogenicity islands of virulent bacteria: Structure, function and impact on microbial evolution. *Molecular Microbiology*, 23(6), 1089–1097.

Halling, S.M., Simons, R.W., Way, J.C., Walsh, R.B., Kleckner, N. (1982). DNA sequence organization of IS10-right of Tn10 and comparison with IS10-left. *Proceedings of the National Academy of Sciences of the United States of America*, 79(8), 2608–2612.

Harmer, C.J., Moran, R.A., Hall, R.M. (2014). Movement of IS26-associated antibiotic resistance genes occurs via a translocatable unit that includes a single IS26 and preferentially inserts adjacent to another IS26. *mBio*, 5(5), e01801–e01814.

Harmer, C.J., Pong, C.H., Hall, R.M. (2020). Structures bounded by directly-oriented members of the IS26 family are pseudo-compound transposons. *Plasmid*, 111, 102530.

Hayes, F. and Van Melderen, L. (2011). Toxins-antitoxins: Diversity, evolution and function. *Critical Reviews in Biochemistry and Molecular Biology*, 46(5), 386–408.

He, S., Corneloup, A., Guynet, C., Lavatine, L., Caumont-Sarcos, A., Siguier, P., Marty, B., Dyda, F., Chandler, M., Ton Hoang, B. (2015). The IS200/IS605 family and "peel and paste" single-strand transposition mechanism. *Microbiology Spectrum*, 3(4).

Hedges, R.W. and Jacob, A.E. (1974). Transposition of ampicillin resistance from RP4 to other replicons. *Molecular & General Genetics: MGG*, 132(1), 31–40.

Hedges, R.W., Jacob, A.E., Smith, J.T. (1974). Properties of an R factor from *Bordetella bronchiseptica*. *Journal of General Microbiology*, 84(1), 199–204.

Heffron, F., McCarthy, B.J., Ohtsubo, H., Ohtsubo, E. (1979). DNA sequence analysis of the transposon Tn3: Three genes and three sites involved in transposition of Tn3. *Cell*, 18(4), 1153–1163.

Hickman, A.B. and Dyda, F. (2015a). Mechanisms of DNA transposition. *Microbiology Spectrum*, 3(2), MDNA3–0034.

Hickman, A.B. and Dyda, F. (2015b). The casposon-encoded Cas1 protein from *Aciduliprofundum boonei* is a DNA integrase that generates target site duplications. *Nucleic Acids Research*, 43(22), 10576–10587.

Hirsch, H.J., Starlinger, P., Brachet, P. (1972). Two kinds of insertions in bacterial genes. *Mol. Gen. Genet.*, 119, 191–206.

Hu, S., Ohtsubo, E., Davidson, N. (1975a). Electron microscopic heteroduplex studies of sequence relations among plasmids of *Escherichia coli*: Structure of F13 and related F-primes. *Journal of Bacteriology*, 122(2), 749–763.

Hu, S., Otsubo, E., Davidson, N., Saedler, H. (1975b). Electron microscope heteroduplex studies of sequence relations among bacterial plasmids: Identification and mapping of the insertion sequences IS1 and IS2 in F and R plasmids. *Journal of Bacteriology*, 122(2), 764–775.

Hu, S., Ptashne, K., Cohen, S.N., Davidson, N. (1975c). Alphabeta sequence of F is IS31. *Journal of Bacteriology*, 123(2), 687–692.

Huang, C.J., Heffron, F., Twu, J.S., Schloemer, R.H., Lee, C.H. (1986). Analysis of Tn3 sequences required for transposition and immunity. *Gene*, 41(1), 23–31.

Huber, M., Faure, G., Laass, S., Kolbe, E., Seitz, K., Wehrheim, C., Wolf, Y.I., Koonin, E.V., Soppa, J. (2019). Translational coupling via termination-reinitiation in archaea and bacteria. *Nature Communications*, 10(1), 4006.

Jordan, E., Saedler, H., Starlinger, P. (1968). O0 and strong-polar mutations in the gal operon are insertions. *Molecular & General Genetics: MGG*, 102(4), 353–363.

Kamali-Moghaddam, M. and Sundström, L. (2000). Transposon targeting determined by resolvase. *FEMS Microbiology Letters*, 186(1), 55–59.

Karim, A., Poirel, L., Nagarajan, S., Nordmann, P. (2001). Plasmid-mediated extended-spectrum beta-lactamase (CTX-M-3 like) from India and gene association with insertion sequence ISEcp1. *FEMS Microbiology Letters*, 201(2), 237–241.

Kehrenberg, C. and Schwarz, S. (2005). Florfenicol-chloramphenicol exporter gene fexA is part of the novel transposon Tn558. *Antimicrobial Agents and Chemotherapy*, 49(2), 813–815.

Kennedy, A.K., Guhathakurta, A., Kleckner, N., Haniford, D.B. (1998). Tn10 transposition via a DNA hairpin intermediate. *Cell*, 95(1), 125–134.

Khan, E., Mack, J.P., Katz, R.A., Kulkosky, J., Skalka, A.M. (1991). Retroviral integrase domains: DNA binding and the recognition of LTR sequences. *Nucleic Acids Res.*, 19, 851–860.

Kholodii, G.Y. (2001). The shuffling function of resolvases. *Gene*, 269(1–2), 121–130.

Kholodii, G.Y., Mindlin, S.Z., Bass, I.A., Yurieva, O.V., Minakhina, S.V., Nikiforov, V.G. (1995). Four genes, two ends, and a res region are involved in transposition of Tn5053: A paradigm for a novel family of transposons carrying either a mer operon or an integron. *Molecular Microbiology*, 17(6), 1189–1200.

Kleckner, N., Barker, D.F., Ross, D.G., Botstein, D. (1978). Properties of the translocatable tetracycline-resistance element Tn10 in *Escherichia coli* and bacteriophage lambda. *Genetics*, 90(3), 427–461.

Klompe, S.E., Vo, P.L.H., Halpin-Healy, T.S., Sternberg, S.H. (2019). Transposon-encoded CRISPR-Cas systems direct RNA-guided DNA integration. *Nature*, 571(7764), 219–225.

Kosek, D., Hickman, A.B., Ghirlando, R., He, S., Dyda, F. (2021). Structures of ISCth4 transpososomes reveal the role of asymmetry in copy-out/paste-in DNA transposition. *The EMBO Journal*, 40(1), e105666.

Krupovic, M., Makarova, K.S., Forterre, P., Prangishvili, D., Koonin, E.V. (2014). Casposons: A new superfamily of self-synthesizing DNA transposons at the origin of prokaryotic CRISPR-Cas immunity. *BMC Biology*, 12, 36.

Lima-Mendez, G., Oliveira Alvarenga, D., Ross, K., Hallet, B., Van Melderen, L., Varani, A.M., Chandler, M. (2020). Toxin-antitoxin gene pairs found in Tn3 family transposons appear to be an integral part of the transposition module. *mBio*, 11(2), e00452-20.

Liu, Y.-Y., Wang, Y., Walsh, T.R., Yi, L.-X., Zhang, R., Spencer, J., Doi, Y., Tian, G., Dong, B., Huang, X. et al. (2016). Emergence of plasmid-mediated colistin resistance mechanism MCR-1 in animals and human beings in China: A microbiological and molecular biological study. *The Lancet Infectious Diseases*, 16(2), 161–168.

Machida, Y., Machida, C., Ohtsubo, E. (1982). A novel type of transposon generated by insertion element IS102 present in a pSC101 derivative. *Cell*, 30(1), 29–36.

Mahillon, J. and Chandler, M. (1998). Insertion sequences. *Microbiology and Molecular Biology Reviews*, 62(3), 725–774.

Mahillon, J. and Lereclus, D. (1988). Structural and functional analysis of Tn4430: Identification of an integrase-like protein involved in the co-integrate-resolution process. *The EMBO Journal*, 7(5), 1515–1526.

Malamy, M.H., Fiandt, M., Szybalski, W. (1972). Electron microscopy of polar insertions in the lac operon of *Escherichia coli*. *Mol Gen Genet.*, 119, 207–222.

Martin, C., Timm, J., Rauzier, J., Gomez-Lus, R., Davies, J., Gicquel, B. (1990). Transposition of an antibiotic resistance element in mycobacteria. *Nature*, 345(6277), 739–743.

May, E.W. and Craig, N.L. (1996). Switching from cut-and-paste to replicative Tn7 transposition. *Science*, 272(5260), 401–404.

McClintock, B. (1950). The origin and behavior of mutable loci in maize. *Proceedings of the National Academy of Sciences of the United States of America*, 36(6), 344–355.

McClintock, B. (1956). Controlling elements and the gene. *Cold Spring Harbor Symposia on Quantitative Biology*, 21, 197–216.

Mindlin, S., Beletsky, A., Mardanov, A., Petrova, M. (2019). Adaptive dif modules in permafrost strains of *Acinetobacter lwoffii* and their distribution and abundance among present day Acinetobacter strains. *Frontiers in Microbiology*, 10, 632.

Mizuuchi, K. and Baker, T.A. (2002). Chemical mechanisms for mobilizing DNA. In *Mobile DNA*, Craig, N.L., Craigie, R., Gellert, M., Lambowitz, A. (eds). American Society for Microbiology Publications, Washington, DC.

Mollet, B., Clerget, M., Meyer, J., Iida, S. (1985). Organization of the Tn6-related kanamycin resistance transposon Tn2680 carrying two copies of IS26 and an IS903 variant, IS903. B. *Journal of Bacteriology*, 163(1), 55–60.

Murphy, R.A. and Boyd, E.F. (2008). Three pathogenicity islands of *Vibrio cholerae* can excise from the chromosome and form circular intermediates. *Journal of Bacteriology*, 190(2), 636–647.

Murphy, E. and Löfdahl, S. (1984). Transposition of Tn554 does not generate a target duplication. *Nature*, 307(5948), 292–294.

Nagy, Z. and Chandler, M. (2004). Regulation of transposition in bacteria. *Research in Microbiology*, 155(5), 387–398.

Nicolas, E., Lambin, M., Dandoy, D., Galloy, C., Nguyen, N., Oger, C.A., Hallet, B. (2015). The Tn3-family of replicative transposons. *Microbiology Spectrum*, 3(4).

Partridge, S.R., Kwong, S.M., Firth, N., Jensen, S.O. (2018). Mobile genetic elements associated with antimicrobial resistance. *Clinical Microbiology Reviews*, 31(4).

Peters, J.E. (2014). Tn7. *Microbiology Spectrum*, 2(5).

Peters, J.E. (2019). Targeted transposition with Tn7 elements: Safe sites, mobile plasmids, CRISPR/Cas and beyond. *Molecular Microbiology*, 112(6), 1635–1644.

Peters, J.E., Makarova, K.S., Shmakov, S., Koonin, E.V. (2017). Recruitment of CRISPR-Cas systems by Tn7-like transposons. *Proceedings of the National Academy of Sciences of the United States of America*, 114(35), E7358–E7366.

Petrovski, S. and Stanisich, V.A. (2010). Tn502 and Tn512 are res site hunters that provide evidence of resolvase-independent transposition to random sites. *Journal of Bacteriology*, 192(7), 1865–1874.

Phillips, S. and Novick, R. (1979). Repressor-controlled transposon in *Staphylococcus aureus*. *Nature*, 278, 476–478.

Poirel, L., Decousser, J.-W., Nordmann, P. (2003). Insertion sequence ISEcp1B is involved in expression and mobilization of a bla(CTX-M) beta-lactamase gene. *Antimicrobial Agents and Chemotherapy*, 47(9), 2938–2945.

Reed, R.R. (1981). Resolution of cointegrates between transposons gamma delta and Tn3 defines the recombination site. *Proceedings of the National Academy of Sciences of the United States of America*, 78(6), 3428–3432.

Reif, H.J. and Saedler, H. (1974). IS1 is involved in deletion formation in the gal region of *E. coli* K12. *Mol. Gen. Genet.*, 137, 17–28.

Rezsöhazy, R., Hallet, B., Delcour, J., Mahillon, J. (1993). The IS4 family of insertion sequences: Evidence for a conserved transposase motif. *Molecular Microbiology*, 9(6), 1283–1295.

Rice, P.A. (2015). Serine resolvases. *Microbiology Spectrum*, 3(2), MDNA3–0045.

Rothstein, S.J. and Reznikoff, W.S. (1981). The functional differences in the inverted repeats of Tn5 are caused by a single base pair nonhomology. *Cell*, 23(1), 191–199.

Saedler, H. and Heiss, B. (1973). Multiple copies of the insertion-DNA sequences IS1 and IS2 in the chromosome of *E. coli* K-12. *Mol. Gen. Genet.*, 122, 267–277.

Saedler, H., Reif, H.J., Hu, S., Davidson, N. (1974). IS2, a genetic element for turn-off and turn-on of gene activity in *E. coli*. *Molecular & General Genetics: MGG*, 132(4), 265–289.

Shapiro, J.A. (1979). Molecular model for the transposition and replication of bacteriophage Mu and other transposable elements. *Proceedings of the National Academy of Sciences of the United States of America*, 76(4), 1933–1937.

Shapiro, J.A. and Sporn, P. (1977). Tn402: A new transposable element determining trimethoprim resistance that inserts in bacteriophage lambda. *Journal of Bacteriology*, 129, 1632–1635.

Sharma, V., Firth, A.E., Antonov, I., Fayet, O., Atkins, J.F., Borodovsky, M., Baranov, P.V. (2011). A pilot study of bacterial genes with disrupted ORFs reveals a surprising profusion of protein sequence recoding mediated by ribosomal frameshifting and transcriptional realignment. *Molecular Biology and Evolution*, 28(11), 3195–3211.

Siguier, P., Gagnevin, L., Chandler, M. (2009). The new IS1595 family, its relation to IS1 and the frontier between insertion sequences and transposons. *Research in Microbiology*, 160(3), 232–241.

Siguier, P., Gourbeyre, E., Chandler, M. (2014). Bacterial insertion sequences: Their genomic impact and diversity. *FEMS Microbiology Reviews*, 38(5), 865–891.

Siguier, P., Gourbeyre, E., Varani, A., Ton-Hoang, B., Chandler, M. (2015). Everyman's guide to bacterial insertion sequences. *Microbiology Spectrum*, 3(2), MDNA3–0030.

Smith, M.C.M. and Thorpe, H.M. (2002). Diversity in the serine recombinases. *Molecular Microbiology*, 44(2), 299–307.

Smyshlyaev, G., Bateman, A., Barabas, O. (2021). Sequence analysis of tyrosine recombinases allows annotation of mobile genetic elements in prokaryotic genomes. *Molecular Systems Biology*, 17(5), e9880.

Snesrud, E., McGann, P., Chandler, M. (2018). The birth and demise of the ISApl1-mcr-1-ISApl1 composite transposon: The vehicle for transferable colistin resistance. *mBio*, 9(1).

So, M., Heffron, F., McCarthy, B.J. (1979). The *E. coli* gene encoding heat stable toxin is a bacterial transposon flanked by inverted repeats of IS1. *Nature*, 277(5696), 453–456.

Stark, W.M., Boocock, M.R., Sherratt, D.J. (1989a). Site-specific recombination by Tn3 resolvase. *Trends in Genetics*, 5(9), 304–309.

Stark, W.M., Sherratt, D.J., Boocock, M.R. (1989b). Site-specific recombination by Tn3 resolvase: Topological changes in the forward and reverse reactions. *Cell*, 58(4), 779–790.

Szuplewska, M., Ludwiczak, M., Lyzwa, K., Czarnecki, J., Bartosik, D. (2014). Mobility and generation of mosaic non-autonomous transposons by Tn3-derived inverted-repeat miniature elements (TIMEs). *PLoS One*, 9(8), e105010.

Toleman, M.A., Bennett, P.M., Walsh, T.R. (2006). ISCR elements: Novel gene-capturing systems of the 21st century? *Microbiology and Molecular Biology Reviews*, 70(2), 296–316.

Trieu-Cuot, P. and Courvalin, P. (1985). Transposition behavior of IS15 and its progenitor IS15-delta: Are cointegrates exclusive end products? *Plasmid*, 14(1), 80–89.

Val, M.-E., Bouvier, M., Campos, J., Sherratt, D., Cornet, F., Mazel, D., Barre, F.-X. (2005). The single-stranded genome of phage CTX is the form used for integration into the genome of *Vibrio cholerae*. *Molecular Cell*, 19(4), 559–566.

Vandecraen, J., Chandler, M., Aertsen, A., Van Houdt, R. (2017). The impact of insertion sequences on bacterial genome plasticity and adaptability. *Critical Reviews in Microbiology*, 43(6), 709–730.

Vanhooff, V., Normand, C., Galloy, C., Segall, A.M., Hallet, B. (2010). Control of directionality in the DNA strand-exchange reaction catalysed by the tyrosine recombinase TnpI. *Nucleic Acids Research*, 38(6), 2044–2056.

Varani, A., He, S., Siguier, P., Ross, K., Chandler, M. (2021). The IS6 family, a clinically important group of insertion sequences including IS26. *Mobile DNA*, 12(1), 11.

Yano, H., Genka, H., Ohtsubo, Y., Nagata, Y., Top, E.M., Tsuda, M. (2013). Cointegrate-resolution of toluene-catabolic transposon Tn4651: Determination of crossover site and the segment required for full resolution activity. *Plasmid*, 69(1), 24–35.

3

Transposable Elements and Human Diseases

Benoît CHÉNAIS
BiOSSE (Biologie des Organismes : Stress, Santé, Environnement), Université du Mans, France

3.1. The moving parts of the human genome

Like many large and structurally complex eukaryotic genomes, the human genome is strewn with ancestral traces of invasion by mobile DNA elements or transposable elements (TEs), which today account for almost half of the human genome; more precisely, the value ranges from 45%, according to the most current data (Lander et al. 2001), to 66% according to a more in-depth analysis (de Koning et al. 2011). This represents not only a quantitatively significant fraction of the human genome, but also an important part of its evolutionary history. While most of these TE insertions have probably been lost, others have become fixed by genetic drift and remained trapped at the heart of our DNA for millions of years after the loss of their ability to propagate through the genome.

As detailed in previous chapters, the two main characteristics of these DNA sequences are (1) their capacity for mobility within a given genome and between different genomes, and (2) their ability to increase in number within the genome, both properties leading TEs to play an important role in genome plasticity and evolution. TEs are diverse in nature, and two main classes have been identified on the basis of their transposition mechanism in eukaryotes (see Chapter 1). Class I elements, retroelements or retrotransposons, are propagated by reverse transcription of an RNA intermediate. Class II elements, or DNA transposons, move mainly by a

Transposable Elements and Genome Evolution,
coordinated by Aurélie HUA-VAN and Pierre CAPY. © ISTE Ltd 2024.

"cut-and-paste" mechanism involving excision and reinsertion of the elements' DNA sequences. In addition, a large number of non-autonomous repeat sequences are affiliated to TEs but are dependent on the enzyme(s) encoded by an auxiliary element, which is itself an autonomous TE, to ensure their mobility and amplification in the genome. *SINE*s (short interspersed nuclear elements) are the main class of non-autonomous TEs that have widely invaded the human genome (de Koning et al. 2011).

Class I elements are by far the most numerous in the human genome. They are subdivided into retrotransposons with long terminal repeats (LTRs) and retrotransposons without LTRs. The latter are highly predominant, and 76% of human TEs are LTR-less retrotransposons, accounting for at least 33.6% of the entire human genome (Lander et al. 2001) (Figure 3.1). Among these LTR-less retrotransposons, the *LINE-1* element (or L1, for long interspersed nuclear element-1) has undergone exceptional expansion and now counts around 500,000 copies, most of which are fragmented. This represents 38% of human TEs and 16.9% of the entire genome (Figure 3.1). However, only a very small proportion of these *LINE-1*s are active, around a hundred sequences among the L1PA1, L1PA2 subfamilies or *Homosapiens*-specific elements (L1Hs).

The second most frequent category of TEs are non-autonomous *SINE*s, in particular *Alu* repeats, which account for 24% of TEs and 10.6% of the genome (Figure 3.1). These *Alu* repeats are present in nearly 1 million copies in the entire human reference genome and derive from a few young *Alu* subfamilies (AluY), with traces of an explosion of *Alu* retrotransposition in ancestral primates dating back 35–40 million years (Konkel et al. 2010). *Alu* sequences depend on the endonuclease and reverse transcriptase activities of the *LINE-1* element for their propagation in genomes. Other *SINE*s found in humans are the *MIR* (mammalian-wide interspersed repeat) elements, notably *MIR3*. Other non-autonomous elements are made up of a composite assembly of sequences homologous to *SINE*s, variable number of tandem repeats (VNTR) and *Alu* sequences, hence their name SVA (for SINE, VNTR, Alu) can be mobilized, that is, moved, by the enzymes of *LINE-1* elements. There are a limited number of subfamilies of *Alu* and SVA elements capable of mobilization in modern humans: for example, *Alu*Ya5, *Alu*Yb8, *Alu*Yb9, SVA_E and SVA_F. *LINE-1* elements are active in the germline, making *LINE-1*, *Alu* and SVA integrations common and segregating DNA markers in populations. Each of us thus inherits a unique assortment of these mobile element insertions, distinguishing us from other individuals by several hundred different sites (loci) in our DNA. It is estimated that one in every 100–200 people carries a de novo germline integration of a *LINE-1* or SVA element, that is, an insertion that is not inherited (Ewing and Kazazian 2010; Huang et al. 2010). The expected frequency for *Alu* insertions is higher, estimated at around one in twenty individuals carrying a new insertion (Xing et al. 2009).

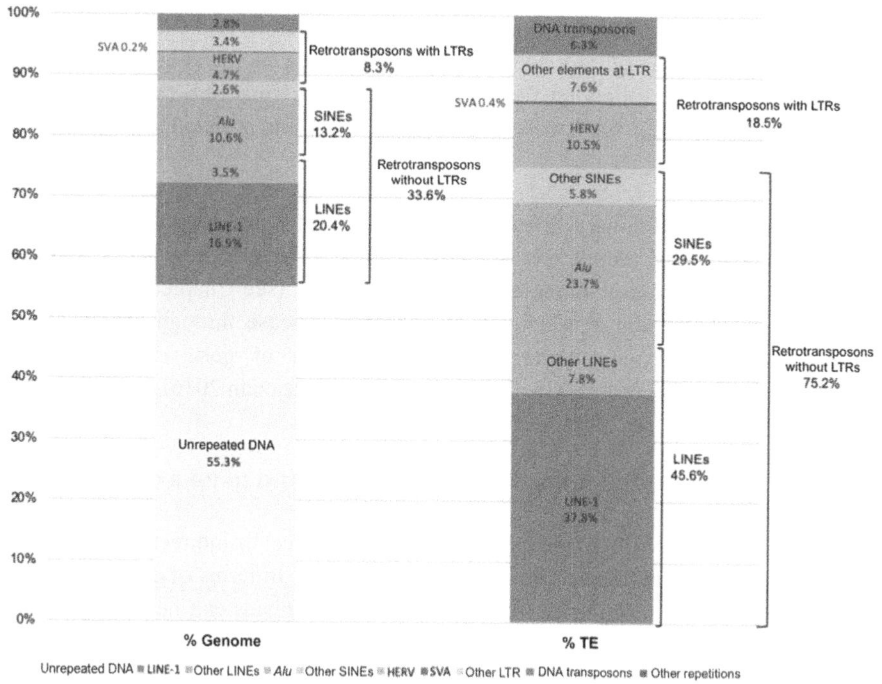

Figure 3.1. *Proportions of TEs in the human genome and distribution in different families and subfamilies. For a color version of this figure, see www.iste.co.uk/huavan/transposable.zip*

LTR retrotransposons are present in smaller numbers in the human genome and are mainly represented by endogenous retroviruses (human endogenous retro virus (HERV)), which resemble retroviruses in both structure and motility, but lack functional envelope genes. In humans, LTR retrotransposon activity has become extinct (or nearly so) over the last few million years (Dewannieux et al. 2006), although their sequences still account for around 8% of our genome (Figure 3.1). The various HERVs are named after the transfer RNA that initiates their reverse transcription reaction, and the most recent active subfamily is HERV-K (initiated by lysine transfer RNA). There are also HERV-I (initiated by isoleucine tRNA) and HERV-L (initiated by leucine tRNA) sequences. The current human genome still contains insertion polymorphisms for these HERV elements. LTRs flanking full proviral insertions tend to recombine, deleting intervening sequences and reducing the insertion to a solo LTR (Thomas et al. 2018).

Class II elements, or transposons, are much less numerous in the human genome, accounting for just 6% of TEs and 2.8% of the whole genome (Figure 3.1). They mainly belong to the following three superfamilies: Tc1/mariner, that is, mariner, MER2-Tigger and Tc2; hAT, that is, MER-1-Charlie and Zaphod; and PiggyBac (Lander et al. 2001).

TEs are an evolutionary force that contributes to the genetic diversity of organisms. TEs can be recruited by the host genome to become new genes, and this domestication of TEs is a source of genetic innovation (see Chapter 6). However, transposition activity can also contribute to human disease through chromosomal rearrangements, insertional mutagenesis, deregulation of gene expression and epigenetic changes (Chénais 2015, 2022; Hancks and Kazazian 2016; Burns 2020).

3.2. TE insertion and its impact on the genome and gene expression

The insertion of a TE into the genome has several direct or indirect consequences on genome structure and the control of gene expression. In terms of events likely to lead to disease in humans, three main groups of mechanisms can be distinguished: (1) genomic rearrangements due to TE abundance; (2) modification of gene structure and regulatory regions by TE insertion; and (3) alteration of epigenetic controls (Chénais 2015, 2022; Hancks and Kazazian 2016; Burns 2020).

3.2.1. *Chromosomal rearrangements*

The high abundance of *LINE-1* and *Alu* repeats (around 17% and 11% of the human genome, respectively) favors ectopic recombination, that is, recombination between nonhomologous loci. Such recombination often leads to major chromosomal rearrangements, such as gene deletions, duplications, translocations or chromosomal inversions, which are the source of pathologies (Tables 3.1 and 3.2). An example of this type of recombination between two nonallelic insertions is shown in Figure 3.2. In this case, *Alu-Alu* recombination occurred between tandem *Alu*Ya5 insertions in the ubiquitin E2 T conjugating enzyme (UBE2T) gene, causing Fanconi anemia. The rearrangement caused an interstitial deletion encompassing exons 2–6 of the paternal allele, with a corresponding duplication of the maternal allele in the patient (Rickman et al. 2015).

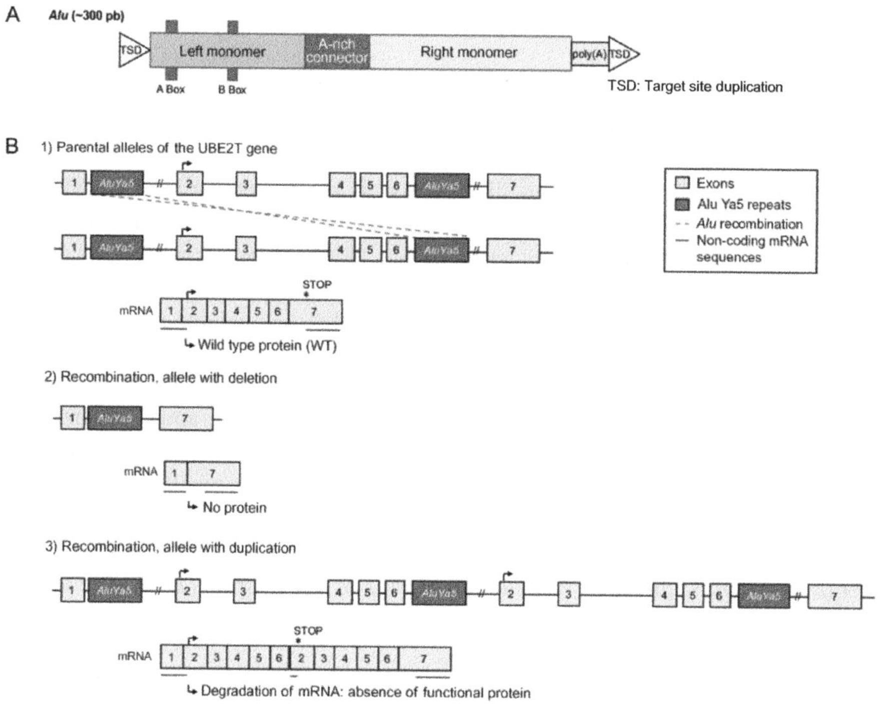

Figure 3.2. *Alu element insertion and recombination. (A) Schematic structure of an Alu element. (B) Example of recombination between two AluYa5 insertions in the UBE2T gene leading to one allele with deletion and one allele with duplication. Figure inspired from Rickman et al. (2015). For a color version of this figure, see www.iste.co.uk/huavan/transposable.zip*

LINE-1 and *Alu* elements are frequently observed in or near chromosomal rearrangement breakpoints, and *LINE-1* alone accounts for 19% of the 2081 breakpoints analyzed in a dataset of 17 entire human genomes (Kidd et al. 2010). Although present in smaller numbers in the human genome, DNA transposons can also be a source of chromosomal rearrangement.

The *MER20* transposon is implicated in 47% of the 1,174 translocations observed at the *TCF3* locus on chromosome 19 and appears to be more particularly linked to the development of acute lymphoblastic leukemia among the dozen or so different leukemias or lymphomas studied (Rodić et al. 2013).

3.2.2. *Modification of gene structure and expression by TE insertion*

Another way in which TEs can have an impact on the human genome and induce disease is by modifying the structure of genes and their regulatory regions. Indeed, de novo insertion, that is, acquired directly from the patient or a mosaic parent, or an insertion inherited from ascendants like a classical allele (in this case referred to as an insertion allele) can be responsible for diseases that behave like classical monogenic diseases. For this to happen, the TE insertion must appear in the germline and critically affect gene function. The TE insertion will therefore create a new allele, which may be either a loss-of-function allele or a gain-of-function allele.

The insertion of a TE into a gene can have several consequences, including the creation of new exons or introns, known as exonization and intronization, respectively. The creation of a new exon causes a reading frame shift and/or creates premature stop codons in around 79% of cases. As a result, the resulting mRNA will either be prematurely degraded or aberrantly spliced, and if it manages to be translated into protein, the latter will be truncated and nonfunctional. Thus, exonization events following insertion of *LINE-1* or *Alu* elements can cause human diseases (Tables 3.1 and 3.2).

TE insertion can also modify mRNA splicing and cause aberrant alternative splicing leading to exon skipping or intron retention (Tables 3.1 and 3.2). For example, the insertion of an SVA element in an intron of the TATA box-binding protein (TAF1)-associated factor 1 gene causes X-linked dystonia with parkinsonism (XDP), a movement disorder endemic to the Philippines. The presence of the SVA element prevents normal splicing of TAF1 mRNA and results in retention of intron 32 (Aneichyk et al. 2018) (Figure 3.3). The situation is more complex, however, and a range of alleles exist in the population with variable-sized expansions of the hexanucleotide repeat domain (CCCTCT)n within the SVA. The length of this repeat is strongly correlated with disease onset, and longer repeat sequences are associated with earlier symptom manifestation (Bragg et al. 2017). Although the precise role of the length of this hexanucleotide repeat is not entirely clear, it affects the efficiency of retroelement transcription and promotes guanine stacking, a DNA structure known as G-quadruplex (Bragg et al. 2017).

LINE-1 elements include a polyadenylation signal in their own sequence, and the polyadenylation signal (i.e. the AATAAA sequence) is frequently created in the A-rich tails of *SINE* and *LINE* elements. Consequently, human retroelements, such as *LINE-1*, *Alu*, or even HERVs that contain a polyadenylation signal in their LTR, can introduce intragenic polyadenylation signals, creating new mRNA isoforms. These alternative mRNAs are often truncated by the loss of one or more exons. This modification of polyadenylation sites has been observed in several human diseases (Tables 3.1 and 3.2). Figure 3.4 shows the example of the insertion of a *LINE-1* element (truncated and partially inverted) in the last exon of the APC gene, causing colon cancer and desmoid tumors (or aggressive fibromatoses) (Miki et al. 1992).

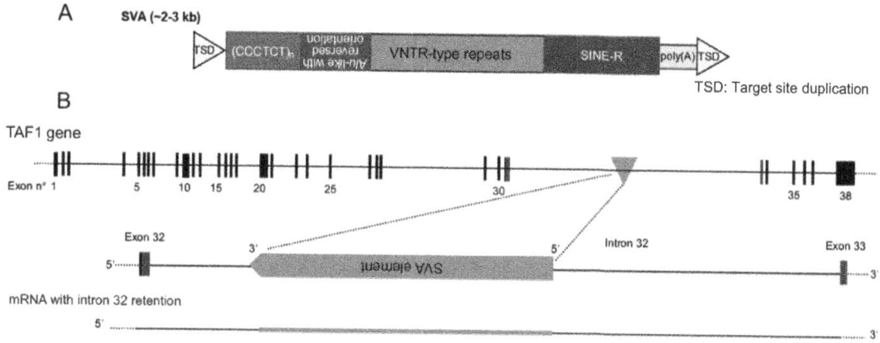

Figure 3.3. *Example of an SVA element insertion at the origin of X-linked dystonia with parkinsonism. (A) Schematic structure of an SVA element. Note the hexanucleotide (CCCTCT)n, whose variable number of repeats affects the disease. (B) Schematic structure of the TAF1 gene, made up of 38 exons (vertical bars) and position of the SVA insertion in reverse orientation. This insertion leads to the retention of intron 32 in the mRNA. For a color version of this figure, see www.iste.co.uk/huavan/transposable.zip*

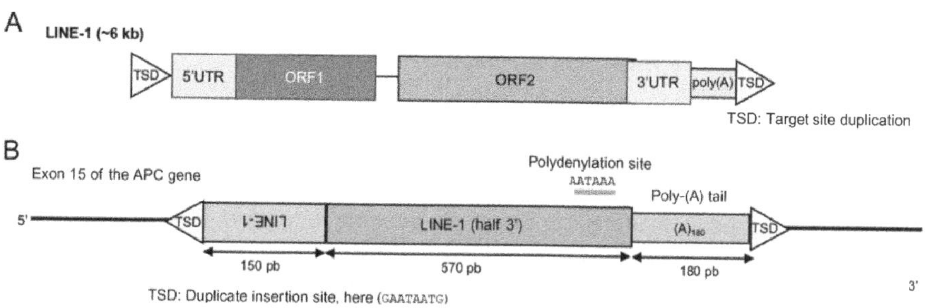

Figure 3.4. *Insertion of a LINE-1 element in the APC gene, resulting in a new polyadenylation site. (A) Schematic structure of a LINE-1 element. (B) Representation of the insertion of a truncated and partially inverted LINE-1 element in the last exon of the APC gene. For a color version of this figure, see www.iste.co.uk/huavan/transposable.zip*

Finally, the use of regulatory elements brought about by the insertion of a TE is a means of modifying gene expression. In particular, the enhancer/silencer elements of HERV LTRs are particularly powerful. For example, the X-linked form of Opitz syndrome is associated with the presence of a HERV-E element whose LTR acts as a tissue-specific promoter and activator of the *MID1* gene (Landry et al. 2002) (Figure 3.5).

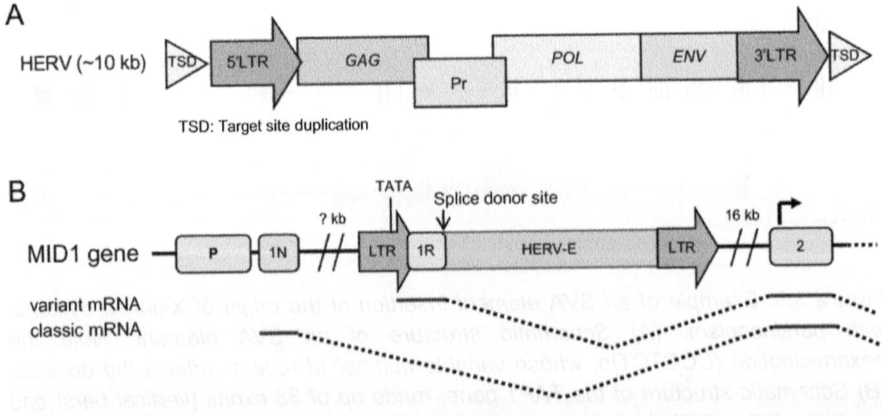

Figure 3.5. *Example of an inserted HERV-E element at the origin of X-linked Opitz syndrome. (A) Schematic structure of a HERV element. (B) Schematic structure of the 5' part of the MID1 gene, comprising the promoter (P) and the first two introns (1N and 2). Insertion of a HERV-E element provides a new transcription start site (TATA) and a new intron 1 (1R, in yellow), leading to an alternative transcript (mRNA). For a color version of this figure, see www.iste.co.uk/huavan/transposable.zip*

3.2.3. *Escaping epigenetic control*

As TEs can be considered a threat to genome integrity, host species have developed defense mechanisms to silence them. These mechanisms rely on epigenetic regulatory pathways such as DNA methylation and inhibitory histone modifications, which, in the case of retrotransposons, are achieved via piRNAs (for PIWI-interacting RNA) and various protein complexes that will not be detailed here (see Chapter 4). However, these silencing mechanisms are imperfect, and some TEs manage to bypass this regulation and are transcribed. For the oldest and/or noncoding TEs, the consequences of this situation may not be significant, but for active TEs, that is, those able to move, transcription of the TE can lead to new insertions. Nevertheless, even transcripts of a priori inactive elements can have an impact on genome expression in the form of long noncoding RNAs (lncRNAs). What is more, the relationship between TEs and epigenetic control is a two-way street: if DNA hypomethylation is a cause of TE activation, the presence of TEs in a genomic region is also a cause of chromatin hypermethylation. Thus, epigenetic modification of TE regulatory sequences can have an impact on the expression of neighboring genes (Grégoire et al. 2016).

A correlation between TE activity and chromatin regulation has been observed, mainly at the level of DNA methylation, and DNA hypomethylation is considered a facilitator of TE mobility. As chromatin methylation occurs preferentially at CG dinucleotides known as CpG (cytosine-phosphate-guanine) islands, the relationship between pathological markers and the methylation status of whole-genome CpG islands, on the one hand, and *LINE-1* elements, on the other hand, has been particularly studied in the case of cancers (Ross et al. 2010). Thus, by analyzing the whole genome of cancer cells of diverse tissue origins, it is possible to observe a preference for insertion of *LINE-1* elements in DNA regions showing hypomethylation and hypomethylation of the ETs themselves and thus their de-repression (Ferreira et al. 2014; Tubio et al. 2014).

3.3. TE involvement in human cancers

3.3.1. *Inserting LINE-1*

Around 50% of human tumors contain somatic retrotranspositions of *LINE-1* elements (Tubio et al. 2014). The first example of a *LINE-1* element insertion was reported as early as 1988 in the article by Morse et al. (1988), which shows the insertion of a *LINE-1* retrotransposon into the human proto-oncogene *c-MYC* from human breast carcinoma cells (Table 3.1). Comparison of the *c-MYC* gene structure of cancerous and noncancerous breast cells from the same patient revealed, in tumor cells, a specific rearrangement of one of the *MYC* loci and amplification of the second *MYC* locus, located on the homologous chromosome. A *LINE-1*-like sequence was found in the second intron of the rearranged *MYC* locus. In this case, the *LINE-1* sequence functioned as a mobile genetic element to induce somatic mutation (Morse et al. 1988). Another example of somatic insertion of the *LINE-1* element was observed in the *APC* tumor suppressor gene (Figure 3.4) in several colon cancer patients and in association with desmoid tumors (or aggressive fibromatoses) (Miki et al. 1992). Genomic (whole genome sequencing) and transcriptomic (mRNA sequencing) data are very useful for observing TE insertions and transcription, respectively. In this way, a high level of *LINE-1* transcription as well as an increase in *LINE-1* insertion activity was found in three cancer types, namely primary tumor samples of invasive breast carcinoma, head and neck squamous cell carcinoma and lung adenocarcinoma (Clayton et al. 2016). More recently, a genomics study on a set of almost 3,000 tumor samples covering 38 different cancer types highlighted the importance of *LINE-1* elements and the genomic rearrangements they induce (Rodriguez-Martin et al. 2020). Indeed, out of almost 20,000 retrotransposition events observed in these cancer samples, 98% are due to a *LINE-1* element, while *Alu* and SVA elements are only a very small minority (Rodriguez-Martin et al. 2020).

Gene	TE type	Type of event(s)	Cancer type
APC	*LINE-1*	Insertion, new polyA site	Colon cancer, desmoid tumors
BCR-ABL	*Alu*	Translocation t(9;22)(q34;q11)	Chronic myeloid leukemia
BRCA1	*Alu*	Insertion, deletion, duplication	Breast cancer, ovarian cancer
BRCA2	*Alu*	Insertion, duplication, aberrant splicing	Breast cancer, ovarian cancer
CAD	*Alu*	Deletion	Hepatoma
CDH1	*Alu*	Deletion	Hereditary diffuse gastric cancer
EWSR1-ETV	*Alu*	Translocation t(5;18)(q23-31;q12)	Ewing's sarcoma
MLL1	*Alu*	Duplication	Acute myeloid leukemia
MYB	*Alu*	Duplication	T-cell lymphoblastic leukemia
MYC	*LINE-1*	Insertion	Breast cancer
TCF3-PBX1	MER20	Translocation t(1;19)(q23;p13.3)	Acute lymphoblastic leukemia

Table 3.1. *Examples of cancers linked to transposable elements*

3.3.2. *Insertion of Alu sequences and chromosomal recombination*

As previously mentioned, *Alu* repeats account for 11% of the human genome. These repeats are at the origin of nonallelic homologous recombination events, or ectopic recombination, leading to several cancer diseases. These recombinations have several consequences: they can lead to chromosome translocations, deletions or duplications. Insertion of an *Alu* sequence, or a deletion resulting from recombination between *Alu* sequences, can themselves give rise to exon skipping.

The *BRCA1* and *BRCA2* genes, which are associated with breast and ovarian cancers, are important insertion sites for *Alu* elements, and an exceptionally high density of *Alu* repeats is found at these loci. Numerous deletion events have also been reported at the *BRCA1* locus and, to a lesser extent, for the *BRCA2* gene (Table 3.1). In addition, one of the first germline mutations caused by TE to be identified is the insertion of an *Alu* element in exon 22 of the *BRCA2* gene. This insertion leads to alternative splicing and skipping of exon 22 in the transcribed RNA (Miki et al. 1996).

Numerous other cases of deletions caused by recombination between *Alu* elements have been described (Table 3.2). For example, homologous recombination between *Alu* sequences in the *CAD* (*Caspase-activated DNase*) gene is frequently detected in human hepatoma tissue. This leads to deletion of exon-3, which in turn leads to skipping of this exon or its replacement by a partial *Alu* sequence, resulting in both cases in a C-terminally truncated CAD protein (Hsieh et al. 2005). A statistically significant over-representation of *Alu* elements around breakpoints can also be observed in deletions of the *cadherin-1* (*CDH1*) gene. These large deletions are found in 4% of families with hereditary diffuse gastric cancer, and therefore appear to originate from homologous recombination between *Alu* repeat sequences (Oliveira et al. 2009). It should also be noted that the insertion of *Alu* elements in the *VHL* (*vonHippel–Lindau*) gene is involved in the deletions that cause vonHippel–Lindau disease, an autosomal dominant syndrome predisposing to the development of benign or malignant tumors (Franke et al. 2009).

Homologous recombination between *Alu* elements sometimes leads to gene duplication, as has been reported for the *BRCA1* gene, the *MLL-1* gene involved in acute myeloid leukemia, and the *MYB* proto-oncogene, duplication of which leads to acute lymphoblastic cell leukemia (Table 3.1). Duplication of the *MYB* locus, which encodes an essential transcription factor, is common in human cancers. The human *MYB* is flanked by *Alu* repeats that are responsible for its duplication by homologous somatic recombination between adjacent *Alu* elements located on sister chromatids, as observed in T-cell acute lymphoblastic leukemia (O'Neil et al. 2007).

Recombination between *Alu* repeats is also involved in the genesis of the Philadelphia chromosome, that is, the translocation between chromosomes 9q34 and 22q11, leading to *BCR/ABL* gene fusion in chronic myeloid leukemia (Jeffs et al. 1998). Similarly, translocation between chromosomes 5q23-31 and 18q12, involving the *TRE-2* oncogene, is due to recombination between *Alu* elements and plays an important role in the development of Ewing's sarcoma (Onno et al. 1992).

These data highlight the role of *Alu* repeats in mediating genome instability in human cancers (Table 3.1).

3.3.3. *Epigenetic alterations due to TE and cancer*

As mentioned above, the frequent hypomethylation of chromatin in tumor cells is considered a facilitator of TE mobility. It is thus possible to observe a correlation between the level of chromatin methylation and the rate of insertion and transcription of *LINE-1* elements in cancer cells (Ross et al. 2010). *LINE-1* element retrotransposition events are frequent in cancers (53% of patients in the study by Tubio et al. 2014, for example) and *LINE-1* activity, which fluctuates with tumor progression, correlates with hypomethylation of the *LINE-1* element promoter (Tubio et al. 2014). Furthermore, these *LINE-1* retrotransposition events are found to have disseminated genes, exons and regulatory elements to new loci, most often in heterochromatin regions of the genome (Tubio et al. 2014).

Escape from epigenetic control of a *LINE-1* element can thus lead to cancer, as shown by the following example, observed in a patient with high-grade mucosal adenocarcinoma of the colon (Scott et al. 2016). Lack of promoter methylation of the specific *LINE-1* locus (*L1Hs*) involved, located on chromosome 17, enabled RNA transcription in samples of the patient's normal tissue. This active element then triggered a de novo insertion of the retrotransposon, assumed to have occurred in the individual's precancerous colonic epithelium. This new *L1Hs* insertion in chromosome 5 disrupted an exon of the adenomatous polyposis colon (*APC*) tumor suppressor gene, producing a trigger mutation that is likely to have caused the subsequent colon cancer. The *L1Hs* source element on chromosome 17 is a polymorphic structural variant in human populations, and methylation studies in other individuals indicate that this specific locus can be effectively silenced. In the patient, the locus was one of around ten *L1Hs* expressed, and all other loci appear to have been silenced. The reason for this specific escape from epigenetic regulation remains unknown (Burns 2020).

3.4. Involvement of TEs in noncancerous pathologies

3.4.1. *Implications of TE in hemoglobinopathies*

The α-globin and β-globin gene clusters, located on chromosomes 16 and 11, respectively, are the target of numerous rearrangements leading to a range of pathologies grouped together under the name of hemoglobinopathies, of which thalassemias are a part.

Deletions involving *Alu* repeats have been observed in the β-globin gene cluster (Table 3.2), which comprises the γ-, δ- and β-globin genes (Harteveld et al. 1997; Yu et al. 2014). For example, in an individual with hereditary persistence of fetal hemoglobin (γ-globin), the β-globin cluster has a large 48.5-kb deletion downstream of the β-globin gene, and the 5′ breakpoint of the deletion is located in a 3′ *Alu* element of the Aγ-globin gene[1], one of two (Aγ and Gγ) fetal globin genes. However, in this particular case, the recombination mechanism involves nonhomologous recombination, as the 3′ breakpoint was located in a region that contains various repeats, including part of a *LINE-1* repeat, a 160-bp perfect palindrome and a 41-bp set of direct repeats found elsewhere in the human genome (Henthorn et al. 1986).

With regard to the α-globin gene cluster, recombination events involving *Alu*- and/or *LINE-1* sequences are numerous and lead globally to various forms of thalassemia (Table 3.2). Here are a few examples of recombination events leading to thalassemia-α, an inherited hemoglobin disease characterized by a quantitative reduction in the α-globin chain and anemia. In one family of patients with thalassemia-α, there is a 28.5 kb deletion involving *Alu* repeats that eliminated one of the duplicated α-globin genes (Yu et al. 2014). In another patient family, an 8.2 kb deletion in the α-globin gene cluster was found. This deletion involves both α-globin genes in *cis* and is caused by a nonhomologous recombination event between an *Alu* element and a *LINE-1* element (Phylipsen et al. 2010). A large 33-kb deletion encompassing the α- and ζ-globin genes and pseudogenes has been observed in another family and causes thalassemia-α^0, that is, the complete absence of the α-globin protein chain (Harteveld et al. 1997).

As described above for cancer, chromatin structure and/or epigenetic regulation associated with TE insertion can lead to noncancerous human diseases, but examples are rare to date. However, the β-globin-L1 allele results from the insertion of an entire *LINE-1* element into intron 2 of the β-globin gene. The decreased transcription rate of this β-globin-L1 allele is due to the hypermethylated profile of its promoter and enhancer sequences, resulting in down-regulation of transcription and thus the thalassemia-β^+ phenotype (Lanikova et al. 2013).

3.4.2. *Implications of TEs in metabolic diseases or diseases linked to metabolic genes*

Deletions involving *Alu* repeats account for numerous mutations in several metabolic genes (Table 3.2), including the apolipoprotein B (*APOB*) gene and the

---

1 In humans, the β-globin gene cluster is located on chromosome 11 and comprises the embryonic chain gene ε followed by the two fetal chain genes γ (Aγ and Gγ), the pseudogene β and then the two adult chain genes δ and β. The two globins Aγ and Gγ differ only in the amino acid at position 136, which is an alanine or a glutamic acid, respectively.

LDL receptor (*LDLR*) gene. For example, in a patient with homozygous hypobetalipoproteinemia, the deletion of exon 21 of the *APOB* gene is the consequence of a recombination event between the *Alu* sequences of introns 20 and 21, resulting in a nonreciprocal exchange between two chromosomes (Huang et al. 1989). The *LDLR* gene is also a site of large deletions (or duplications) by homologous recombination between *Alu* repeats, which are numerous in almost all introns of the gene and in the 3′ untranslated region (3'-UTR) of the mRNA (Hobbs et al. 1992). Thus, analysis of autosomal dominant hypercalcemia in two patient families showed the presence of a large deletion in the *LDLR* gene due to nonallelic intra-chromosomal homologous recombination between two homologous *Alu* sequences located away from the splice sites in the intronic sequences (Jelassi et al. 2012). Furthermore, the *LINE-1* element can also be the source of deletions (Table 3.2), as in the case of the pyruvate dehydrogenase E1 (*PDHX*) gene located on the X chromosome. In this case, a large 46 kb deletion linked to the insertion of a complete *LINE-1* element leads to pathological deficiency of the pyruvate dehydrogenase complex (Miné et al. 2007). However, insertion of *LINE-1* into the *PDHX* gene can also lead to an aberrantly spliced isoform resulting from the use of two cryptic splice sites (Miné et al. 2007).

Insertion of *Alu* or *LINE-1* elements can also lead to the creation of a new exon, or "exonization" (Table 3.2). For example, a G to C mutation in an *Alu* element present in intron 3 of the ornithine delta-aminotransferase (*OAT*) gene results in the creation of a new splice donor site and potentially a new exon (Mitchell et al. 1991). This insertion of an *Alu* element also provides a premature transcription termination site and leads to an OAT deficiency. In both cases, this *Alu* insertion is involved in mitochondrial OAT deficiency, which causes hyperornithinemia leading to gyrate atrophy of the choroid and the retina, a disease characterized by myopia and night blindness progressing to progressive vision loss. Similarly, exon creation due to insertion of an *Alu* element has been observed in the β-glucuronidase gene (*GUSB*) causing Sly syndrome or mucopolysaccharidosis type VII (Vervoort et al. 1998). The *LINE-1* retroelement is also a potential source of exon creation, implicated in at least one case of chronic granulomatous disease, a pathology favoring bacterial and fungal infections (Meischl et al. 2000). In this case, the *CYBB* gene, linked to the X chromosome and encoding the beta chain of cytochrome B245, has an insertion of *LINE-1* in intron 5, resulting in internal rearrangements and new splice sites, leading to a highly heterogeneous splicing pattern with the introduction of two *LINE-1* fragments as new exons in the mRNAs and exon skipping (Meischl et al. 2000).

Gene	TE type	Type of event(s)	Pathology type(s)
APOB	*Alu*	Deletion	Hypo-betalipoproteinemia
CASR	*Alu*	New polyA site	Hypocalciuric hypercalcemia and severe neonatal hyperparathyroidism
CD58	*Alu*	Exon skipping, change of reading frame	Multiple sclerosis
COL4A3	*Alu*	Exonization	Alport syndrome
CYBB	*LINE-1*	Exonization, intron retention	Chronic granulomatous disease
DMD	*LINE-1*	Exon skipping	Duchenne muscular dystrophy
DMD	*Alu*, *LINE-1*	Aberrant splicing, new polyA site	X-linked dilated cardiomyopathy
F8	*Alu*	Exon skipping	Hemophilia A
F9	*Alu*, *LINE-1*	New polyA site	HemophiliaB
FAS	*Alu*	Exon skipping, new polyA site	Autoimmune lymphoproliferative syndrome
FGFR2	*Alu*	Aberrant splicing, exon skipping	Apert's syndrome
FKTN	*LINE-1*	Exon skipping	Fukuyama-type congenital muscular dystrophy
a-GALA	*Alu*	Deletion	Fabry's disease
GUSB	*Alu*	Exonisation	Sly's syndrome
HBB	*Alu*	Deletion	Thalassemia
HEXB	*Alu*	Deletion	Sandhoff's disease
IDS	*Alu*	Exon skipping	Hunter's syndrome
LDLR	*Alu*	Deletion	Familial hypercholesterolemia
MID1	HERV-E	New regulatoryregion	Opitz's syndrome
NF1	*Alu*	Insertion, aberrant splicing, change of reading frame	Neurofibroma, neurofibromatosis type 1
OAT	*Alu*	Exonization	Ornithine delta-aminotransferase deficiency
OPA1	*Alu*	Exon skipping	Autosomal dominant opticatrophy
PDHX	*LINE-1*	Aberrant splicing	Pyruvate dehydrogenase complex deficiency
RP2	*LINE-1*	Aberrant splicing, change of reading frame	X-linked retinitispigmentosa
RSK2	*LINE-1*	Exon skipping, change of reading frame	Coffin-Lowry's syndrome
TAF1	SVA	Intron retention	X-linked dystonia with parkinsonism
UBE2T	*Alu*	Deletion, duplication	Fanconi Anemia
VHL	*Alu*	Deletion	Von Hippel–Lindau disease

Table 3.2. *Examples of noncancerous pathologies linked to transposable elements*

3.4.3. *Implications of TEs in neurological diseases*

TE insertion can be the cause of neurological diseases or syndromes (Ahmadi et al. 2020) (Table 3.2). The insertion of a *LINE-1* element in the ribosomal S6 kinase 2 (*RSK2*) gene was reported in a patient with Coffin–Lowry syndrome, a disease characterized by psychomotor and growth retardation, facial dysmorphia and skeletal abnormalities. This insertion of *LINE-1* at position 8 of intron 3 leads to a skipping of exon 4, followed by a reading frame shift and the appearance of a premature stop codon (Martínez-Garay et al. 2003).

Fabry disease, characterized by central nervous system neurodegeneration, is an X-linked recessive disorder of glycosphingolipid catabolism, resulting from deficient activity of the lysosomal hydrolase α-galactosidase A encoded by a gene rich in *Alu* repeats. This gene has 12 *Alu* elements in its 12-kb sequence (Kornreich et al. 1990). Several mutations and gene rearrangements have been described as causing the disease, of which a single deletion was originally identified as a recombination between *Alu* repeats and ten others as rearrangements mediated by *Alu* repeats (Kornreich et al. 1990). Subsequently, a large deletion of 3.1 kb due to *Alu-Alu* recombination and including the entire exon 2 was identified among 50 new Fabry disease mutations (Shabbeer et al. 2006).

Sandhoff disease, a recessive lysosomal storage disease, is another example of an TE-related neurodegenerative disorder. Deficiency of β-hexosaminidase (*HEXB*) activity is due to deletion alleles of the *HEXB* gene in 27% of all data examined (Neote et al. 1990). The *HEXB* gene contains two *Alu* repeats which are involved in the deletion event, which removes around 16 kb, including the *HEXB* gene promoter, exons 1 to 5, and part of intron 5. The deletion also created a reconstituted *Alu* sequence, the left half of which comes from the 5′-*Alu* sequence and the right half from the 3′-*Alu* sequence (Neote et al. 1990).

Neurofibromatosis type I is a common genetic disease linked to the *NF-1* tumor suppressor gene and characterized by changes in skin coloration and the growth of benign tumors along the nerves, that is, neurofibromas. Insertion of an *Alu* sequence into the intron of the *NF-1* gene results in deletion and frameshifting in the downstream exon during splicing, which may be a cause of the disease (Wallace et al. 1991).

High TE expression has been observed in several human neurodegenerative diseases, including sporadic Creutzfeldt-Jakob disease, age-related macular degeneration and amyotrophic lateral sclerosis (ALS) (Ahmadi et al. 2020). For example, the RNA-binding protein TDP-43, which is implicated in ALS and frontotemporal lobar degeneration (FTLD), interacts with many TE transcripts and this association of TDP-43 with its target TE RNAs is reduced in FTLD patients (Li et al. 2012). As previously mentioned, insertion of an SVA element into an intron of the *TAF1* gene causes XPD (Bragg et al. 2017) (Figure 3.3). The insertion prevents normal mRNA splicing and leads to retention of intron 32 (Aneichyk et al. 2018). Furthermore, this SVA insertion specifically reduces acetylated histone H3 binding on exon 32, closest to the insertion. Conversely, excision of the SVA by CRISPR/Cas9 technology restores acetylated histone H3 binding (Petrozziello et al. 2020).

Finally, a link between TEs and psychiatric disorders has also been demonstrated (Guffanti et al. 2014; Misiak et al. 2019). In particular, whole-genome sequencing revealed brain tissue-specific *LINE-1* insertions in patients, preferentially localized in genes linked to synapses and schizophrenia, suggesting a causal role for *LINE-1* in the onset of the disease (Bundo et al. 2014).

3.4.4. *Implications of TE in various other diseases*

Insertions of *Alu* elements in the *VHL* gene cause a major alteration leading to VHL disease, an autosomal dominant familial syndrome predisposing to the development of benign and malignant tumors (Franke et al. 2009).

The creation of an exon following the insertion of an *Alu* element in the gene for the alpha-3 chain of type IV collagen (*COL4A3*) could be at the origin of Alport syndrome, a nephropathy (glomerulonephritis) due to a defect in collagen IV (Knebelmann et al. 1995).

Alternative splicing is often affected by TE insertion and, for example, insertion of the *Alu* element into an intron of the human factor VIII (*F8*) gene results in exon skipping and, consequently, the development of hemophilia A (Ganguly et al. 2003). Another example is the insertion of an *Alu* element in exon 9, or just at the start of exon 9, of the fibroblast growth factor receptor 2 (*FGFR2*) gene, which could be a cause of Apert syndrome (Oldridge et al. 1999). The latter mutation affects the use of the 3′ splice site, leading to the genesis of various nonfunctional splice forms of *FGFR2* gene transcripts.

The dystrophin gene (*DMD*), located on the X chromosome, is a giant gene of around 2.5 Mb in size, with over 60 exons. Numerous TE insertions into the *DMD* gene have resulted in deletions or exon skipping. For example, insertion of *LINE-1* induces skipping of exon 44, resulting in Duchenne muscular dystrophy (Narita et al. 1993). Insertion of an *Alu*-type sequence downstream of intron 11 of the *DMD* gene has been observed in individuals with X-linked dilated cardiomyopathy (Ferlini et al. 1998). The insertion induces a rearrangement that activates a cryptic splice site in intron 11 and produces an alternative transcript containing the *Alu*-like sequence and part of the adjacent intron 11, spliced between exons 11 and 12, but untranslated due to the presence of numerous stop codons.

In addition, several diseases have been linked to the introduction of a new polyadenylation site by the insertion of *LINE-1* or *Alu* elements into human genes, such as hemophilia A and B, autoimmune lympho-proliferative syndrome, X-linked dilated cardiomyopathy, hypocalciuric hypercalcemia and severe hyperparathyroidism of the newborn (Table 3.2) (Chénais 2015, 2022).

3.4.5. *Link between TE and common diseases*

The increase in genomic studies, and in particular genome-wide association studies (GWASs), strengthens the case for a genetic basis for common diseases, that is, diseases that have a high incidence in the population (> 1:2,000) and are not usually classified as hereditary. However, these diseases often have a genetic basis, and a family aggregation can then be observed. We speak of multifactorial diseases, since in addition to genetic predisposition factors, a complex set of environmental factors (diet, viral and bacterial infections, other diseases, exposure to pollutants, smoking, alcohol, drugs, etc.) will play a part in the development of the pathology. These common diseases are highly varied and include cardiovascular diseases (hypertension, myocardial infarction, etc.), endocrine diseases (type 2 diabetes, obesity, hypercholesterolemia, etc.), osteoarticular pathologies (osteoarthritis, rheumatoid arthritis, osteoporosis, etc.), neurological diseases (senile dementia), dysimmune and autoimmune diseases (e.g. multiple sclerosis (MS)), psychiatric diseases (schizophrenia, bipolar disorder, autism, etc.) and allergy/asthma problems. Hundreds of different *Alu* element insertions have been identified in the best-defined predisposition loci or genes for these diseases (Payer et al. 2017). Furthermore, in many cases, the mobile element insertion is in linkage disequilibrium with disease haplotype markers, that is, the *Alu* element appears linked to disease-associated single nucleotide polymorphisms (SNPs). This suggests that the TE insertion is probably associated with, or even the cause of, the genetic risk factor for developing the disease in question.

One example is the insertion of an *Alu* element at the *CD58* locus on a haplotype that affects susceptibility to MS. Among the many risk alleles identified by GWAS for this disease, the *CD58* genotype is one of the strongest genetic predictors of MS risk. The risk allele reduces *CD58* gene expression, and this reduction in *CD58* expression is associated with the risk of developing MS and relapsing. An in-depth study of the sequence of this locus showed a very strong genetic link between a

302 bp *Alu* insertion and a particular SNP (rs2300747 for those in the know). Only this allele of the *CD58* locus containing both the *Alu* insertion and the rs2300747 SNP is associated with an increased risk of developing the disease. This *Alu* insertion is relatively frequent, being present in 66% of individuals. Moreover, this *Alu* sequence insertion alters *CD58* mRNA splicing, promoting exon 3 skipping and a reading frame shift leading to a nonfunctional mRNA (Payer et al. 2019). These observations therefore show that routine insertion of an *Alu* sequence compromises *CD58* gene expression, creating an allele that increases susceptibility to MS. More generally, we can therefore conclude that inherited TE insertion alleles, which are common to all our genomes, can modify the risk of developing many common types of disease.

3.5. The role of stress and environmental pollution in TE mobility

In plants and animals, various forms of stress are known to act as triggers or facilitators of TE mobility (Lanciano and Mirouze 2018) (see Chapters 4 and 5). Several plant studies report TE overexpression following abiotic or biotic stress conditions such as temperature, nitrate deprivation, wounding, etc. (Hou et al. 2019; Fambrini et al. 2020). Activation of TEs in response to various stresses has also been demonstrated in insects, notably Drosophila (Capy et al. 2000; García Guerreiro 2012), but also in the context of insecticide resistance (Chénais et al. 2012). This suggests that human exposure to pesticides and other environmental pollutants could affect the mobility of TEs in humans, and consequently be the cause of certain pathologies (Chénais 2022).

To date, there is no direct demonstration of a causal relationship between exposure to a pollutant, the mobility of TEs and the development of a pathology. However, the arguments in favor of this hypothesis are of two types: (1) it has been shown experimentally or in population studies that pesticides and other environmental contaminants can modify the various levels of epigenetic regulation and thus create a context favorable to TE mobility; (2) mobility tests carried out on human or rodent cell models have demonstrated that exposure to various contaminants or stresses can promote the mobility of TEs, notably *LINE-1* elements (Del Re and Giorgi 2020).

3.5.1. *Epigenetic alterations caused by environmental pollutants*

Pesticides of various classes can induce epigenetic modifications such as DNA demethylation, histone modification and microRNA synthesis. Several animal studies have shown hypomethylation of certain promoters after exposure to high doses of pesticides (organochlorines, biphenyl-polychlorines, methyl-mercury). In addition, dichlorodiphenyltrichloroethane (DDT) is known to alter DNA methylation profiles, particularly in the brains of young rats, where DDT-induced hypomethylation suggests a role for this pesticide in neurodegenerative diseases (Shutoh et al. 2009). Metoxychlor, an organochlorine pesticide acting as an endocrine disruptor, also alters methylation imprinting leading to reproductive system defects in female offspring (Stouder and Paoloni-Giacobino 2011). The methylation of all of the genes tested is affected and these changes are passed on to the offspring as only the gametes are affected, but they gradually disappear between the first and third generation. Vinclozolin, a dicarboximide fungicide known to be an endocrine disruptor, also alters methylation patterns in mouse embryos (Kang et al. 2011).

In humans, global DNA hypomethylation, measured by % 5-methyl-cytosine at *Alu* and *LINE-1* sequences in adipocytes where xenobiotics are concentrated, has been linked to low-dose exposure to organochlorine pesticides in healthy individuals (Kim et al. 2010). The same correlation between blood levels of persistent organic pollutants (DDT, DDE, β-BHC, oxychlordane, mirex) or polychlorobenzenes and global DNA hypomethylation was observed in another population (Rusiecki et al. 2008). Arsenic is also a common pollutant in the waters of some countries and its impact on methylation, particularly at the level of *LINE-1* elements, has been evaluated in numerous, sometimes contradictory studies (Paul et al. 2017).

The modification of DNA methylation levels, and in particular the hypomethylation of *Alu* and *LINE-1* sequences by pesticides and other pollutants, represents a highly probable mechanism for the reactivation of TEs and consequent induction of pathologies, particularly cancers (Chénais 2022).

3.5.2. *Methylation and mobility of LINE-1 elements in response to environmental stresses*

The use of *LINE-1* element mobility assays using cells transfected with modified *LINE-1* elements containing a reporter gene, and the estimation of the copy number and methylation level of these TEs are valuable approaches for assessing the impact of stress or environmental pollutants on *LINE-1* activity. These approaches, which are still not widely used, are being developed and enable us to conclude that environmental pollutants have a strong impact on the activity of *LINE-1* elements and, by extension, on all TEs. Numerous pollutants have been tested for their ability to activate the mobility of *LINE-1* elements in various human and animal cell models (Del Re and Giorgi 2020). Metals such as aluminum, arsenic, cadmium, mercury and nickel were found to increase *LINE-1* retrotransposition events, while cobalt, lead, magnesium and zinc had no effect. Copper even appears to have an inhibitory effect (Karimi et al. 2014). It is important to note, however, that these metals were tested alone, and that it might be interesting to analyze the effects of mixtures, since this is what humans are most often exposed to. Carcinogens such as benzo[a]pyrene, phorbol esters and heterocyclic amines also appear to induce the mobility of *LINE-1* elements, although the results obtained in mouse models are not necessarily generalizable to humans (Del Re and Giorgi 2020). Certain psychoactive substances such as cocaine and methamphetamine are also capable of inducing the displacement of *LINE-1* elements in human neuronal cells. This has not been tested for morphine, but it has been shown to induce hypomethylation of the *LINE-1* promoter and an increase in its expression (mRNA levels) (Trivedi et al. 2014; Del Re and Giorgi 2020).

On the other hand, ionizing radiation, gamma radiation (from the decay of radioactive elements) and X-rays increase the mobility of *LINE-1* elements in cellular models, but at doses much higher than the usual exposure conditions. Ultraviolet (UV) radiation also appears to be able to increase *LINE-1* element expression and mobility but, curiously, these results were obtained in cell types not naturally exposed to UV (Teneng et al. 2007; Del Re and Giorgi 2020). On the other hand, very low-frequency magnetic fields, generated by power lines and ordinary electrical and electronic appliances, alter neither the methylation level nor the activity of *LINE-1* elements (Del Re and Giorgi 2020).

3.5.3. *Influence of occupational and psychosocial stresses on the mobility of LINE-1 elements*

Psychosocial stresses, such as isolation or lack of maternal care, can also affect *LINE-1* elements. For example, it is possible to observe an increase in *LINE-1* copy number in the hippocampus of 8-week-old mice compared with 3-week-old mice. This increase in copy number was not found in mice placed alone in an isolated cage immediately after weaning. This social isolation after weaning, which is a stress classically used in rodents, leads to a decrease in the retrotransposition rate of *LINE-1* in 8-week-old mice (Ueno et al. 2016). Conversely, a decrease in maternal care will cause an increase in the transcription and mobility of *LINE-1* elements in the hippocampus of the mice. Thus, the time mothers spend on maternal care is significantly and inversely correlated with their offspring's L1 copy number (Bedrosian et al. 2018). Physical exercise has also been tried as a source of stress and potential inducer of retrotransposition of *LINE-1* elements. Using transgenic mice that will express the EGFP (*enhanced green fluorescent protein*) reporter gene only after a somatic retrotransposition event, it has been shown that the number of *EGFP*-positive cells, thus reflecting the number of *LINE-1* retrotranspositions, is three times higher in the brains of running mice compared with non-running control mice (Muotri et al. 2005).

However, since all of these results were obtained in mice, we must remain cautious about generalizing them to humans. Indeed, particularly in the case of transgenic mice with the 5'UTR region of a human *LINE-1* element as promoter, it is not certain that this element is subject to the same regulations in the mouse genome as in the human genome from which it is derived.

3.6. Conclusion

Initially regarded as junk DNA, TEs are now recognized as major players in genome evolution, and their role in genome plasticity has now been clearly demonstrated. The high abundance of TEs in the human genome, in particular the *Alu* and *LINE-1* repeats, makes them responsible for the molecular origin of many diseases. This involves several molecular mechanisms, in addition to TE insertion per se, including DNA recombination and chromosomal rearrangements (particularly deletion); modification of gene expression through the introduction or alteration of poly-A sites, splice sites, exons or introns and regulatory elements, as well as alteration of epigenetic regulations. Consequently, improving our knowledge of TEs through in-depth genomic approaches may lead to potential new diagnostic markers of disease and prenatal markers of genetic disease. Thus, the development of combined molecular and bioinformatics tools is of great interest, but this has so far mainly been used in the field of oncology (Clayton et al. 2016; Tang et al. 2017; Rodriguez-Martin et al. 2020). Finally, the exposure of individuals to numerous stresses and in particular to environmental pollutants and contaminants seems to have a nonnegligible impact on epigenetic derepression and TE mobility, potentially leading in this way to the development of certain pathologies such as cancers or neurological or autoimmune diseases.

3.7. References

Ahmadi, A., De Toma, I., Vilor-Tejedor, N., EftekhariyanGhamsari, M.R., Sadeghi, I. (2020). Transposable elements in brain health and disease. *Ageing Research Reviews*, 64, 101153.

Aneichyk, T., Hendriks, W.T., Yadav, R., Shin, D., Gao, D., Vaine, C.A., Collins, R.L., Domingo, A., Currall, B., Stortchevoi, A. et al. (2018). Dissecting the causal mechanism of X-linked dystonia-Parkinsonism by integrating genome and transcriptome assembly. *Cell*, 172(5), 897–909, e21.

Bedrosian, T.A., Quayle, C., Novaresi, N., Gage, F.H. (2018). Early life experience drives structural variation of neural genomes in mice. *Science*, 359(6382), 1395–1399.

Bragg, D.C., Mangkalaphiban, K., Vaine, C.A., Kulkarni, N.J., Shin, D., Yadav, R., Dhakal, J., Ton, M.-L., Cheng, A., Russo, C.T. et al. (2017). Disease onset in X-linked dystonia-Parkinsonism correlates with expansion of a hexameric repeat within an SVA retrotransposon in TAF1. *Proceedings of the National Academy of Sciences of the United States of America*, 114(51), E11020–E11028.

Bundo, M., Toyoshima, M., Okada, Y., Akamatsu, W., Ueda, J., Nemoto-Miyauchi, T., Sunaga, F., Toritsuka, M., Ikawa, D., Kakita, A. et al. (2014). Increased L1 retrotransposition in the neuronal genome in schizophrenia. *Neuron*, 81(2), 306–313.

Burns, K.H. (2020). Our conflict with transposable elements and its implications for human disease. *Annual Review of Pathology*, 15, 51–70.

Capy, P., Gasperi, G., Biémont, C., Bazin, C. (2000). Stress and transposable elements: Co-evolution or useful parasites? *Heredity*, 85(Pt 2), 101–106.

Chénais, B. (2015). Transposable elements in cancer and other human diseases. *Current Cancer Drug Targets*, 15(3), 227–242.

Chénais, B. (2022). Transposable elements and human diseases: Mechanisms and implication in the response to environmental pollutants. *International Journal of Molecular Sciences*, 23, 2551.

Chénais, B., Caruso, A., Hiard, S., Casse, N. (2012). The impact of transposable elements on eukaryotic genomes: From genome size increase to genetic adaptation to stressful environments. *Gene*, 509(1), 7–15.

Clayton, E.A., Wang, L., Rishishwar, L., Wang, J., McDonald, J.F., Jordan, I.K. (2016). Patterns of transposable element expression and insertion in cancer. *Frontiers in Molecular Biosciences*, 3, 76.

Del Re, B. and Giorgi, G. (2020). Long INterspersed element-1 mobility as a sensor of environmental stresses. *Environmental and Molecular Mutagenesis*, 61(4), 465–493.

Ewing, A.D. and Kazazian, H.H. (2010). High-throughput sequencing reveals extensive variation in human-specific L1 content in individual human genomes. *Genome Research*, 20(9), 1262–1270.

Fambrini, M., Usai, G., Vangelisti, A., Mascagni, F., Pugliesi, C. (2020). The plastic genome: The impact of transposable elements on gene functionality and genomic structural variations. *Genesis*, 58(12), e23399.

Ferlini, A., Galié, N., Merlini, L., Sewry, C., Branzi, A., Muntoni, F. (1998). A novel Alu-like element rearranged in the dystrophin gene causes a splicing mutation in a family with X-linked dilated cardiomyopathy. *American Journal of Human Genetics*, 63(2), 436–446.

Ferreira, P.G., Jares, P., Rico, D., Gómez-López, G., Martínez-Trillos, A., Villamor, N., Ecker, S., González-Pérez, A., Knowles, D.G., Monlong, J. et al. (2014). Transcriptome characterization by RNA sequencing identifies a major molecular and clinical subdivision in chronic lymphocytic leukemia. *Genome Research*, 24(2), 212–226.

Franke, G., Bausch, B., Hoffmann, M.M., Cybulla, M., Wilhelm, C., Kohlhase, J., Scherer, G., Neumann, H.P.H. (2009). Alu-Alu recombination underlies the vast majority of large VHL germline deletions: Molecular characterization and genotype-phenotype correlations in VHL patients. *Human Mutation*, 30(5), 776–786.

Ganguly, A., Dunbar, T., Chen, P., Godmilow, L., Ganguly, T. (2003). Exon skipping caused by an intronic insertion of a young Alu Yb9 element leads to severe hemophilia A. *Human Genetics*, 113(4), 348–352.

García Guerreiro, M.P. (2012). What makes transposable elements move in the *Drosophila* genome? *Heredity*, 108(5), 461–468.

Grégoire, L., Haudry, A., Lerat, E. (2016). The transposable element environment of human genes is associated with histone and expression changes in cancer. *BMC Genomics*, 17, 588.

Guffanti, G., Gaudi, S., Fallon, J.H., Sobell, J., Potkin, S.G., Pato, C., Macciardi, F. (2014). Transposable elements and psychiatric disorders. *American Journal of Medical Genetics. Part B, Neuropsychiatric Genetics: The Official Publication of the International Society of Psychiatric Genetics*, 165B(3), 201–216.

Hancks, D.C. and Kazazian, H.H. (2016). Roles for retrotransposon insertions in human disease. *Mobile DNA*, 7, 9.

Harteveld, K.L., Losekoot, M., Fodde, R., Giordano, P.C., Bernini, L.F. (1997). The involvement of Alu repeats in recombination events at the alpha-globin gene cluster: Characterization of two alphazero-thalassaemia deletion breakpoints. *Human Genetics*, 99(4), 528–534.

Henthorn, P.S., Mager, D.L., Huisman, T.H., Smithies, O. (1986). A gene deletion ending within a complex array of repeated sequences 3' to the human beta-globin gene cluster. *Proceedings of the National Academy of Sciences of the United States of America*, 83(14), 5194–5198.

Hobbs, H.H., Brown, M.S., Goldstein, J.L. (1992). Molecular genetics of the LDL receptor gene in familial hypercholesterolemia. *Human Mutation*, 1(6), 445–466.

Hou, J., Lu, D., Mason, A.S., Li, B., Xiao, M., An, S., Fu, D. (2019). Non-coding RNAs and transposable elements in plant genomes: Emergence, regulatory mechanisms and roles in plant development and stress responses. *Planta*, 250(1), 23–40.

Hsieh, S.-Y., Chen, W.-Y., Yeh, T.-S., Sheen, I.-S., Huang, S.-F. (2005). High-frequency Alu-mediated genomic recombination/deletion within the caspase-activated DNase gene in human hepatoma. *Oncogene*, 24(43), 6584–6589.

Huang, L.S., Ripps, M.E., Korman, S.H., Deckelbaum, R.J., Breslow, J.L. (1989). Hypobetalipoproteinemia due to an apolipoprotein B gene exon 21 deletion derived by Alu-Alu recombination. *The Journal of Biological Chemistry*, 264(19), 11394–11400.

Huang, C.R.L., Schneider, A.M., Lu, Y., Niranjan, T., Shen, P., Robinson, M.A., Steranka, J.P., Valle, D., Civin, C.I., Wang, T. et al. (2010). Mobile interspersed repeats are major structural variants in the human genome. *Cell*, 141(7), 1171–1182.

Jeffs, A.R., Benjes, S.M., Smith, T.L., Sowerby, S.J., Morris, C.M. (1998). The BCR gene recombines preferentially with Alu elements in complex BCR-ABL translocations of chronic myeloid leukaemia. *Human Molecular Genetics*, 7(5), 767–776.

Jelassi, A., Slimani, A., Rabès, J.P., Jguirim, I., Abifadel, M., Boileau, C., Najah, M., M'rabet, S., Mzid, J., Slimane, M.N. et al. (2012). Genomic characterization of two deletions in the LDLR gene in Tunisian patients with familial hypercholesterolemia. *Clinica Chimica Acta; International Journal of Clinical Chemistry*, 414, 146–151.

Kang, E.-R., Iqbal, K., Tran, D.A., Rivas, G.E., Singh, P., Pfeifer, G.P., Szabó, P.E. (2011). Effects of endocrine disruptors on imprinted gene expression in the mouse embryo. *Epigenetics*, 6(7), 937–950.

Karimi, A., Madjd, Z., Habibi, L., Akrami, S.M. (2014). Evaluating the extent of *LINE-1* mobility following exposure to heavy metals in HepG2 cells. *Biological Trace Element Research*, 160(1), 143–151.

Kidd, J.M., Graves, T., Newman, T.L., Fulton, R., Hayden, H.S., Malig, M., Kallicki, J., Kaul, R., Wilson, R.K., Eichler, E.E. (2010). A human genome structural variation sequencing resource reveals insights into mutational mechanisms. *Cell*, 143(5), 837–847.

Kim, K.-Y., Kim, D.-S., Lee, S.-K., Lee, I.-K., Kang, J.-H., Chang, Y.-S., Jacobs, D.R., Steffes, M., Lee, D.-H. (2010). Association of low-dose exposure to persistent organic pollutants with global DNA hypomethylation in healthy Koreans. *Environmental Health Perspectives*, 118(3), 370–374.

Knebelmann, B., Forestier, L., Drouot, L., Quinones, S., Chuet, C., Benessy, F., Saus, J., Antignac, C. (1995). Splice-mediated insertion of an Alu sequence in the COL4A3 mRNA causing autosomal recessive Alport syndrome. *Human Molecular Genetics*, 4(4), 675–679.

de Koning, A.P.J., Gu, W., Castoe, T.A., Batzer, M.A., Pollock, D.D. (2011). Repetitive elements may comprise over two-thirds of the human genome. *PLoS Genetics*, 7(12), e1002384.

Konkel, M.K., Walker, J.A., Batzer, M.A. (2010). LINEs and SINEs of primate evolution. *Evolutionary Anthropology: Issues, News, and Reviews*, 19(6), 236–249.

Kornreich, R., Bishop, D.F., Desnick, R.J. (1990). Alpha-galactosidase A gene rearrangements causing Fabry disease. Identification of short direct repeats at breakpoints in an Alu-rich gene. *The Journal of Biological Chemistry*, 265(16), 9319–9326.

Lanciano, S. and Mirouze, M. (2018). Transposable elements: All mobile, all different, some stress responsive, some adaptive? *Current Opinion in Genetics & Development*, 49, 106–114.

Lander, E.S., Linton, L.M., Birren, B., Nusbaum, C., Zody, M.C., Baldwin, J., Devon, K., Dewar, K., Doyle, M., FitzHugh, W. et al. (2001). Initial sequencing and analysis of the human genome. *Nature*, 409(6822), 860–921.

Landry, J.-R., Rouhi, A., Medstrand, P., Mager, D.L. (2002). The Opitz syndrome gene Mid1 is transcribed from a human endogenous retroviral promoter. *Molecular Biology and Evolution*, 19(11), 1934–1942.

Lanikova, L., Kucerova, J., Indrak, K., Divoka, M., Issa, J.-P., Papayannopoulou, T., Prchal, J.T., Divoky, V. (2013). β-Thalassemia due to intronic *LINE-1* insertion in the β-globin gene (HBB): Molecular mechanisms underlying reduced transcript levels of the β-globin(L1) allele. *Human Mutation*, 34(10), 1361–1365.

Li, W., Jin, Y., Prazak, L., Hammell, M., Dubnau, J. (2012). Transposable elements in TDP-43-mediated neurodegenerative disorders. *PLoS One*, 7(9), e44099.

Makino, S., Kaji, R., Ando, S., Tomizawa, M., Yasuno, K., Goto, S., Matsumoto, S., Tabuena, M.D., Maranon, E., Dantes, M. et al. (2007). Reduced neuron-specific expression of the TAF1 gene is associated with X-linked dystonia-parkinsonism. *American Journal of Human Genetics*, 80(3), 393–406.

Martínez-Garay, I., Ballesta, M.J., Oltra, S., Orellana, C., Palomeque, A., Moltó, M.D., Prieto, F., Martínez, F. (2003). Intronic L1 insertion and F268S, novel mutations in RPS6KA3 (RSK2) causing Coffin-Lowry syndrome. *Clinical Genetics*, 64(6), 491–496.

Meischl, C., Boer, M., Ahlin, A., Roos, D. (2000). A new exon created by intronic insertion of a rearranged *LINE-1* element as the cause of chronic granulomatous disease. *European Journal of Human Genetics: EJHG*, 8(9), 697–703.

Miki, Y., Nishisho, I., Horii, A., Miyoshi, Y., Utsunomiya, J., Kinzler, K.W., Vogelstein, B., Nakamura, Y. (1992). Disruption of the APC gene by a retrotransposal insertion of L1 sequence in a colon cancer. *Cancer Research*, 52(3), 643–645.

Miki, Y., Katagiri, T., Kasumi, F., Yoshimoto, T., Nakamura, Y. (1996). Mutation analysis in the BRCA2 gene in primary breast cancers. *Nature Genetics*, 13(2), 245–247.

Miné, M., Chen, J.-M., Brivet, M., Desguerre, I., Marchant, D., de Lonlay, P., Bernard, A., Férec, C., Abitbol, M., Ricquier, D. et al. (2007). A large genomic deletion in the PDHX gene caused by the retrotranspositional insertion of a full-length *LINE-1* element. *Human Mutation*, 28(2), 137–142.

Misiak, B., Ricceri, L., Sąsiadek, M.M. (2019). Transposable elements and their epigenetic regulation in mental disorders: Current evidence in the field. *Frontiers in Genetics*, 10, 580.

Mitchell, G.A., Labuda, D., Fontaine, G., Saudubray, J.M., Bonnefont, J.P., Lyonnet, S., Brody, L.C., Steel, G., Obie, C., Valle, D. (1991). Splice-mediated insertion of an Alu sequence inactivates ornithine delta-aminotransferase: A role for Alu elements in human mutation. *Proceedings of the National Academy of Sciences of the United States of America*, 88(3), 815–819.

Morse, B., Rotherg, P.G., South, V.J., Spandorfer, J.M., Astrin, S.M. (1988). Insertional mutagenesis of the myc locus by a *LINE-1* sequence in a human breast carcinoma. *Nature*, 333(6168), 87–90.

Muotri, A.R., Chu, V.T., Marchetto, M.C.N., Deng, W., Moran, J.V., Gage, F.H. (2005). Somatic mosaicism in neuronal precursor cells mediated by L1 retrotransposition. *Nature*, 435(7044), 903–910.

Narita, N., Nishio, H., Kitoh, Y., Ishikawa, Y., Ishikawa, Y., Minami, R., Nakamura, H., Matsuo, M. (1993). Insertion of a 5' truncated L1 element into the 3' end of exon 44 of the dystrophin gene resulted in skipping of the exon during splicing in a case of Duchenne muscular dystrophy. *The Journal of Clinical Investigation*, 91(5), 1862–1867.

Neote, K., McInnes, B., Mahuran, D.J., Gravel, R.A. (1990). Structure and distribution of an Alu-type deletion mutation in Sandhoff disease. *The Journal of Clinical Investigation*, 86(5), 1524–1531.

Oldridge, M., Zackai, E.H., McDonald-McGinn, D.M., Iseki, S., Morriss-Kay, G.M., Twigg, S.R., Johnson, D., Wall, S.A., Jiang, W., Theda, C. et al. (1999). De novo alu-element insertions in FGFR2 identify a distinct pathological basis for Apert syndrome. *American Journal of Human Genetics*, 64(2), 446–461.

Oliveira, C., Senz, J., Kaurah, P., Pinheiro, H., Sanges, R., Haegert, A., Corso, G., Schouten, J., Fitzgerald, R., Vogelsang, H. et al. (2009). Germline CDH1 deletions in hereditary diffuse gastric cancer families. *Human Molecular Genetics*, 18(9), 1545–1555.

O'Neil, J., Tchinda, J., Gutierrez, A., Moreau, L., Maser, R.S., Wong, K.-K., Li, W., McKenna, K., Liu, X.S., Feng, B. et al. (2007). Alu elements mediate MYB gene tandem duplication in human T-ALL. *The Journal of Experimental Medicine*, 204(13), 3059–3066.

Onno, M., Nakamura, T., Hillova, J., Hill, M. (1992). Rearrangement of the human tre oncogene by homologous recombination between Alu repeats of nucleotide sequences from two different chromosomes. *Oncogene*, 7(12), 2519–2523.

Paul, S., Bhattacharjee, P., Giri, A.K., Bhattacharjee, P. (2017). Arsenic toxicity and epimutagenecity: The new LINEage. *Biometals: An International Journal on the Role of Metal Ions in Biology, Biochemistry, and Medicine*, 30(4), 505–515.

Payer, L.M., Steranka, J.P., Yang, W.R., Kryatova, M., Medabalimi, S., Ardeljan, D., Liu, C., Boeke, J.D., Avramopoulos, D., Burns, K.H. (2017). Structural variants caused by Alu insertions are associated with risks for many human diseases. *Proceedings of the National Academy of Sciences of the United States of America*, 114(20), E3984–E3992.

Payer, L.M., Steranka, J.P., Ardeljan, D., Walker, J., Fitzgerald, K.C., Calabresi, P.A., Cooper, T.A., Burns, K.H. (2019). Alu insertion variants alter mRNA splicing. *Nucleic Acids Research*, 47(1), 421–431.

Petrozziello, T., Dios, A.M., Mueller, K.A., Vaine, C.A., Hendriks, W.T., Glajch, K.E., Mills, A.N., Mangkalaphiban, K., Penney, E.B., Ito, N. et al. (2020). SVA insertion in X-linked dystonia Parkinsonism alters histone H3 acetylation associated with TAF1 gene. *PLoS One*, 15(12), e0243655.

Phylipsen, M., Vogelaar, I.P., Schaap, R.A.C., Arkesteijn, S.G.J., Boxma, G.L., van Helden, W.C.H., Wildschut, I.C.M., de Bruin-Roest, A.C., Giordano, P.C., Harteveld, C.L. (2010). A new alpha(0)-thalassemia deletion found in a Dutch family (–(AW)). *Blood Cells, Molecules & Diseases*, 45(2), 133–135.

Rickman, K.A., Lach, F.P., Abhyankar, A., Donovan, F.X., Sanborn, E.M., Kennedy, J.A., Sougnez, C., Gabriel, S.B., Elemento, O., Chandrasekharappa, S.C. et al. (2015). Deficiency of UBE2T, the E2 ubiquitin ligase necessary for FANCD2 and FANCI ubiquitination, causes FA-T subtype of Fanconi anemia. *Cell Reports*, 12(1), 35–41.

Rodić, N., Zampella, J.G., Cornish, T.C., Wheelan, S.J., Burns, K.H. (2013). Translocation junctions in TCF3-PBX1 acute lymphoblastic leukemia/lymphoma cluster near transposable elements. *Mobile DNA*, 4(1), 22.

Rodriguez-Martin, B., Alvarez, E.G., Baez-Ortega, A., Zamora, J., Supek, F., Demeulemeester, J., Santamarina, M., Ju, Y.S., Temes, J., Garcia-Souto, D. et al. (2020). Pan-cancer analysis of whole genomes identifies driver rearrangements promoted by *LINE-1* retrotransposition. *Nature Genetics*, 52(3), 306–319.

Ross, J.P., Rand, K.N., Molloy, P.L. (2010). Hypomethylation of repeated DNA sequences in cancer. *Epigenomics*, 2(2), 245–269.

Rusiecki, J.A., Baccarelli, A., Bollati, V., Tarantini, L., Moore, L.E., Bonefeld-Jorgensen, E.C. (2008). Global DNA hypomethylation is associated with high serum-persistent organic pollutants in Greenlandic Inuit. *Environmental Health Perspectives*, 116(11), 1547–1552.

Scott, E.C., Gardner, E.J., Masood, A., Chuang, N.T., Vertino, P.M., Devine, S.E. (2016). A hot L1 retrotransposon evades somatic repression and initiates human colorectal cancer. *Genome Research*, 26(6), 745–755.

Shabbeer, J., Yasuda, M., Benson, S.D., Desnick, R.J. (2006). Fabry disease: Identification of 50 novel alpha-galactosidase A mutations causing the classic phenotype and three-dimensional structural analysis of 29 missense mutations. *Human Genomics*, 2(5), 297–309.

Shutoh, Y., Takeda, M., Ohtsuka, R., Haishima, A., Yamaguchi, S., Fujie, H., Komatsu, Y., Maita, K., Harada, T. (2009). Low dose effects of dichlorodiphenyltrichloroethane (DDT) on gene transcription and DNA methylation in the hypothalamus of young male rats: Implication of hormesis-like effects. *The Journal of Toxicological Sciences*, 34(5), 469–482.

Stouder, C. and Paoloni-Giacobino, A. (2011). Specific transgenerational imprinting effects of the endocrine disruptor methoxychlor on male gametes. *Reproduction*, 141(2), 207–216.

Teneng, I., Stribinskis, V., Ramos, K.S. (2007). Context-specific regulation of *LINE-1*. *Genes to Cells: Devoted to Molecular & Cellular Mechanisms*, 12(10), 1101–1110.

Thomas, J., Perron, H., Feschotte, C. (2018). Variation in proviral content among human genomes mediated by LTR recombination. *Mobile DNA*, 9, 36.

Trivedi, M., Shah, J., Hodgson, N., Byun, H.-M., Deth, R. (2014). Morphine induces redox-based changes in global DNA methylation and retrotransposon transcription by inhibition of excitatory amino acid transporter type 3-mediated cysteine uptake. *Molecular Pharmacology*, 85(5), 747–757.

Tubio, J.M.C, Li, Y., Ju, Y.S., Martincorena, I., Cooke, S.L., Tojo, M., Gundem, G., Pipinikas, C.P., Zamora, J., Raine, K. et al. (2014). Mobile DNA in cancer. Extensive transduction of nonrepetitive DNA mediated by L1 retrotransposition in cancer genomes. *Science*, 345(6196), 1251343.

Ueno, M., Okamura, T., Mishina, M., Ishizaka, Y. (2016). Modulation of long interspersed nuclear element-1 in the mouse hippocampus during maturation. *Mobile Genetic Elements*, 6(4), e1211980.

Vervoort, R., Gitzelmann, R., Lissens, W., Liebaers, I. (1998). A mutation (IVS8+0.6kbdelTC) creating a new donor splice site activates a cryptic exon in an Alu-element in intron 8 of the human beta-glucuronidase gene. *Human Genetics*, 103(6), 686–693.

Wallace, M.R., Andersen, L.B., Saulino, A.M., Gregory, P.E., Glover, T.W., Collins, F.S. (1991). A de novo Alu insertion results in neurofibromatosis type 1. *Nature*, 353(6347), 864–866.

Xing, J., Zhang, Y., Han, K., Salem, A.H., Sen, S.K., Huff, C.D., Zhou, Q., Kirkness, E.F., Levy, S., Batzer, M.A. et al. (2009). Mobile elements create structural variation: Analysis of a complete human genome. *Genome Research*, 19(9), 1516–1526.

Yu, J., Xie, J., Luo, L., Li, Z. (2014). An Alu element-mediated 28.5 kb α-thalassemia deletion found in a Chinese family. *Hemoglobin*, 38(6), 427–430.

4

The Silencing Mechanisms Inhibiting Transposable Element Activity in Somatic and Germ Cells

Chantal VAURY
Institute of Genetics, Reproduction and Development (iGRed), CNRS, Université Clermont Auvergne, Clermont-Ferrand, France

4.1. Introduction

Two fundamental cell types are found in multicellular organisms: somatic cells and germ cells. In animals, somatic cells constitute all of the tissues from organs to skin or bones. Any mutation occurring within a somatic cell will be transmitted through mitosis to the descendant cell. Its deleterious effect is then restricted to the tissue affected. Germ cells produce gametes: spermatozoa and ova, whose primary function is to transmit genetic information to future embryos, resulting from fertilization. Since gametes transmit genetic information through meiosis, any mutation occurring in a germ cell will be transmitted from parents to offspring. In response to their harmful impact, mechanisms exist in both types of cells to ensure genomic protection against mutations. A lower frequency of spontaneous mutations, indicative of a higher degree of protection, has been observed in the germ line in which any mutation can have deleterious effects on the entire progeny (Murphey et al. 2013).

Among agents that may induce mutations are transposable elements (TEs) (Bourque et al. 2018). TEs are major constituents of genomes and, through diverse

Transposable Elements and Genome Evolution,
coordinated by Aurélie HUA-VAN and Pierre CAPY. © ISTE Ltd 2024.

strategies, have the capacity to move from one chromosome site to another. These mobilizations may then affect genomic expression and be a threat to genetic stability. In somatic tissues of humans and animals, they have been identified as pathogenic factors implicated in diseases such as cancers or neurodegenerative disorders (see Chapter 3). In the germline, a high mutation rate due to a non-controlled mobilization may not affect the parents themselves but the progeny in a trans-generation transmission. Harmful impact of new TE insertions may then jeopardize the whole species' fitness (Choi and Lee 2020). Dedicated systems have evolved to constrain TE activity. In somatic and in germ line cells, the mechanisms involved differ, although they both entail ancient and conserved mechanisms based on DNA, chromatin modification and small RNA (sRNA)-mediated silencing. This chapter updates the highly connected, dynamic set of mechanisms through which TE control is assured or sometimes lost in both types of cells.

4.2. Silencing of transposable elements in somatic tissues

4.2.1. *Mechanisms that ensure deposition of epigenetic hallmarks at TEs*

In the soma, TE regulation depends on the tissue specificity of the TE promoter and on an active transcriptional and post-transcriptional repression. Several studies have been carried out to accurately investigate the tissue-specific expression of each TE because of these multiple layers of control exerted on the elements (Desset et al. 2008). However, we have to keep in mind that the first barriers defining the restricted transcriptional competence of TEs are the cell-type transcription factors (TFs) exploited (or not) by TEs for their own expression (Desset et al. 2008; Ghildiyal et al. 2008; Kuwabara et al. 2009).

In cells competent for TE transcription, the survey of TE activity hinges on three critical actors: small RNAs, DNA methylation and chromatin compaction. Endogenous small interfering RNAs (endo-siRNAs) target TE RNAs by sequence complementarity and initiate silencing. Endo-siRNAs have been reported in many organisms, from plants to mammals (Ghildiyal et al. 2010; Hamilton et al. 2015). Their defining properties are produced from long double-strand RNAs (dsRNAs), processed by RNase III enzymes Dicer, and are 2′-O-methylated at their 3′ ends. They can induce a post-transcriptional silencing (PTGS), leading to TE mRNA destruction in the cytoplasm, or a co-transcriptional silencing (TGS) on nascent transcripts in the nucleus. DNA methylation and/or chromatin modifications are also hallmarks of chromatin structures deposited on TE loci to control their repression. Although it is not yet fully elucidated how RNAi machinery, DNA methylation and chromatin modifications are coordinated to ensure proper heterochromatin assembly, or how DNA methyl transferases (DMTs) and/or histone methyl transferases (HMTs) are recruited to their target sites, numerous studies have identified many factors involved, reported their specificities and assessed similarities and differences between species. The information from studies performed in *Arabidopsis thaliana*, *Drosophila melanogaster* and mammals is reported in the following.

In plants, transcripts from the specific RNA polymerase IV (Pol IV) are converted into double-strand RNAs due to the RNA-dependent RNA Polymerase 2 (RDR2). These double-strand RNAs are cleaved into 24-nucleotide (nt) siRNAs by DICER-LIKE3 (DCL3) and exported to the cytoplasm, where one of the strands is incorporated onto Argonaute 4 or 6 (AGO4, AGO6) (Matzke and Mosher 2014). This complex goes back to the nucleus and, by sequence complementarity, targets nascent transcripts from Pol V, mostly corresponding to TEs. A DNA methyltransferase activity is recruited to the site and mediates DNA methylation and transcriptional silencing. This transcriptionally repressive DNA methylation occurs through the methyl transferase DRM1 and 2 and is named RNA-directed DNA methylation or RdDM (Figure 4.1). This occurs at CG, CHG and CHH cytosines (H represents A, C or T) (Cao and Jacobsen 2002). Non-canonical RdDM pathways involving other types of siRNAs have also been implicated in the RdDM process such as an RdDM using 21–22 nts TE-siRNAs incorporated onto AGO6, or another using 21 to 24 siRNAs incorporated onto AGO4 (McCue et al. 2015; Ye et al. 2016). The DNA methylated sites are often marked by H3-K9 methylation; however, it has been shown that DNA methylation does not control the methylation of histone H3-K9 (Johnson et al. 2002).

In *D. melanogaster*, double-strand RNAs are cleaved by Dicer-2 into 21-nt siRNAs. An isoform of the double-strand RNA binding protein (dsRBP) Loquacious, Loqs-PD, associated with Dcr-2, is required during the nucleolytic processing of hairpin-derived endo-siRNAs (Zhou et al. 2009; Hartig and Förstemann 2011). When exported to the cytoplasm, the dsRBP R2D2 ensures endo-siRNA sorting by localizing Dcr-2 to specific cytoplasmic foci called D2 bodies, where the Argonaute protein, Ago2, is localized (Nishida et al. 2013). The matured endo-siRNAs are loaded onto Ago2 to form the RNA-induced silencing complex (RISC), which silences TEs through RNA degradation in the cytoplasm, and TGS after its transport to the nucleus (Figure 4.1). In *Drosophila* and in some other species like *Sacchamoryces pombe*, the RISC does not induce TGS via DNA methylation but via a range of chromatin modifications that suppress transcription, including histone modifications and altered chromatin packing. Histone modification for TE silencing is primarily a methylation of the amino-terminal tail of histone H3 on Lysine 9 or Lysine 27 (H3K9me or H3K27me). The methylation of H3K27 is a hallmark of the silencing mediated by the polycomb repressive complex 2 (PRC2), since it is carried out by enhancer-of-zeste (E(z)), the catalytic subunit of

PRC2. The chromodomain of the *Drosophila* PC protein from the PRC1 binds to the H3K9me3 and more strongly at K27me2/3. Then, the binding of PRC1 and PRC2 is a hallmark of the chromatin structure associated with the repressed states of TEs in somatic tissues of *Drosophila* (Dufourt et al. 2011) (Figure 4.1).

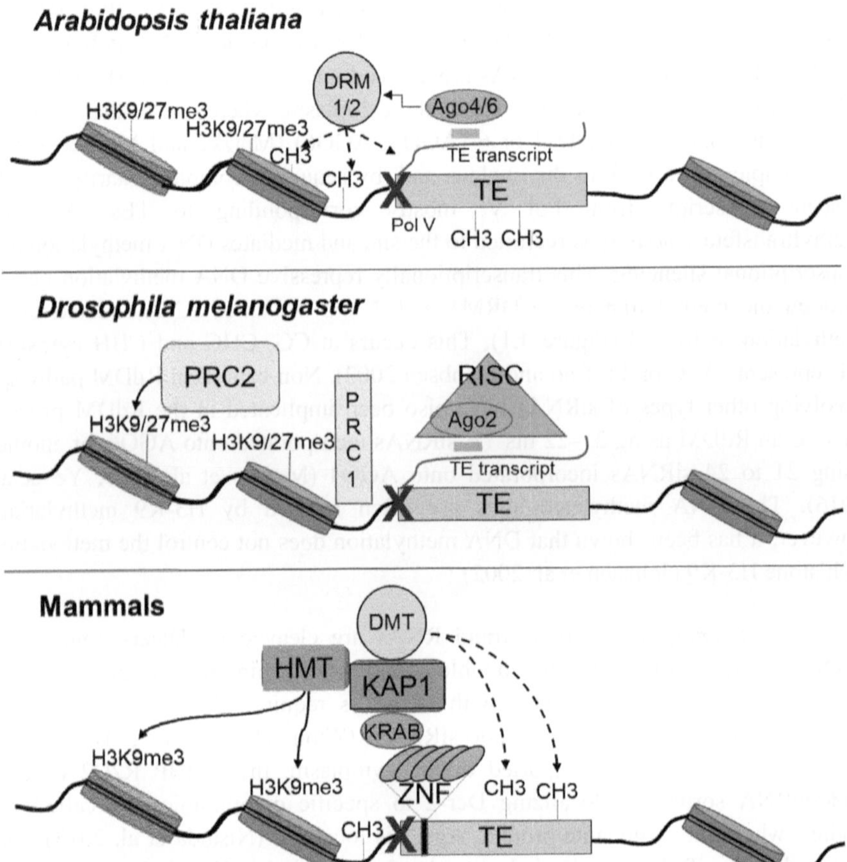

Figure 4.1. *Mechanisms of TE epigenetic silencing in somatic cells. Differences between A. thaliana, D. melanogaster, and mammals are presented. DNA: black line; TE: gray box; nucleosome: purple disk; siRNA: small light blue line; TE transcript: blue line; DRM: domain rearranged methylase (encoding DNA methylase transferase); Pol V: RNA polymerase V; PRC1/2: polycomb repressive complex; RISC: RNA-induced silencing complex; DMT: DNA methyl transferase; HMT: histone methyl transferase; DNA methylation and histone modifications are indicated by the dotted and solid lines, respectively; stop of transcription is indicated by a red cross. For a color version of this figure, see www.iste.co.uk/huavan/transposable.zip*

In mammalian genomes, TEs occupy almost half of the genome. They are the main targets of DNA methylation, which occurs almost exclusively at cytosine-guanine dinucleotides (CpG). In particular, *ERVs* (endogenous retroviruses) are found heavily DNA methylated. If DNA methylation is abolished, TEs are reactivated, which underlines its important role for the establishment of the transcriptional silencing (Walsh et al. 1998). Deposition of histone modifications, H3K9me3 or H3K27me3, ensures proper heterochromatin assembly, the transcriptional silencing hallmarks of these loci. Very little is known about how HMTs are recruited to their target sites (Elgin and Reuter 2013). siRNAs have been found in mammalian somatic cells but it is not clear yet whether they contribute to TE silencing through their sequence complementarity (Czech et al. 2008). Studies on the Krüppel-associated box (KRAB) domain containing zinc-finger proteins (KRAB-ZFPs) helped to understand how TEs may be specifically targeted. KRAB-ZFPs are transcriptional repressors (Ecco et al. 2017). They bind to DNA through their zinc finger domain (ZNF) and recruit the corepressor KAP1 (also called TRIM28). KAP1 then recruits epigenetic modifiers such as DNA methyltransferases and histone modifiers (Wolf et al. 2015) (Figure 4.1). A series of data have illustrated how the KRAB-ZFP targets the VL30 family of proviral genetic elements in mouse in a sequence-specific manner. Zfp809, a member of the KRAB-ZFP family, binds to the primer binding site (PBS-Pro) used by VL30 and some retroviruses. From here, it primes reverse transcription and recruits the complex KAP1 with its associated Setdb1 protein. Setdb1, a H3K9 methyltransferase, contributes to initiate silencing (Wolf et al. 2015). Once TE silencing is established, KRAB-ZFPs and KAP1 are no longer required.

4.2.2. *Inheritance of the epigenetic state of TEs through cell division*

A key feature of such an epigenetic silencing resulting from DNA and/or histone methylation is that, after it is initiated, it must be maintained and transmitted through cell division (Reinberg and Vales 2018).

In yeast (saccharomyces *cerevisiae*), studies have shown that establishment, spreading and maintenance are sequential steps in the assembly of centromeric heterochromatin. At mating type or telomeres, the Sir1-Orc complex (silent information regulator 1-origin recognition complex) is required to initiate heterochromatin. This recruits the Sir complex (Sir2-Sir3-Sir4) through direct interaction with Sir2. This latter desacetylates H3K16ac and the Sir complex expands this desacetylated state, establishing a compacted state. The initiator, Sir1-ORC, is then no longer required for maintaining the heterochromatic state.

The Gasser's lab has shown that in somatic cells of *Caenorabditis elegans*, two pathways distinguish the maintenance of somatic H3K9me3 from its nucleation (Padeken et al. 2021). Worms have two somatic H3K9 HMTs, met-2 and set-25. SET-25 has the unique ability to establish de novo H3K9me-mediated repression. MET-2 catalyzes H3K9me1-2. Together with LIN-61, it is required to maintain heterochromatin integrity through cell division.

In eukaryotes, polycomb group (PcG) proteins are a group of proteins critical for maintaining gene repression through mitosis (Margueron et al. 2008). Within multiprotein complexes PRC1 and PRC2, PcG proteins silence their target genes by remodeling chromatin via histones modifications. In multiple organisms, PcG proteins, more importantly E(z)-like proteins from PRC2, have been found to be associated with TE loci, where the H3K27me3 and H3K9me3 marks may co-occur (Dufourt et al. 2011; Frapporti et al. 2019; Zhao et al. 2019b). These data design PcG proteins as critical actors of epigenetic inheritance in somatic tissues. Found at TEs from diverse organisms, the interplay between small RNAs and PcG repression is proposed to be a widely conserved ancestral phenomenon that underlies cell memory of TE silencing.

With regards to transmission of TE silencing to the next generation of cell, it must be added that DNA methylation may also support the specific deamination at TEs and consequently their irreversible inactivation.

4.2.3. *Influence of environmental stresses on TE control in somatic cells*

As early as the 1950s, Barbara McClintock proposed that transposition might lead to genome restructuring and bringing the required flexibility to face environmental stresses (McClintock 1984). The environmental stresses are diverse. They can come from a natural environment such as temperature changing, ground or space radiation. They can result from human activity such as pollution, radiation and toxicity of nutrient. They can also come from idiosyncratic risks of living such as psychic stress or ageing. When organisms are confronted with these traumatic events, epigenetic hallmarks at TE loci may be erased, resulting in altered DNA methylation, histone modifications or small RNAs production. These failures in the silencing mechanisms have been related to TE transcription and transposition events, underlying human disorders and the onset of diseases, including cancers (Burns 2017; Pappalardo et al. 2021). For example, LINE-1 promoter hypomethylation has been associated with genomic instability coupled with poor prognosis in several cancers: lung (Daskalos et al. 2009), colon (Ogino et al. 2008), esophageal squamous cell carcinoma (Iwagami et al. 2013) and ovaries (Pattamadilok et al. 2008). More examples can be found in Chapter 3.

Environmental stress may also indirectly alter TE activity in genomes. The control of Ty5 retrotransposon in *S. cerevisiae* illustrates one way this control is lost. Ty5 integrates into heterochromatin at the telomeres and silent mating loci. This target specificity is due to the interaction of Ty5 integrase-targeting-domain with the heterochromatin protein Sir4 occurring when Ty5 integrase is phosphorylated (Dai et al. 2007). However, the protein kinase A pathway is weakened during starvation for amino acids, nitrogen, or fermentation carbon. Thus, when yeasts are starved, Ty5 integrase fails to be phosphorylated, which prevents its interaction with Sir4 and leads to mutations due to random integrations throughout the genome.

Beside these few examples, it has been shown that the compaction of chromatin at TE loci not only blocks their transcription but also prevents recombination between these homologous dispersed sequences, limiting the threats posed by genomic rearrangements. When compact chromatin at TEs is lost in response to environmental insults, genome restructuring associated with chromosomal inversions, deletions or duplications occurs between dispersed TE loci. This is notably observed in cancer cells (Burns 2017).

Overall, epigenetic control of TEs in somatic cells helps the host genome to withstand the threats of these repeated sequences. However, under environmental stresses, the epigenetic control can be erased, leading to mutational events and genetic instabilities associated with TE desilencing.

4.3. Silencing of transposable elements in the germline

Similar to somatic cells, epigenetic silencing of TEs in germ cells is achieved through TGS and PTGS mechanisms implicating DNA methylation, chromatin modifications and small RNAs. Here, by contrast to somatic cells, the major small RNA family used is the PIWI-interacting RNA (piRNA) family (Brennecke et al. 2007; Carmell et al. 2007; Ozata et al. 2019).

4.3.1. *The piRNA pathway in ovarian germ cells*

In animal gonads, the piRNA pathway implicates single-strand RNAs of 21–35 nts in length, phosphorylated and 2′-O-methylated at their 5′ and 3′ ends, respectively. piRNAs associated with members of the PIWI subclade of the Argonaute protein family have been found in the vast majority of animals. Studies report that they were likely established before the evolution of bilaterians (Lim et al. 2014). Although piRNA biogenesis has been studied in several organisms, among which we can cite mouse or silkworm, the mechanisms involved have been mostly elucidated in *D. melanogaster* and will be described thereafter.

In *Drosophila*, piRNAs originate from genomic regions rich in TEs, intermingled within each other and called piRNA clusters. These clusters are initially transcribed as long non-coding RNAs. What defines a piRNA cluster remains mostly unclear. The maternal inheritance of piRNAs has been incriminated (Akkouche et al. 2017; Huang et al. 2017) and recent studies have shown that establishment of a piRNA cluster may also result from environmental changes in the absence of maternal inheritance (Casier et al. 2019). In the germline, piRNA clusters are predominantly transcribed from both strands, giving the so-called dual-strand piRNA clusters. Dual-strand piRNA clusters are decorated with three proteins: Rhino, Deadlock, and Cutoff in a so-called RDC complex (Mohn et al. 2014). Transcription of piRNA clusters starts internally in both directions at sites where Moonshiner, a paralog of a TF IIA subunit, is recruited via the heterochromatin protein-1 variant Rhino bound to H3K9me3 marks at piRNA clusters (Andersen et al. 2017). If Moonshiner has been found in *D. melanogaster*, a similar non-canonical transcriptional pathway seems to occur in plants (Law et al. 2013). Cutoff is proposed to bind the 5′ end of these long RNA precursors, inhibiting their splicing and polyadenylation (Zhang et al. 2014; Chen et al. 2016). These long transcripts, unspliced and lacking a poly(A) tail, are exported to the cytoplasm where they are cleaved in a perinuclear granule called nuage (Lim and Kai 2007; Lim et al. 2013). The endonuclease Zucchini localized at the outer membrane of mitochondria is thought to process these RNAs into primary piRNAs that are loaded onto the PIWI proteins Piwi, or Aubergine (Aub). This first production of piRNAs is called the de novo pathway.

The population of piRNAs bound to Piwi (Piwi-piRNAs) leaves the cytoplasm, enters the nucleus and guides Piwi to nascent transcripts with complementary sequences. This results in the establishment of transcriptional repression and heterochromatin formation *in trans* (coTGS) at TE loci throughout the genome. The histone methyl-transferase SetDB1, histone demethylase Lsd1, SUMO E3 ligase Su(Var)2-10 and HP1a are required for this TGS at TE loci. Additional factors such as Maelstrom and Asterix are also required as well as Panoramix, which functions via the Nxf2-Nxt1 heterodimer in a nuclear protein complex called SFinX (Batki et al. 2019) or Pandas (Zhao et al. 2019a).

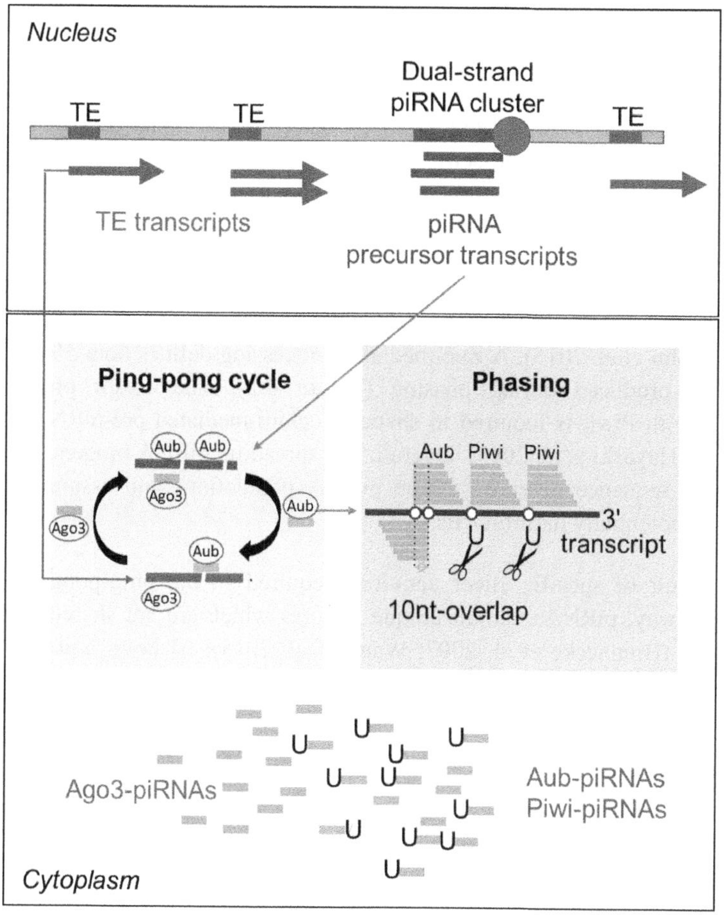

Figure 4.2. *In Drosophila germ cells, piRNAs are amplified through the ping-pong cycle and phasing pathway: precursors of piRNA clusters (blue) and mRNAs expressed from active TEs (orange) are transported from the nucleus to the cytoplasm to be processed into piRNAs. Reciprocal cleavage of piRNA precursor transcripts and TE mRNAs by the cytoplasmic endonucleases Ago3 and Aub produces Aub-piRNAs (light blue) and Ago3-piRNAs (light orange). A piRNA loaded onto Aub and generated by the ping-pong pathway initiates phased piRNAs. The endonuclease Zucchini forms the 5' end of these piRNAs, which are mostly loaded onto Piwi. For a color version of this figure, see www.iste.co.uk/huavan/transposable.zip*

The cytoplasmic Aub-piRNAs produced by the de novo pathway are predominantly anti-sense to TE mRNAs. By sequence complementarity, they target and cleave sense transcripts arising from the transcription of active TEs. The products are proposed to be transferred to Ago3 that will, after processing, give rise to mature Ago3-piRNAs. Ago3-piRNAs are therefore mainly sense-TE sequences. In turn, by sequence complementarity, Ago3-piRNAs are able to target and cleave anti-sense transcripts arising from piRNA clusters. These latter are transferred to Aub and generate new anti-sense Aub-piRNAs. Constant rounds of piRNA production lead to the amplification of sense- and anti-sense piRNAs. This system is called the ping-pong cycle (Brennecke et al. 2007) (Figure 4.2).

A third pathway called phasing then comes into play. PIWI proteins direct a stepwise fragmentation of the sliced precursors, leading to phased piRNAs (Han et al. 2015; Mohn et al. 2015). A Zuc-mediated processing defines both 5′ and 3′ ends of piRNAs produced during phasing (Figure 4.2). The Tudor protein Papi, interacting with Piwi, is required to shape Zucchini-mediated pre-piRNAs to their mature size (Hayashi et al. 2016). Overall, this spreading piRNA biogenesis helps to increase the sequence diversity of the piRNA populations and assures the solid silencing of potentially harmful TEs.

As a result of specific slicer activities required in the ping-pong cycle and phasing pathway, piRNAs exhibit unique features which are not shared with other small RNAs (Brennecke et al. 2007; Wang et al. 2015): (i) Piwi- and Aub-bound piRNAs show a strong bias for uridine (U) nt at their 5′ end; (ii) the 5′ end of a slicer product is shifted by exactly 10 nts relative to the 5′ end of the guide piRNA. Therefore Ago3-bound piRNAs tend to have an Adenine at base 10 from the 5′ end (Brennecke et al. 2007).

This strategy of RNA amplification to enhance RNA-mediated silencing of TEs is found in animals ranging from hydra to human (Gainetdinov et al. 2018).

4.3.2. *Additional strategies used to silence TEs in the germline*

In the germline, like in somatic cells, TE repression employs DNA methylation and chromatin modifications, including sequence-specific transcriptional repression by KRAB complexes. The deposition of these chromatin modifications and the mechanisms involved to target TEs are similar to the one described above in somatic tissues. This will not be detailed again. Details can also be found in several reviews (Molaro and Malik 2016; Maupetit-Mehouas and Vaury 2020). Here, we will only focus on two specific mechanisms of repression in the germline recently described in literature.

First, during germ cell development in mammals, de novo DNA methylation is performed by the DNA methyl transferases DNMT3A and B (Maupetit-Mehouas and Vaury 2020). However, a third DNMT has been discovered in rodents, DNMT3C. This DNMT displays the unique function to methylate the promoter of young retrotransposons and helps to discriminate different evolutionary ages of retrotransposons. The studies support that positive selection of *Dnmt3C* is the result of an ongoing evolutionary arms race with young retrotransposon lineages in muroid genomes. The authors hypothesize that repression of young retrotransposons might be driving the recurrent innovation of a functional domain in the N-termini on germline DNMT3s in mammals (Molaro et al. 2020).

Second, although TE mobilization appears to be constantly under control, several studies have reported transient relaxation of chromatin-based TE silencing during short windows of the germline development (Maupetit-Mehouas and Vaury 2020). This suggests that additional controls might exist. While seeking out new factors which could limit mammalian retrotransposon activity, Chelmicki et al. (2021) found that RNA methylation might play this role. They reported that m^6A RNA methylation affects mRNA abundance of ERVs. m^6A mostly occurs on the 5′ UTR of ERV mRNAs, where the tRNA binding site essential for reverse transcription is localized. Their study suggests that this pathway could prevent m^6A-modified ERV mRNAs from being reverse transcribed, translated or assembled into protective retrotransposition complexes. RNA methylation could then be considered as an additional protective pathway with several layers of control. The authors note that if IAP-derived mRNAs were mostly targeted in their study, other retrotransposon types may also undergo alternative modes of RNA methylation. They also proposed that m^6A-dependent regulation could decrease the immunopotential of ERV-RNAs and then dampen their ability to trigger inflammatory responses. Other studies have shown that RNA methyltransferases are required in the control of TEs. As an example, in Drosophila, TE silencing and genome integrity are affected by mutations of two cytosine-5 RNA methyltransferases, Dnmt2 and NSun2 (Genenncher et al. 2018). These RNA methyltransferases are known to methylate specific tRNAs, suggesting that their tRNA-related functions indirectly impact TE control. Elucidating functions of RNA methylation in TE silencing is only in its early phase of understanding. Hopefully, future studies will soon bring more light to this obscure field of regulation.

Overall, with a sophisticated mechanism leading to piRNA amplification, TGS and PTGS are highly efficient to prevent TE mobilization in the germline. This silencing is coupled with other layers of control involving the specific recognition of active TEs, and a control of retrotransposition via RNA methylation.

4.4. The specific case of somatic cells surrounding the germline

Specific mechanisms controlling TEs have been found in somatic cells located in close proximity to the germline. Characteristics of these mechanisms have been well described in model organisms such as ciliates, plants and *Drosophila*.

Ciliates are unicellular eukaryotes with two separate germinal and somatic nuclei with different functions and genome structures. Silent micronuclei contain the complete germline genome and then transmit the genetic information to the next generation. The somatic macronuclei contain a reduced genome expressing the genetic information. In *Paramecium*, rearrangements of the macronucleus genome include elimination of repeated sequences (IES for internal eliminated sequences). This process involves small RNA-directed heterochromatin formation and DNA excision. It has been hypothesized that ancient domestication of a transposase, PiggyBac, has favored propagation of TEs within the germline, which could be allowed because of the silent state of this lineage in ciliates (Arnaiz et al. 2012). In turn, the subsequent rearrangements eliminating TEs protect the transcriptionally active somatic genome from their mutagenic impact. The first direct evidence that TEs can account for the massive proliferation of IESs in *Paramecium* has been recently identified, after the micronucleus of none Paramecium species was sequenced (Sellis et al. 2021).

In plants, the terminally differentiated pollen vegetative nurse cells surround the sperm cells. Two types of gametophytic small RNAs are required for TE repression. Together with the RNA-directed DNA methylation (RdDM) pathway, 24-nt long RNAs assure TGS. If reactivated, TEs can produce 21/22-nt long RNAs that assure PTGS or, if loaded onto AGO6, also direct TGS. In the mature sperm, vegetative nurse cells undergo a programmed heterochromatin decondensation and transcriptional reactivation of TEs. 21-nt RNAs dramatically increase for several TEs. These TE small RNAs are loaded into the sperm cells, where they can silence TE reporters in a non-cell-autonomous way. It is proposed that this cell-type specific epigenetic reprogramming could serve as a mechanism that reveals TE presence to the germline, ensuring their control after fertilization (Martínez et al. 2016).

In *Drosophila* ovaries, somatic cells surrounding the germline are called follicle cells. TEs are silenced in the follicule cells via TGS and PTGS pathways similar to germline but in a simplified version. piRNA populations are 23–29 nt in length, 70% of them map to annotated TEs, and they preferentially display a uridine at their 5′ end. They originate from the processing of precursor transcripts produced by piRNA clusters that are single strand, contrary to the double-strand piRNA clusters of the germline (Brennecke et al. 2007). While transcription of piRNA clusters in the germline starts internally at sites where Moonshiner is recruited, somatic piRNA clusters are transcribed from a polymerase II promoter as long-single stranded RNAs such as regular genes. A major somatic piRNA cluster, *flamenco*, produces more than 70% of the follicular piRNAs (Brennecke et al. 2007; Goriaux et al. 2014). Thanks to the Exon Junction Complex containing Nxf1 and Nxt1, the somatic piRNA precursors are transferred to a single nuclear focus called Dot COM, where they accumulate before being exported to the cytoplasm (Dennis et al. 2016). In the cytoplasm, they localize within a structure facing Dot COM and called Yb-body where their processing is believed to happen (Qi et al. 2011). Yb-bodies are specific to the follicle cells. Follicle piRNAs are bound to Piwi, the only member of the PIWI clade expressed in these cells. Many factors required in the processing have been reported and are referred to in numerous reviews (Théron et al. 2014; Huang et al. 2017; Sato and Siomi 2020). Similar to germ cells, matured Piwi-piRNAs enter the nucleus, where they target transcripts from active TEs in a homology-dependent base-pairing mechanism. This results in the deposition of repressive marks H3K9me3 and a decrease of TE transcription. Since Ago3 and Aub are absent from the follicle cells, the pool of piRNAs is not amplified by the ping-pong cycle. This specific silencing in a somatic tissue raises the question of whether silencing of TEs expressed in the follicle cells contributes to dampening TE activity in the closely apposed oocyte via the import of TE-derived piRNAs. Indeed, a pool of Piwi-piRNAs produced within the follicle cells could be transferred passively to the oocyte during late stages of oogenesis, when an active uptake of vitellogenin produced partly by the follicle cells happens within the oocyte (Brasset et al. 2006). This is still an open question.

4.5. Transmission of silencing through generations

TE epigenetic silencing, independent of changes in DNA sequence, may be transmitted over multiple generations via the gametes as a transgenerational epigenetic inheritance (Blake and Watson 2016). However, teasing out the mechanisms responsible for this transmission is highly complex. The main reason is that only a few model organisms can survive to mutations affecting epigenetic silencing and produce offspring. Then, most of our current knowledge has been acquired in plants, *C. elegans* and *D. melangaster*.

In plants, DNA methylation tends to be propagated across generations, rather than re-established at each generation. Several studies have revealed the interplay between the mechanisms that enable methylation at TEs within and between generations (Teixeira and Colot 2010). A series of observations reported that reactivation of TEs in specific cells of the gametocyte might reinforce TE silencing in the future embryo. They have pointed out the essential role of siRNAs in this epigenetic contribution to the progeny (Mosher and Melnyk 2010). Additionally, when sequencing the methylome of haploid cell types from developing pollen of *A. thaliana*, Calarco et al. have found that CHH methylation is lost from TEs, mostly retrotransposons, in microspores and sperm cells. However, this methylation is restored by DNA methyltransferase guided by 24 nt small RNAs in the embryo after fertilization. This designated small RNAs act as guides for epigenetic inheritance by specifying cytosine methylation (Calarco et al. 2012).

In the nematode worm *C. elegans*, the silencing of TEs initiated in the germline by the piRNA pathway can be stably maintained over multiple generations via a core set of small RNAs and chromatin factors (Ashe et al. 2012). A multi-step siRNA amplification occurs. A first population of siRNAs is localized close to the piRNA targets. A second population, imported to the nucleus by the Argonaute protein HRDE-1, leads to the production of a third population of siRNAs that map throughout the transcripts. This siRNA amplification confers dominant silencing in the progeny. The third category of siRNAs is sufficient for multigenerational silencing independently of the initial trigger (Sapetschnig et al. 2015).

In *D. melanogaster*, early studies had reported that two TEs, the P- or I-elements, were involved in a genetic syndrome called hybrid dysgenesis. Males with P- or I-elements crossed to females lacking these TEs yielded progeny with germline mutations and temperature-sensitive sterility. By contrast, the offspring from the reciprocal cross was fertile (Bingham et al. 1982; Bucheton 1990). Further studies revealed that P and I were responsible for this syndrome and that the cytoplasmic contribution of the mother in the egg was determinant for the sterility. piRNAs targeting P- or I-elements and deposited by the mother in the embryo were found to serve as vectors for this epigenetically inherited information. The lack of these maternally inherited piRNAs was responsible for the hybrid dysgenic syndromes (Marin et al. 2000; Brennecke et al. 2008).

Another important set of experiments published in 2012 revealed that an active transgenic piRNA cluster could induce piRNA production from another homologous locus, originally incompetent for piRNA production. This phenomenon, already described in plants, is called paramutation. This study highlighted that the epigenetic signals sufficient to activate a new piRNA locus without the inheritance of the active locus are PIWI proteins, and the pool of piRNAs expressed in the maternal germline during oogenesis are deposited into the developing egg (de Vanssay et al. 2012). More recently, it was also reported that the identity of piRNA clusters is acquired in a Piwi-dependent manner. piRNA clusters are transcribed in early embryos before zygotic genome activation. The proposed model is that their transcripts are targeted by PIWI-piRNAs deposited into the developing egg during oogenesis, and thus present in the early embryo. This allows the subsequent H3K9me3 deposition and dual-strand piRNA clusters licensing (Le Thomas et al. 2014; Akkouche et al. 2017).

Overall, current knowledge does not bring a complete picture of the mechanisms involved in the heritability of epigenetic traits, but inherited DNA methylation or histone marks might be compelling mediators. Small RNAs that mediate DNA methylation and histone modification are at the forefront of this research. Since the control of TEs is mediated by these mechanisms in many species, including human, it is tempting to consider TEs as a potential source of epigenetic inheritance.

4.6. Environmental stresses and their influence on TEs

The evolutionary potential of TEs resides in their capacity to bypass the strict germline control preventing their transposition, so that new copies can invade the genomes of the descendants. Several reasons may lay at the heart of new genomic copies. Horizontal transfers are one of these but will not be developed in this chapter (See Chapter 7 this book). Modification of their epigenetic regulation is also invoked in the appearance of new copies in the progeny. Environmental stresses such as heat shock, UV, diet, stress and toxin exposure have been shown to compromise TE silencing in the germline and also be a source of TE transposition. However, it is not clear yet why some TE families are upregulated, whereas others are downregulated.

The environment experienced during lifetime may then impact the health of descendants after harmful TE mutations. Far more dramatic consequences on the progeny have been suggested when bursts of transpositions, due to relaxation of TE silencing, were associated with macroevolution and speciation (Rebollo et al. 2010). New insertions occurring in the progeny may also have a positive impact with acquisition of genomic novelties, inheritance of new traits that will be maintained for organismal function (Casacuberta and González 2013). As a recent example, evidence was presented that DNA transposons promote exon shuffling by inserting transposase domains in new genomic contexts. The authors proposed that this process provides a plausible path for the emergence of several ancient TFs with important developmental functions (Cosby et al. 2021).

Heat-shock is one of the best stress conditions studied so far in its impact on TE control. In plant, numerous studies have reported the influence of heat shock on TE regulation. Interestingly, a work from Mittelsten-Sheid's group pointed out that an LTR-retrotransposon from *A. thaliana*, *ONSEN*, acquired a heat-responsive element recognized by plant-derived heat stress defense factors. Under elevated temperatures, transcription of *ONSEN* increases, leading to the production of full-length extrachromosomal DNA capable of reintegrating the genome (Cavrak et al. 2014). As another example, exposure of worms to high temperatures leads to expressions of heterochromatic silenced repeats associated with altered H3K9me3 chromatin. This transgenerational impact of heat shock is inherited through eggs and sperm, and persists for at least 14 generations of *C. elegans* (Klosin et al. 2017).

Numerous studies have also reported the crucial role of nutrients on the epigenome. Although the application of such findings may be crucial for a better life, studies are not that easy. They require the use of hypomorphic alleles of genes allowing detection of any phenotypic variation that results from diet. In mice, maternal methyl supplement has been reported to affect epigenetic and DNA methylation of offspring. The study was revealed when analyzing the epigenetic regulation of an allele of the *agouti* gene responsible for the coat color. Expression of this allele (A^{vy}) is under the epigenetic regulation of an LTR-retrotransposon, IAP. Due to the IAP epigenetic regulation, A^{vy} is expressed to different degrees in mice with the same genotype. Failure to epigenetically suppress expression of this allele results in the yellow agouti obese syndrome characterized by different coat colors, altered metabolism and obesity (Cooney et al. 2002). It has been shown that the epigenetic phenotypes observed in the progeny of A^{vy} mothers are strongly correlated with altered methylation of the IAP LTR; these significantly vary after maternal methyl supplements. Such studies highlight the importance of TEs as crucial components of the epigenome. They also reveal TEs as useful tools to elucidate the complexity of epigenetic variations resulting from nutrients.

It is interesting to mention studies reporting how TEs may also be used for a beneficial adaptive response in the host. Recent studies highlighted a mechanism by which inheritance of a stress response requiring the retrotransposon Cer1 helps *C. elegans* to survive. From its nutritious food sources, *C. elegans* is exposed to many pathogens such as *Pseudomonas aeruginosa*, which kills *C. elegans* within 2–3 days. However, *C. elegans* has the ability to learn how to avoid this pathogen and to pass this avoidance to its progeny. A single non-coding RNA (PA14) produced by *P. aeruginosa* is sufficient to induce pathogen avoidance in the worms and in four subsequent generations of progeny (Kaletsky et al. 2020). It was recently found that virus-like particles (VLPs) encoded by the Cer1 retrotransposon are required for learned and transgenerationally inherited P14-avoidance behavior. Since *C. elegans* secretes Cer1-VLPs into the environment, the study suggests that Cer1 VLPs function as a form of communication between individuals avoiding the pathogen, and naive individuals carrying Cer1 in their genomes (Moore et al. 2021). The authors propose that *C. elegans* might have coopted a potentially harmful TE to instead protect itself and its progeny to pathogens from their natural environment.

Another potential adaptive response implicating TEs upon stress comes from *D. melanogaster*. Indeed, de novo piRNA clusters in flies can be established at the high temperature of 29°C, instead of 25°C (Casier et al. 2019). This process might be helpful in case of warming to increase the production of specific piRNA populations and protect the host against reactivation of heat-shock sensitive TEs.

Many insights remain to be gained before unraveling the processes involved in the effects caused by environmental stresses mediated by TEs. However, at the level of species, the flexible TE response in the germline to environmental insults might already be considered as a necessary capacity to successfully survive ever-changing environments.

4.7. Conclusion

TEs are genomic components that may be harmful if consistent systems of repression did not exist, both in somatic lineages to preserve the individuals and in the germline to preserve the progeny. However, failures in silencing exist and reveal their complex relationship with genomes – for better or for worse. Considering their capacity to insert new genomic copies, it is now well established that TEs are responsible for harmful mutations; however, they also contribute to genome plasticity and are a major source of genomic diversity. The ubiquitous presence of TEs in living organisms, together with their epigenetic silencing, place them as significant epigenetic components of genomes prone to epigenetic variations under adverse environmental conditions. Since a host's gene epigenetic regulation and its links to health or diseases is far from being elucidated, studies on TEs offer the perspective to better understand and predict disease susceptibility to ultimately improve the health of populations. Tracking inherited epigenetic information and its effects for multiple generations remains a significant challenge to overcome.

4.8. References

Akkouche, A., Mugat, B., Barckmann, B., Varela-Chavez, C., Li, B., Raffel, R., Pélisson, A., Chambeyron, S. (2017). Piwi is required during drosophila embryogenesis to license dual-strand piRNA clusters for transposon repression in adult ovaries. *Mol. Cell*, 66, 411–419.e4.

Andersen, P.R., Tirian, L., Vunjak, M., Brennecke, J. (2017). A heterochromatin-dependent transcription machinery drives piRNA expression. *Nature*, 549, 54–59.

Arnaiz, O., Mathy, N., Baudry, C., Malinsky, S., Aury, J.-M., Denby Wilkes, C., Garnier, O., Labadie, K., Lauderdale, B.E., Le Mouël, A. (2012). The Paramecium germline genome provides a niche for intragenic parasitic DNA: Evolutionary dynamics of internal eliminated sequences. *PLoS Genet.*, 8, e1002984.

Ashe, A., Sapetschnig, A., Weick, E.-M., Mitchell, J., Bagijn, M.P., Cording, A.C., Doebley, A.-L., Goldstein, L.D., Lehrbach, N.J., Le Pen, J. (2012). piRNAs can trigger a multigenerational epigenetic memory in the germline of *C. elegans*. *Cell*, 150, 88–99.

Batki, J., Schnabl, J., Wang, J., Handler, D., Andreev, V.I., Stieger, C.E., Novatchkova, M., Lampersberger, L., Kauneckaite, K., Xie, W. (2019). The nascent RNA binding complex SFiNX licenses piRNA-guided heterochromatin formation. *Nat. Struct. Mol. Biol.*, 26, 720–731.

Bingham, P.M., Kidwell, M.G., Rubin, G.M. (1982). The molecular basis of P-M hybrid dysgenesis: The role of the P element, a P-strain-specific transposon family. *Cell*, 29, 995–1004.

Blake, G.E. and Watson, E.D. (2016). Unravelling the complex mechanisms of transgenerational epigenetic inheritance. *Curr. Opin. Chem. Biol.*, 33, 101–107.

Bourque, G., Burns, K.H., Gehring, M., Gorbunova, V., Seluanov, A., Hammell, M., Imbeault, M., Izsvák, Z., Levin, H.L., Macfarlan, T.S. (2018). Ten things you should know about transposable elements. *Genome Biol.*, 19, 199.

Brasset, E., Taddei, A., Arnaud, F., Faye, B., Fausto, A., Mazzini, M., Giorgi, F., Vaury, C. (2006). Viral particles of the endogenous retrovirus ZAM from *Drosophila melanogaster* use: A pre-existing endosome/exosome pathway for transfer to the oocyte. *Retrovirology*, 3, 25.

Brennecke, J., Aravin, A.A., Stark, A., Dus, M., Kellis, M., Sachidanandam, R., Hannon, G.J. (2007). Discrete small RNA-generating loci as master regulators of transposon activity in Drosophila. *Cell*, 128, 1089–1103.

Brennecke, J., Malone, C.D., Aravin, A.A., Sachidanandam, R., Stark, A., Hannon, G.J. (2008). An epigenetic role for maternally inherited piRNAs in transposon silencing. *Science*, 322, 1387–1392.

Bucheton, A. (1990). I transposable elements and I-R hybrid dysgenesis in Drosophila. *Trends Genet.*, 6, 16–21.

Burns, K.H. (2017). Transposable elements in cancer. *Nat. Rev. Cancer*, 17, 415–424.

Calarco, J.P., Borges, F., Donoghue, M.T.A., Van Ex, F., Jullien, P.E., Lopes, T., Gardner, R., Berger, F., Feijó, J.A., Becker, J.D. (2012). Reprogramming of DNA methylation in pollen guides epigenetic inheritance via small RNA. *Cell*, 151, 194–205.

Cao, X. and Jacobsen, S.E. (2002). Locus-specific control of asymmetric and CpNpG methylation by the DRM and CMT3 methyltransferase genes. *Proc. Natl. Acad. Sci. USA*, 99(Suppl 4), 16491–16498.

Carmell, M.A., Girard, A., van de Kant, H.J.G., Bourc'his, D., Bestor, T.H., de Rooij, D.G., Hannon, G.J. (2007). MIWI2 is essential for spermatogenesis and repression of transposons in the mouse male germline. *Dev. Cell*, 12, 503–514.

Casacuberta, E. and González, J. (2013). The impact of transposable elements in environmental adaptation. *Mol. Ecol.*, 22, 1503–1517.

Casier, K., Delmarre, V., Gueguen, N., Hermant, C., Viodé, E., Vaury, C., Ronsseray, S., Brasset, E., Teysset, L., Boivin, A. (2019). Environmentally-induced epigenetic conversion of a piRNA cluster. *eLife*, 8, e39842.

Cavrak, V.V., Lettner, N., Jamge, S., Kosarewicz, A., Bayer, L.M., Mittelsten Scheid, O. (2014). How a retrotransposon exploits the plant's heat stress response for its activation. *PLoS Genet.*, 10, e1004115.

Chelmicki, T., Roger, E., Teissandier, A., Dura, M., Bonneville, L., Rucli, S., Dossin, F., Fouassier, C., Lameiras, S., Bourc'his, D. (2021). m6A RNA methylation regulates the fate of endogenous retroviruses. *Nature*, 591, 312–316.

Chen, Y.-C.A., Stuwe, E., Luo, Y., Ninova, M., Le Thomas, A., Rozhavskaya, E., Li, S., Vempati, S., Laver, J.D., Patel, D.J. (2016). Cutoff suppresses RNA polymerase II termination to ensure expression of piRNA precursors. *Mol. Cell*, 63, 97–109.

Choi, J.Y. and Lee, Y.C.G. (2020). Double-edged sword: The evolutionary consequences of the epigenetic silencing of transposable elements. *PLOS Genet.*, 16, e1008872.

Cooney, C.A., Dave, A.A., Wolff, G.L. (2002). Maternal methyl supplements in mice affect epigenetic variation and DNA methylation of offspring. *J. Nutr.*, 132, 2393S-2400S.

Cosby, R.L., Judd, J., Zhang, R., Zhong, A., Garry, N., Pritham, E.J., Feschotte, C. (2021). Recurrent evolution of vertebrate transcription factors by transposase capture. *Science*, 371, eabc6405.

Czech, B., Malone, C.D., Zhou, R., Stark, A., Schlingeheyde, C., Dus, M., Perrimon, N., Kellis, M., Wohlschlegel, J.A., Sachidanandam, R. et al. (2008). An endogenous small interfering RNA pathway in Drosophila. *Nature*, 453, 798–802.

Dai, J., Xie, W., Brady, T.L., Gao, J., Voytas, D.F. (2007). Phosphorylation regulates integration of the yeast Ty5 retrotransposon into heterochromatin. *Mol. Cell*, 27, 289–299.

Daskalos, A., Nikolaidis, G., Xinarianos, G., Savvari, P., Cassidy, A., Zakopoulou, R., Kotsinas, A., Gorgoulis, V., Field, J.K., Liloglou, T. (2009). Hypomethylation of retrotransposable elements correlates with genomic instability in non-small cell lung cancer. *Int. J. Cancer*, 124, 81–87.

Dennis, C., Brasset, E., Sarkar, A., Vaury, C. (2016). Export of piRNA precursors by EJC triggers assembly of cytoplasmic Yb-body in Drosophila. *Nat. Commun.*, 7, 13739.

Desset, S., Buchon, N., Meignin, C., Coiffet, M., Vaury, C. (2008). In *Drosophila melanogaster* the COM locus directs the somatic silencing of two retrotransposons through both Piwi-dependent and -independent pathways. *PLoS ONE*, 3, e1526.

Dufourt, J., Brasset, E., Desset, S., Pouchin, P., Vaury, C. (2011). Polycomb group-dependent, heterochromatin protein 1-independent, chromatin structures silence retrotransposons in somatic tissues outside ovaries. *DNA Res. Int. J. Rapid Publ. Rep. Genes Genomes*, 18, 451–461.

Ecco, G., Imbeault, M., Trono, D. (2017). KRAB zinc finger proteins. *Dev. Camb. Engl.*, 144, 2719–2729.

Elgin, S.C.R. and Reuter, G. (2013). Position-effect variegation, heterochromatin formation, and gene silencing in Drosophila. *Cold Spring Harb. Perspect. Biol.*, 5, a017780.

Frapporti, A., Miró Pina, C., Arnaiz, O., Holoch, D., Kawaguchi, T., Humbert, A., Eleftheriou, E., Lombard, B., Loew, D., Sperling, L. (2019). The polycomb protein Ezl1 mediates H3K9 and H3K27 methylation to repress transposable elements in Paramecium. *Nat. Commun.*, 10, 2710.

Gainetdinov, I., Colpan, C., Arif, A., Cecchini, K., Zamore, P.D. (2018). A single mechanism of biogenesis, initiated and directed by PIWI proteins, explains piRNA production in most animals. *Mol. Cell*, 71, 775-790.e5.

Genenncher, B., Durdevic, Z., Hanna, K., Zinkl, D., Mobin, M.B., Senturk, N., Da Silva, B., Legrand, C., Carré, C., Lyko, F. (2018). Mutations in cytosine-5 tRNA methyltransferases impact mobile element expression and genome stability at specific DNA repeats. *Cell Reports*, 22, 1861–1874.

Ghildiyal, M., Seitz, H., Horwich, M.D., Li, C., Du, T., Lee, S., Xu, J., Kittler, E.L.W., Zapp, M.L., Weng, Z. (2008). Endogenous siRNAs derived from transposons and mRNAs in Drosophila somatic cells. *Science*, 320, 1077–1081.

Ghildiyal, M., Xu, J., Seitz, H., Weng, Z., Zamore, P.D. (2010). Sorting of Drosophila small silencing RNAs partitions microRNA* strands into the RNA interference pathway. *RNA*, 16, 43–56.

Goriaux, C., Théron, E., Brasset, E., Vaury, C. (2014). History of the discovery of a master locus producing piRNAs: The flamenco/COM locus in *Drosophila melanogaster*. *Front. Genet.*, 5, 257.

Hamilton, A., Voinnet, O., Chappell, L., Baulcombe, D. (2015). Two classes of short interfering RNA in RNA silencing. *EMBO J.*, 34, 2590.

Han, B.W., Wang, W., Li, C., Weng, Z., Zamore, P.D. (2015). piRNA-guided transposon cleavage initiates Zucchini-dependent, phased piRNA production. *Science*, 348, 817–821.

Hartig, J.V. and Förstemann, K. (2011). Loqs-PD and R2D2 define independent pathways for RISC generation in Drosophila. *Nucleic Acids Res.*, 39, 3836–3851.

Hayashi, R., Schnabl, J., Handler, D., Mohn, F., Ameres, S.L., Brennecke, J. (2016). Genetic and mechanistic diversity of piRNA 3'-end formation. *Nature*, 539, 588–592.

Huang, X., Fejes Tóth, K., Aravin, A.A. (2017). piRNA biogenesis in *Drosophila melanogaster*. *Trends Genet.*, 33, 882–894.

Iwagami, S., Baba, Y., Watanabe, M., Shigaki, H., Miyake, K., Ishimoto, T., Iwatsuki, M., Sakamaki, K., Ohashi, Y., Baba, H. (2013). LINE-1 hypomethylation is associated with a poor prognosis among patients with curatively resected esophageal squamous cell carcinoma. *Ann. Surg.*, 257, 449–455.

Johnson, L., Cao, X., Jacobsen, S. (2002). Interplay between two epigenetic marks. DNA methylation and histone H3 lysine 9 methylation. *Curr. Biol.*, 12, 1360–1367.

Kaletsky, R., Moore, R.S., Vrla, G.D., Parsons, L.R., Gitai, Z., Murphy, C.T. (2020). *C. elegans* interprets bacterial non-coding RNAs to learn pathogenic avoidance. *Nature*, 586, 445–451.

Klosin, A., Casas, E., Hidalgo-Carcedo, C., Vavouri, T., Lehner, B. (2017). Transgenerational transmission of environmental information in *C. elegans*. *Science*, 356, 320–323.

Kuwabara, T., Hsieh, J., Muotri, A., Yeo, G., Warashina, M., Lie, D.C., Moore, L., Nakashima, K., Asashima, M., Gage, F.H. (2009). Wnt-mediated activation of NeuroD1 and retro-elements during adult neurogenesis. *Nat. Neurosci.*, 12, 1097–1105.

Law, J.A., Du, J., Hale, C.J., Feng, S., Krajewski, K., Palanca, A.M.S., Strahl, B.D., Patel, D.J., Jacobsen, S.E. (2013). Polymerase IV occupancy at RNA-directed DNA methylation sites requires SHH1. *Nature*, 498, 385–389.

Le Thomas, A., Stuwe, E., Li, S., Du, J., Marinov, G., Rozhkov, N., Chen, Y.-C.A., Luo, Y., Sachidanandam, R., Toth, K.F. (2014). Transgenerationally inherited piRNAs trigger piRNA biogenesis by changing the chromatin of piRNA clusters and inducing precursor processing. *Genes Dev.*, 28, 1667–1680.

Lim, A.K. and Kai, T. (2007). Unique germ-line organelle, nuage, functions to repress selfish genetic elements in *Drosophila melanogaster*. *Proc. Natl. Acad. Sci. USA*, 104, 6714–6719.

Lim, A.K., Lorthongpanich, C., Chew, T.G., Tan, C.W.G., Shue, Y.T., Balu, S., Gounko, N., Kuramochi-Miyagawa, S., Matzuk, M.M., Chuma, S. (2013). The nuage mediates retrotransposon silencing in mouse primordial ovarian follicles. *Development*, 140, 3819–3825.

Lim, R.S.M., Anand, A., Nishimiya-Fujisawa, C., Kobayashi, S., Kai, T. (2014). Analysis of hydra PIWI proteins and piRNAs uncover early evolutionary origins of the piRNA pathway. *Dev. Biol.*, 386, 237–251.

Margueron, R., Li, G., Sarma, K., Blais, A., Zavadil, J., Woodcock, C.L., Dynlacht, B.D., Reinberg, D. (2008). Ezh1 and Ezh2 maintain repressive chromatin through different mechanisms. *Mol. Cell*, 32, 503–518.

Marin, L., Lehmann, M., Nouaud, D., Izaabel, H., Anxolabéhère, D., Ronsseray, S. (2000). P-element repression in *Drosophila melanogaster* by a naturally occurring defective telomeric P copy. *Genetics*, 155, 1841–1854.

Martínez, G., Panda, K., Köhler, C., Slotkin, R.K. (2016). Silencing in sperm cells is directed by RNA movement from the surrounding nurse cell. *Nat. Plants*, 2, 16030.

Matzke, M.A. and Mosher, R.A. (2014). RNA-directed DNA methylation: An epigenetic pathway of increasing complexity. *Nat. Rev. Genet.*, 15, 394–408.

Maupetit-Mehouas, S. and Vaury, C. (2020). Transposon reactivation in the germline may be useful for both transposons and their host genomes. *Cells*, 9, 1172.

McClintock, B. (1984). The significance of responses of the genome to challenge. *Science*, 226, 792–801.

McCue, A.D., Panda, K., Nuthikattu, S., Choudury, S.G., Thomas, E.N., Slotkin, R.K. (2015). ARGONAUTE 6 bridges transposable element mRNA-derived siRNAs to the establishment of DNA methylation, *EMBO*, 34(1), 20–35. doi: 0.15252/embj.201489499.

Mohn, F., Sienski, G., Handler, D., Brennecke, J. (2014). The Rhino-Deadlock-Cutoff complex licenses noncanonical transcription of dual-strand piRNA clusters in Drosophila. *Cell*, 157, 1364–1379.

Mohn, F., Handler, D., Brennecke, J. (2015). piRNA-guided slicing specifies transcripts for Zucchini-dependent, phased piRNA biogenesis. *Science*, 348, 812–817.

Molaro, A. and Malik, H.S. (2016). Hide and seek: How chromatin-based pathways silence retroelements in the mammalian germline. *Curr. Opin. Genet. Dev.*, 37, 51–58.

Molaro, A., Malik, H.S., Bourc'his, D. (2020). Dynamic evolution of de novo DNA methyltransferases in rodent and primate genomes. *Mol. Biol. Evol.*, 37, 1882–1892.

Moore, R.S., Kaletsky, R., Lesnik, C., Cota, V., Blackman, E., Parsons, L.R., Gitai, Z., Murphy, C.T. (2021). The role of the Cer1 transposon in horizontal transfer of transgenerational memory. *Cell*, 184, 4697–4712.e18.

Mosher, R.A. and Melnyk, C.W. (2010). siRNAs and DNA methylation: Seedy epigenetics. *Trends Plant Sci.*, 15, 204–210.

Murphey, P., McLean, D.J., McMahan, C.A., Walter, C.A., McCarrey, J.R. (2013). Enhanced genetic integrity in mouse germ cells. *Biol. Reprod.*, 88, 6.

Nishida, K.M., Miyoshi, K., Ogino, A., Miyoshi, T., Siomi, H., Siomi, M.C. (2013). Roles of R2D2, a cytoplasmic D2 body component, in the endogenous siRNA pathway in Drosophila. *Mol. Cell*, 49, 680–691.

Ogino, S., Nosho, K., Kirkner, G.J., Kawasaki, T., Chan, A.T., Schernhammer, E.S., Giovannucci, E.L., Fuchs, C.S. (2008). A cohort study of tumoral LINE-1 hypomethylation and prognosis in colon cancer. *J. Natl. Cancer Inst.*, 100, 1734–1738.

Ozata, D.M., Gainetdinov, I., Zoch, A., O'Carroll, D., Zamore, P.D. (2019). PIWI-interacting RNAs: Small RNAs with big functions. *Nat. Rev. Genet.*, 20, 89–108.

Padeken, J., Methot, S., Zeller, P., Delaney, C.E., Kalck, V., Gasser, S.M. (2021). Argonaute NRDE-3 and MBT domain protein LIN-61 redundantly recruit an H3K9me3 HMT to prevent embryonic lethality and transposon expression. *Genes Dev.*, 35, 82–101.

Pappalardo, A.M., Ferrito, V., Biscotti, M.A., Canapa, A., Capriglione, T. (2021). Transposable elements and stress in vertebrates: An overview. *Int. J. Mol. Sci.*, 22, 1970.

Pattamadilok, J., Huapai, N., Rattanatanyong, P., Vasurattana, A., Triratanachat, S., Tresukosol, D., Mutirangura, A. (2008). LINE-1 hypomethylation level as a potential prognostic factor for epithelial ovarian cancer. *Int. J. Gynecol. Cancer Off. J. Int. Gynecol. Cancer Soc.*, 18, 711–717.

Qi, H., Watanabe, T., Ku, H.-Y., Liu, N., Zhong, M., Lin, H. (2011). The Yb body, a major site for Piwi-associated RNA biogenesis and a gateway for Piwi expression and transport to the nucleus in somatic cells. *J. Biol. Chem.*, 286, 3789–3797.

Rebollo, R., Horard, B., Hubert, B., Vieira, C. (2010). Jumping genes and epigenetics: Towards new species. *Gene*, 454, 1–7.

Reinberg, D. and Vales, L.D. (2018). Chromatin domains rich in inheritance. *Science*, 361, 33–34.

Sapetschnig, A., Sarkies, P., Lehrbach, N.J., Miska, E.A. (2015). Tertiary siRNAs mediate paramutation in *C. elegans*. *PLOS Genet.*, 11, e1005078.

Sato, K. and Siomi, M.C. (2020). The piRNA pathway in *Drosophila* ovarian germ and somatic cells. *Proc. Jpn. Acad. Ser. B*, 96, 32–42.

Sellis, D., Guérin, F., Arnaiz, O., Pett, W., Lerat, E., Boggetto, N., Krenek, S., Berendonk, T., Couloux, A., Aury, J.-M. et al. (2021). Massive colonization of protein-coding exons by selfish genetic elements in Paramecium germline genomes. *PLOS Biol.*, 19, e3001309.

Teixeira, F.K. and Colot, V. (2010). Repeat elements and the Arabidopsis DNA methylation landscape. *Heredity*, 105, 14–23.

Théron, E., Dennis, C., Brasset, E., Vaury, C. (2014). Distinct features of the piRNA pathway in somatic and germ cells: From piRNA cluster transcription to piRNA processing and amplification. *Mob. DNA*, 5, 28.

de Vanssay, A., Bougé, A.-L., Boivin, A., Hermant, C., Teysset, L., Delmarre, V., Antoniewski, C., Ronsseray, S. (2012). Paramutation in *Drosophila* linked to emergence of a piRNA-producing locus. *Nature*, 490, 112–115.

Walsh, C.P., Chaillet, J.R., Bestor, T.H. (1998). Transcription of IAP endogenous retroviruses is constrained by cytosine methylation. *Nat. Genet.*, 20, 116–117.

Wang, W., Han, B.W., Tipping, C., Ge, D.T., Zhang, Z., Weng, Z., Zamore, P.D. (2015). Slicing and binding by Ago3 or Aub trigger Piwi-bound piRNA production by distinct mechanisms. *Mol. Cell*, 59, 819–830.

Wolf, G., Yang, P., Füchtbauer, A.C., Füchtbauer, E.-M., Silva, A.M., Park, C., Wu, W., Nielsen, A.L., Pedersen, F.S., Macfarlan, T.S. (2015). The KRAB zinc finger protein ZFP809 is required to initiate epigenetic silencing of endogenous retroviruses. *Genes Dev.*, 29, 538–554.

Ye, R., Chen, Z., Lian, B., Jordan Rowley, M., Xia, N., Chai, J., Li, Y., He, X.-J., Wierzbicki, A.T., Qi, Y. (2016). A dicer-independent route for biogenesis of siRNAs that direct DNA methylation in arabidopsis, *Molecular Cell*, 61(2), 222-35. doi: 10.1016/j.molcel. 2015.11.015.

Zhang, Z., Wang, J., Schultz, N., Zhang, F., Parhad, S.S., Tu, S., Vreven, T., Zamore, P.D., Weng, Z., Theurkauf, W.E. (2014). The HP1 homolog rhino anchors a nuclear complex that suppresses piRNA precursor splicing. *Cell*, 157, 1353–1363.

Zhao, K., Cheng, S., Miao, N., Xu, P., Lu, X., Zhang, Y., Wang, M., Ouyang, X., Yuan, X., Liu, W. et al. (2019a). A Pandas complex adapted for piRNA-guided transcriptional silencing and heterochromatin formation. *Nat. Cell Biol.*, 21, 1261–1272.

Zhao, X., Xiong, J., Mao, F., Sheng, Y., Chen, X., Feng, L., Dui, W., Yang, W., Kapusta, A., Feschotte, C. et al. (2019b). RNAi-dependent polycomb repression controls transposable elements in tetrahymena. *Genes Dev.*, 33, 348–364.

Zhou, R., Czech, B., Brennecke, J., Sachidanandam, R., Wohlschlegel, J.A., Perrimon, N., Hannon, G.J., (2009). Processing of *Drosophila* endo-siRNAs depends on a specific Loquacious isoform. *RNA*, 15, 1886–1895.

5

Transposable Elements and Adaptation

Marta CORONADO-ZAMORA and Josefa GONZÁLEZ
Institute of Evolutionary Biology, CSIC, UPF, Barcelona, Spain

Transposable elements (TEs) are mobile genomic sequences present in virtually all of the organisms analyzed to-date. TEs cause many mutations, and in this chapter, we have summarized evidence for the role of some of these mutations in the ability of organisms to respond to environmental stress.

5.1. Transposable elements are mobile genomic sequences

The genome, that is, the complete set of DNA of an organism, contains different types of DNA sequences, including genes and gene regulatory regions. Genes are DNA sequences that contain the instructions to construct a specific molecule (a protein or an RNA); whereas, gene regulatory regions are DNA sequences that contain information about when, where and how much of these molecules are going to be constructed. Besides genes and gene regulatory sequences, there is another component present in virtually all of the genomes studied to-date: TEs (Hua-Van et al. 2011; Wells and Feschotte 2020). TEs are DNA sequences that can move and multiply across the genome. TEs contain genes and gene regulatory regions that allow them to make copies of themselves and insert those copies throughout the genome. Due to this capacity to move around in the genome, TEs generate mutations. Not all TEs in a genome are active; in fact, the majority of TEs have lost the capacity to move from one genomic position to another. However, TEs cause mutations not only because they move around in the genome, but also because they

Transposable Elements and Genome Evolution,
coordinated by Aurélie HUA-VAN and Pierre CAPY. © ISTE Ltd 2024.

are repetitive sequences that can be involved in the generation of chromosomal rearrangements.

TEs were discovered by Barbara McClintock in the 1940s when she was trying to understand the changes in the genome that were responsible for the colors of kernels on a cob of maize. Her discovery was rejected at the time, as genomes were thought to be a collection of genes laid out in a linear sequence, like "beads on a string", and the existence of TEs suggested that genomes were dynamic (Keller 1983). Dr. McClintock not only described the ability of TEs to move around in the genome and affect the expression of genes, referring to them as "controlling elements", but she also acknowledged that TEs could take part in the cellular stress response. She inferred that the activation of TEs could lead to gene modifications and genome restructuring, and this variation could help the cell adapt to stressful conditions (McClintock 1984). Later, TEs were described in organisms other than corn. After the ability to sequence whole genomes was developed, TEs were found to be present in virtually all species across the tree of life (Wells and Feschotte 2020). The proportion of TEs in a genome varies, from as low as <1% in some yeasts to >66% in human genome and >85% in some plants (Hua-Van et al. 2011). The ability to study TEs in a variety of organisms has also led to the discovery of the diversity within these genomic components, along with the diversity of their effects on genome function and genome evolution.

5.1.1. *Transposable elements are diverse genomic sequences*

TEs are very diverse. Some of them are rather simple and contain just a single gene flanked by short sequences that are exactly the same on both ends. On the other hand, other TEs contain multiple genes, and some of these TEs are also flanked by repeated sequences, while others are not. If we compare the sequence and the structure (e.g., number and type of genes, and presence or absence of repeat sequences flanking these genes) of the TEs that have been described so far, it is clear that TEs have evolved by recurrent gain and loss of genes from other TE sequences, and from the genome where they are inserted. All TE insertions can be classified into two large classes, based on the transposition mechanism. These two classes are further subclassified into several orders, mainly based on the TE structure; and each one of these orders is then further classified into multiple families, based on TE sequence similarities (Wicker et al. 2007). The family level therefore groups TEs that are from the same class and share the same overall structure, and are very similar at the sequence level. Despite this great diversity of TEs, TEs share one thing in common, as shown in the following sections: the regulatory regions they contain can have a significant impact on the genome where they are inserted.

5.1.2. *Transposable elements generate different types of mutations*

Due to their ability to insert anywhere in the genome, TEs can create different types of mutations. The effects of these mutations will vary depending on where the TE is inserted and which type of sequence the TE contains. Broadly, we can distinguish four different mutational effects due to TE insertions (Figure 5.1): (i) disruption of gene or gene regulatory sequences; (ii) source of new regulatory regions (*cis*-regulatory sequences) or epigenetic modifications; (iii) source of new protein domains; and (iv) chromosome rearrangements due to ectopic recombination.

Disruption of gene or gene regulatory sequences: The most evident way in which a TE can alter the genome is by inserting into gene or gene regulatory sequences and disrupting them. If a TE jumps into the gene itself, it can completely disrupt its function: the instructions to construct the specific molecule can be broken apart by the insertion of a TE (Figure 5.1(A)). If a TE inserts into a gene regulatory region, it can disrupt its regulation by abolishing the gene expression in either the tissue or developmental stage that is controlled by that particular region (Figure 5.1(A)).

Source of regulatory changes: TEs can be a source of new regulatory sequences that can affect the way genes are expressed; either modifying the amount, the timing, or the location of expression of the gene. There are two main mechanisms by which TEs can affect gene regulation. First, when TEs are inserted close to a gene, they can supply new regulatory sequences, called *cis*-regulatory elements; these new regulatory regions could then have an impact on the nearby gene expression (Figure 5.1(B)). Besides affecting individual genes, TEs have also been shown to affect gene regulatory networks that coordinate gene expression. Second, TEs can also affect gene expression without modifying the gene sequence: through epigenetic changes. Silencing epigenetic marks – such as DNA methylation and repressive modifications of the proteins that condense DNA into chromatin (histones) – are found throughout the genome as a general mechanism to regulate gene expression, and they are also found in TE sequences. These TEs silencing marks can spread to neighbor genes, leading to gene inactivation (Figure 5.1(B)). TEs can also contain active chromatin marks that can lead to gene overexpression (Guio et al. 2018).

Source of new protein domains: Genes contain different building blocks – domains – that are used together to construct a protein. When TEs are inserted within a gene, they can contribute to the formation of new proteins by providing a new functional domain. This process of generating a novel protein with new functions by incorporating a domain present in the TE insertion is known as exaptation (Figure 5.1(C)).

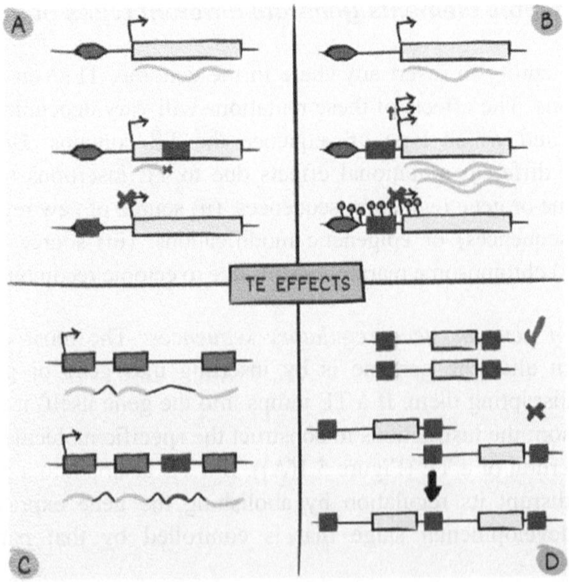

Figure 5.1. *Transposable elements can generate mutations through different molecular mechanisms. For a color version of this figure, see www.iste.co.uk/huavan/transposable.zip*

COMMENT ON FIGURE 5.1.– *Hexagons represent cis-regulatory elements, light boxes represent full gene sequences, dark boxes represent exons, non-outlined boxes represent TEs and arrows indicate where the gene starts to get expressed. (A) Disruption of gene or gene regulatory sequences. A TE inserting into the gene sequence can disrupt it by, for example, introducing a stop codon. A TE can disrupt an existing regulatory region, leading to gene inactivation in a particular tissue or developmental stage. (B) Source of regulatory changes. A TE inserting near the regulatory region of a gene can add new regulatory regions (cis-regulatory elements) affecting the gene expression, for example, increasing the level of expression. Silencing epigenetics marks in a TE sequence can spread to the neighbor gene, leading to the gene inactivation. (C) Source of new protein domains. A TE inserting between two exons of a gene (intron) can be incorporated in the final sequence, providing a new functional domain to the protein. (D) Chromosome rearrangements due to ectopic recombination. Sequences of chromosomes align through similar (homologous) regions to exchange genetic information. However, due to their repetitive nature, TEs that belong to the same family also have very similar sequences and are often located in different regions of the genome, TEs can therefore induce misaligned pairing leading to aberrant chromosomal rearrangements such as duplications, as represented in the figure. This erroneous process is known as ectopic recombination.*

Chromosome rearrangements due to ectopic recombination: Due to their repetitive nature, TEs can lead to chromosomal rearrangements through a mechanism known as ectopic recombination. During cell division (meiosis), homologous chromosomes align and similar DNA sequences from the paired chromosomes cross over each other, leading to the exchange of chromosome segments through a process known as recombination. The high sequence similarity between insertions of the same type of TEs across the genome can also induce chromosome pairing, leading to chromosome rearrangements, such as duplications, deletions and inversions (Figure 5.1(D)).

As for other types of mutations, mutations caused by TEs can be beneficial, harmful, or have no effect on the organism (neutral mutations). The discovery of the first active TE in the human genome was an insertion of an L1 element in an exon of the *Factor VIII* (*F8*) that disrupted this gene's function (Figure 5.1(A)), causing the disorder known as hemophilia A (Kazazian et al. 1988). Several other TE insertions related to human disease, such as cancer and neurological diseases, have since been described in the human genome (Payer and Burns 2019; Saleh et al. 2019). Evidence for TE mutations having beneficial effects on an organism has also started to accumulate (Casacuberta and González 2013; Schrader and Schmitz 2019) and several works point towards the relevance of the effects of TE insertions in environmental adaptation (Dubin et al. 2018; Rech et al. 2019; Hunter 2020; Baduel and Quadrana 2021). In the next three sections, we will focus on the role of TEs in three relevant environmental stress situations: (i) resistance to insecticides; (ii) response to immune stress; and (iii) response to environmental shock.

5.2. Transposable elements and insecticide resistance

TEs have long been hypothesized to play a role in insecticide resistance due to their capacity to create many types of mutations, as exemplified in Figure 5.1 (Wilson 1993; Rostant et al. 2012). Resistance to an insecticide occurs when a genetic change allows insects that were otherwise susceptible to that particular insecticide to survive. These resistant individuals would have a higher chance of going on to reproduce and, hence, a higher chance of transmitting their mutation to their offspring, ultimately increasing its frequency in the population. There are two main modes of insecticide resistance. One is through a faster detoxification of the insecticide toxic compounds (metabolic resistance). This can be achieved by an increased abundance of the proteins implicated in the metabolism of the insecticide, for example, due to changes in the regulatory regions of these genes (Ffrench-Constant et al. 2005). The other mode of insecticide resistance is through impeding the insecticide to bind to its target receptor or molecule (target resistance)

(Ffrench-Constant et al. 2005). The most common insecticides act on the central nervous system of insects, by binding neural receptors or other important molecules, and blocking them from developing their normal function. Thus, a mutation in the gene that encodes the receptor targeted by the insecticide can prevent the insecticide from properly binding to it, leading to resistance to this insecticide. In the two examples described below, we provide details on the molecular mechanisms underlying a TE insertion, leading to overexpression of a detoxifying gene that generates insecticide resistance; and another TE insertion involved in the duplication of a receptor gene that fine-tunes the resistance generated by a previous mutation.

5.2.1. *Transposable elements and metabolic resistance*

Metabolic resistance occurs through a faster metabolization of insecticides, usually due to an increased abundance of detoxification enzymes. The family of cytochrome P450s enzymes plays a crucial role in the detoxification process in many pest and non-pest insect species (Lu et al. 2021). In *Drosophila melanogaster*, the overexpression of one particular P450 gene, the *Cyp6g1* gene, has been associated with an increased resistance to different types of insecticides, including DDT, which was massively used between the 1940s and 1970s (Daborn et al. 2002; Le Goff and Hilliou 2016). Daborn et al. (2002) discovered that the expression level of *Cyp6g1* was 10 to 100 times higher in insecticide-resistant flies, compared to susceptible ones (Figure 5.2(A)). In order to assess whether the overexpression of *Cyp6g1* alone was responsible for the increase in insecticide resistance, they created a transgenic fly that overexpressed the gene 100-fold compared with a susceptible non-transgenic fly. They exposed both flies to DDT and found that the transgenic flies that overexpressed *Cyp6g1* survived after exposing them to insecticide, while the flies lacking *Cyp6g1* overexpression did not (Figure 5.2(B)). With this experiment, they demonstrated that the overexpression of *Cyp6g1* alone was sufficient to confer insecticide resistance.

To identify the mutation responsible for the increased *Cyp6g1* expression, Daborn et al. analyzed the regulatory region of this gene. As mentioned above (section 5.1), a gene regulatory region controls when, where and how much a gene is expressed. Mutations in gene regulatory regions can therefore alter the expression of genes, for example, by increasing gene expression levels or by altering where a gene is expressed. The authors found that an *Accord* TE insertion was present in all resistant flies, very close to the location of *Cyp6g1*: only 287 base pairs upstream of the transcription start site. On the contrary, no TE insertion was present in the susceptible flies. These results suggest that the TE is the mutation causing the increased *Cyp6g1* expression and therefore the resistance to insecticides.

The *Accord* insertion was absent in flies collected in the 1930s, before the commercialization of DDT; while after 1960, 37% of flies were resistant, and this percentage is even higher in other populations all over the world (Daborn et al. 2002; Catania et al. 2004; Chung et al. 2007; Le Goff and Hilliou 2016); this therefore demonstrates the relevance of TEs for adaptation to insecticides.

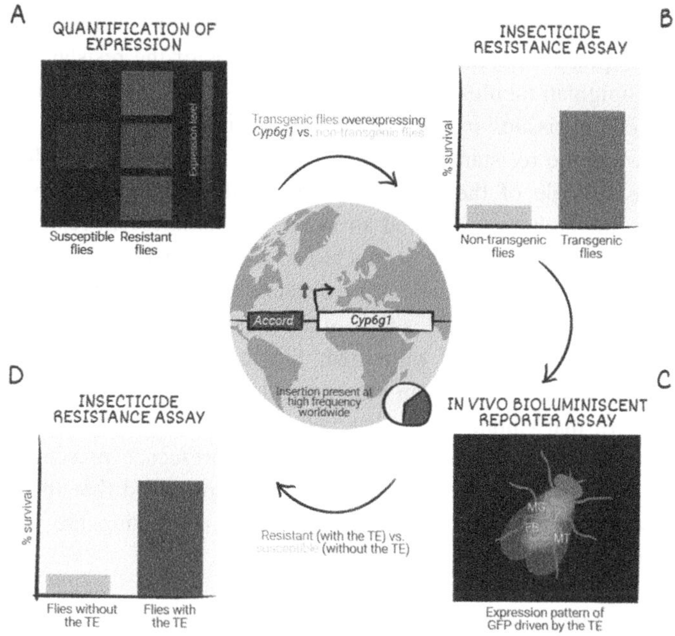

Figure 5.2. *Expression and insecticide resistance assays to understand the role of* Cyp6g1 *and* Accord *TE insertion in insecticide resistance. The insertion of an* Accord *TE in the upstream region of* Cyp6g1 *gene is found at high frequency in worldwide populations of D. melanogaster. (A) Daborn et al. (2002) discovered that the expression level of* Cyp6g1 *was 10–100 higher in insecticide-resistant flies, compared to susceptible ones. Each square represents one strain. (B) The authors created transgenic flies that overexpressed* Cyp6g1. *An insecticide resistance assay comparing the survival of transgenic versus non-transgenic flies showed that the overexpression of* Cyp6g1 *conferred insecticide resistance. (C) Chung et al. (2007) discovered with an in vivo bioluminescent reporter assay that the overexpression of* Cyp6g1 *was restricted to three tissues (the midgut (MG), the Malpighian tubules (MT) and the fat body (FB)), suggesting that the TE provides tissue-specific regulatory regions. (D) An insecticide resistance assay comparing flies with the TE insertion and flies without it, showed that tissue-specific overexpression due to the TE was enough to confer insecticide resistance. For a color version of this figure, see www.iste.co.uk/huavan/transposable.zip*

Although the authors found a clear correlation between the overexpression of *Cyp6g1* and the presence of the *Accord* TE insertion, how exactly the TE was responsible for the increased expression was still not known (Daborn et al. 2002). The answer to this question was found some years later by Chung et al. (2007). Chung et al. confirmed that the *Cyp6g1* expression level was higher in resistant flies (flies with the TE insertion) compared to susceptible flies (flies without the TE insertion). Furthermore, they discovered, by performing in situ hybridization, that this increased expression occurred in particular tissues of the resistant flies: in the midgut, the Malpighian tubules, and in the fat body, which, not surprisingly, are the major detoxification tissues in insects. As the TE insertion was the only genetic difference between the resistant and the susceptible flies, the authors focused on a possible regulatory role of the TE. To determine if the TE had a tissue-specific regulatory role, the authors performed an in vivo bioluminescent reporter assay. In this kind of assay, a reporter gene is attached to a regulatory region of interest – in this case, the *Accord* TE sequence – to evaluate if the TE alone can drive the expression of the reporter gene. This type of assay is called a reporter assay because it can be easily visualized and measured: the reporter gene expresses a green fluorescent protein (GFP) that can be quantified by fluorescence. The spatial expression pattern that Chung et al. observed with their *Accord* reporter gene was the same as reported in the resistant flies: with fluorescence present only in the midgut, Malpighian tubules and the fat body. This suggested that the *Accord* TE carries tissue-specific regulatory regions capable of expressing the gene only in these three detoxification tissues (Figure 5.2(C)).

Finally, the authors wanted to prove that the TE insertion was linked to the insecticide resistance. Although the initial experiment comparing the survival of transgenic flies overexpressing *Cyp6g1* and non-transgenic flies showed that the increased expression of the gene conferred insecticide resistance, the transgenic flies overexpressed the gene ubiquitously in the whole organism, and not in the three tissues controlled by the TE regulatory region. Therefore, the authors wanted to test if the overexpression of *Cyp6g1* driven by the TE only in the midgut, the Malpighian tubules, and in the fat body was enough to confer insecticide resistance. They found this to be the case: flies with the *Accord* TE insertion had a higher survival compared to flies without it (Figure 5.2(D)). Therefore, the tissue-specific increase in *Cyp6g1* expression due to the *Accord* TE insertion is sufficient to confer insecticide resistance.

5.2.2. *Target resistance and transposable elements*

In this mode of insecticide resistance, even a subtle disruption of the protein structure targeted by the insecticide, is enough to confer high levels of resistance. This is the case with the *Rdl* (*Resistance to dieldrin*) gene, which in *D. melanogaster*, encodes a nervous system receptor that is the primary target of many types of insecticides, such as cyclodienes. A single mutation which changes the amino acid alanine (Ala) to serine (Ser) or glycine (Gly) in this gene is responsible for a 4,000-fold cyclodiene resistance in homozygous individuals – insects carrying both resistant alleles, Rld^{R} – compared to susceptible (wild-type) individuals – non-resistant insects carrying both susceptible alleles, Rld^{WT} – (Ffrench-Constant et al. 1993a). Although this mutation does not abolish the primary function of the receptor, it has a negative effect on the insect, the so-called cost of resistance. When insecticide-resistant flies are exposed to high temperatures – 38°C for 10 min – they need a longer time to recover compared to non-resistant flies that are able to recover the ability to fly after a few minutes (Ffrench-Constant et al. 1993b). Therefore, while the mutation confers a high resistance against the insecticide, there is also a temperature-tolerance cost.

The Ala to Ser or Gly mutation has been documented in other species (Thompson et al. 1993); and interestingly some insect genomes contain multiple copies of the *Rdl* gene (Anthony et al. 1998; Del Villar and Jones 2018; Sheng et al. 2018; Hazzouri et al. 2020). Analyses of the genomes of a *D. melanogaster* population revealed that the *Rdl* region was also duplicated in some of the individuals, with one of the alleles carrying the known resistant mutation (Rld^{R}) and the other the wild-type allele (Rld^{WT}) (Remnant et al. 2013). Remnant et al. performed a resistance to cyclodiene insecticide assay with the three types of flies: wild-type (both Rld^{WT} copies), resistant (both Rld^{R} copies) and carrying the duplication (Rld^{WT}-Rld^{R}). They showed that flies with the duplication were also resistant over the wild type, although with an intermediate resistance compared to the resistant Rld^{R} flies – 27-fold versus >45,000-fold survival (Figure 5.3(A)). Interestingly, a temperature-tolerance assay in which flies were exposed to high temperature (38°C) for 10 min showed that the flies with the duplication recovered faster than the Rld^{R} resistant ones: while resistant individuals needed, on average, 10 min to recover, the individuals with the duplication needed on average 7 min, closer to the 4 min that the average wild-type individual needed (Figure 5.3(B)).

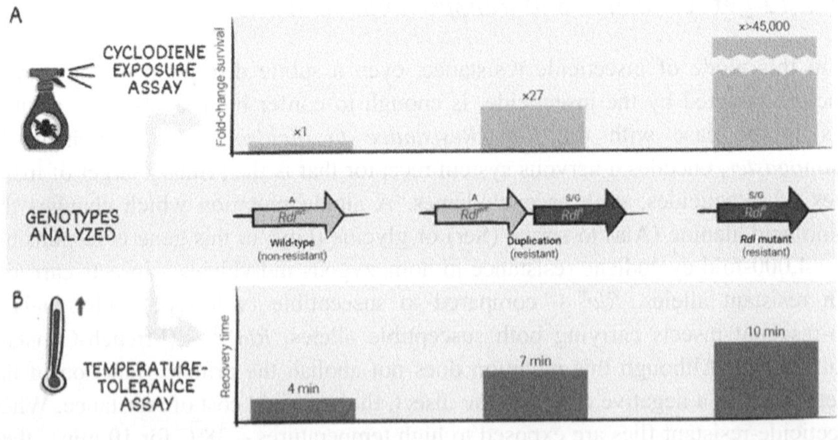

Figure 5.3. *Insecticide-exposure and temperature-tolerance assays in* Rld *wild-type flies (*Rld^{WT} *genotype), flies carrying the* Rld *duplication (*Rld^{WT}-Rld^{R} *genotype) and* Rld *resistant (*Rld^{R} *genotype) flies (Remnant et al. 2013). For a color version of this figure, see www.iste.co.uk/huavan/transposable.zip*

COMMENT ON FIGURE 5.3.– *(A) In the cyclodiene insecticide exposure assay, flies were exposed to toxic levels of the insecticide and the survival of the mutant flies was measured, compared with the wild-type flies. Flies carrying the duplication had a 27-fold survival, compared with the wild-type flies, and resistant flies showed a >45,000-fold survival compared with the wild-type flies. (B) In the temperature-tolerance assay, flies were exposed to 38°C for 10 min and the recovery time was recorded. Wild-type flies recovered after 4 min, flies carrying the duplication after 7 min and resistant flies needed 10 min to recover.*

These experiments exemplify how gene duplication can be a powerful mechanism for fine-tuning the resistance to insecticides by providing an intermediate insecticide resistance while reducing the temperature-tolerance cost. However, we are missing the final question in this story: what was the origin of this duplication? The authors found that two *roo* TEs inserted in the introns of *Rdl*'s neighboring genes were the mutations causing the duplication (Remnant et al. 2013). One of the *roo* insertions is located inside an intron of the *nwk* gene, and the second one is located 113 kb downstream in an intron of the *Glu-RIB* gene (Figure 5.4(A)). The ectopic recombination between the two *roo* insertions in a heterozygous individual (Figure 5.4(B)) led to a 113 kb duplication rearrangement. The rearrangement contains two copies of the *Rdl* gene: a wild-type allele (Rld^{WT}) and a resistant allele (Rld^{R}). When these two copies are expressed, flies have an intermediate resistance to the insecticide while the associated temperature-tolerance cost is reduced.

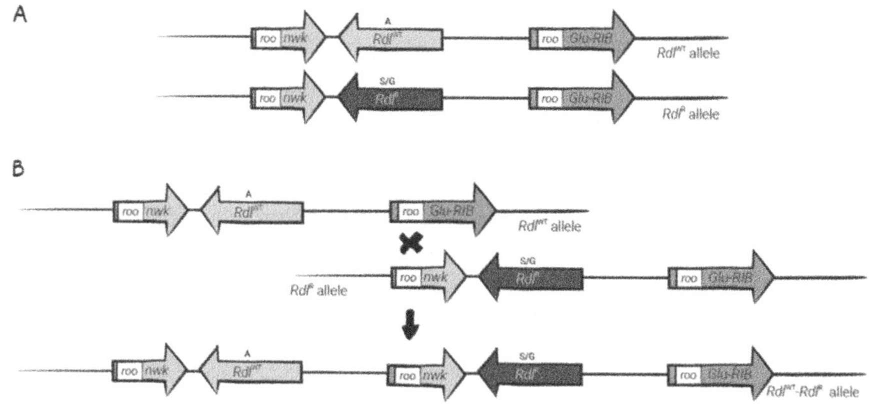

Figure 5.4. *Genomic structure of the* Rdl *region and molecular mechanism generating the* Rdl *duplication. (A) Normal sequence alignment between* Rdl^{WT} *and* Rdl^{R} *in a heterozygous individual. (B) Abnormal sequence alignment leading to ectopic recombination in a heterozygous individual. The ectopic recombination between the two* roo *TEs led to a 113 kb duplication rearrangement with the two Rdl alleles, creating the Rdl^{WT}-*Rdl^{R} *allele. The arrow-shaped boxes represent the orientation of the genes (figure adapted and modified from Remnant et al. 2013). For a color version of this figure, see www.iste.co.uk/huavan/transposable.zip*

5.3. Transposable elements and the immune response

There is evidence across organisms suggesting that TEs play a role in the expression of immune-related genes and immune-related gene networks. In *D. melanogaster*, a genome-wide screen aimed at identifying adaptive TE insertions reported several insertions that affect the expression of immune-related genes (Ullastres et al. 2021). For one of these insertions, the CRISPR/Cas9 editing technique was used to delete the TE insertion from the genome. These editing experiments showed that the TE insertion was responsible for the increased resistance to bacterial infection detected in flies containing that particular TE insertion. In humans, several insertions of TEs from the *MER41* family have been shown to contribute to the regulation of essential immune functions (Chuong et al. 2016). Besides animals, there is also evidence that TEs play a role in immune response in plants. Below, we describe the findings that show how copies from the *ATCOPIA93* family in the thale cress *Arabidopsis thaliana* have been recruited to activate the response of a disease resistance gene (Zervudacki et al. 2018).

In *A. thaliana*, TEs from the *ATCOPIA93* family are methylated and contain repressive histone marks. DNA methylation and the presence of histone marks are two of the mechanisms that cells use to silence TEs and thus prevent any possible effect of the insertions on gene expression (González and Petrov 2012). Zervudacki et al. identified an *ATCOPIA93* insertion located nearby *RPP4*, a disease response gene, that was unmethylated and lacked repressive histone marks. They showed that this particular *ATCOPIA93* insertion led to increased *RPP4* gene expression in response to pathogens (Figure 5.5(A)). They further proved that activation of the nearby disease gene was due to the presence of a regulatory sequence, a binding site for a transcription factor, in the *ATCOPIA93* insertion (Figure 5.5(B)). Furthermore, they also showed that other copies of the *ATCOPIA93* family that were very similar in sequence but were methylated and contained histone marks were not associated with increased expression of a reporter gene (Figure 5.5(C)).

As *ATCOPIA93* insertions have very similar sequences, it is likely that insertions of this family would be activated in response to pathogens but that the presence of both methylation and repressive histone marks ensures their inactive state. Indeed, Zervudacki et al. showed increased expression of *ATCOPIA93* copies when methylation and enzymes responsible for histone methylation were impaired (Figure 5.5(D)). Overall, these results provide a link between TE activation and stress response as first hypothesized by Dr. Barbara McClintock.

5.4. Transposable elements and environmental shock response

Besides specific roles in insecticide resistance and immune response, some TE insertions have been shown to affect gene expression in response to multiple stress conditions. This is the case of the highly active *Tf1* family in the yeast *Schizosaccharomyces pombe* (Chatterjee et al. 2014; Esnault and Levin 2015). Although most TEs do not have an insertion preference for certain locations in the genome, the analysis of the distribution of *Tf1* integrations has shown a clear preference for gene promoters, affecting approximately 1,000 of them (Chatterjee et al. 2014). This selective integration of *Tf1* in gene promoters is mediated by a protein, called Sap1, that interacts with the Tf1 integrase protein, the protein that facilitates the integration of the TE in the genome. One interesting feature of this TE is that it contains a regulatory element; and, as we have already seen (sections 5.2.1 and 5.3), regulatory elements have the ability to induce the expression of adjacent genes under certain scenarios, such as stress conditions. The fact that *Tf1* has been detected in many different gene promoters opens the possibility that TEs can orchestrate complex genetic programs for responding to different stress conditions. To test the implication of *Tf1* integrations in the environmental stress response, Feng et al. (2013), and later Esnault et al. (2019) studied the relationship between this

TE and different types of stresses, including temperature stress, oxidative stress, heavy metals and pollutants.

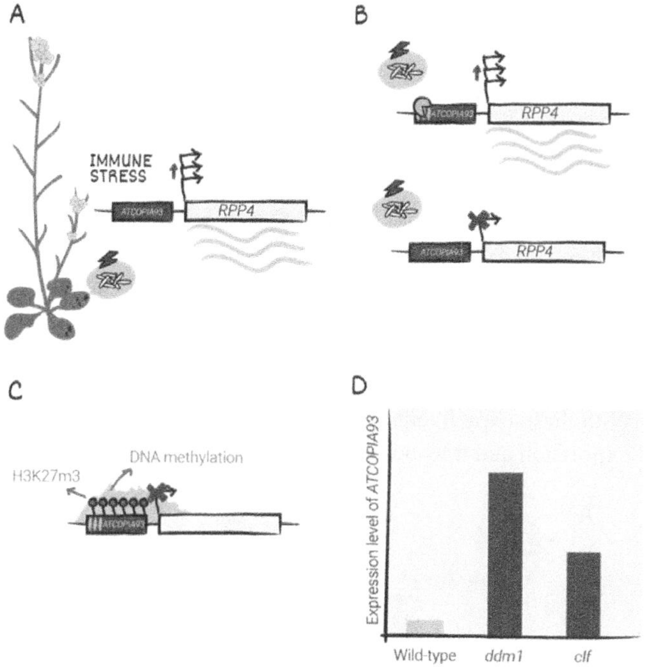

Figure 5.5. *Unraveling the role of* ATCOPIA93 *in immune stress response. For a color version of this figure, see www.iste.co.uk/huavan/transposable.zip*

COMMENT ON FIGURE 5.5.– *(A) The* ATCOPIA93 *insertion located upstream of the* RPP4 *disease response gene leads to increased gene expression under immune stress. (B)* ATCOPIA93 *insertion contains a transcription factor binding site. Under biotic stress, the binding site recruits a transcription factor that is responsible for the increased expression of* RPP4 *gene. The removal of this regulatory region results in the lack of overexpression of* RPP4 *gene under immune stress. (C)* ATCOPIA93 *copies methylated and with histone marks do not induce the expression of a reporter gene. Both DNA methylation and H3K27m3 histone marks targeting the* ATCOPIA93 *copy results in gene silencing. (D) Quantification of the expression level of* ATCOPIA93 *in wild-type plants, and in two mutants with impaired copies of* ddm1 *and* clf *genes. Plants mutant for ddm1 gene are methylation-defective, and plants mutant for clf show reduced levels of H3K27m3 histone marks. In a wild-type plant (with normal methylation and epigenetic mechanisms), the levels of* ATCOPIA93 *are low, but mutant plants lacking ddm1 or* clf1 *show increased expression of* ATCOPIA93.

Feng et al. (2013) created a collection of 32 strains that each had a single insertion of *Tf1* in a gene promoter. To test if any of these insertions were related to high-temperature stress (39°C) and oxidative stress (exposure to H_2O_2), they analyzed the gene expression levels after the strains were exposed to both stresses. They found that in four out of the 32 strains, the genes targeted by *Tf1* were overexpressed. Moreover, the expression level of the *Tf1* TE itself was also higher, suggesting that *Tf1* expression is induced by these two stresses.

Some years later, Esnault et al. (2019) checked whether the *Tf1* expression was also higher in these four strains after exposing yeast cells to other types of stresses: two heavy metals – $CoCl_2$ and $ZnCl_2$ – caffeine and the pollutant phthalate. They found that in three out of four strains, there was also a higher expression of *Tf1* in response to these stresses (Figure 5.6). Interestingly, they also found that the higher transcription of this TE resulted in an increase in transposition events: *Tf1* insertion in other places of the genome was triggered by these environmental stress conditions (Figure 5.6). With these experiments, they demonstrated that a range of stresses can stimulate *Tf1* expression and transposition.

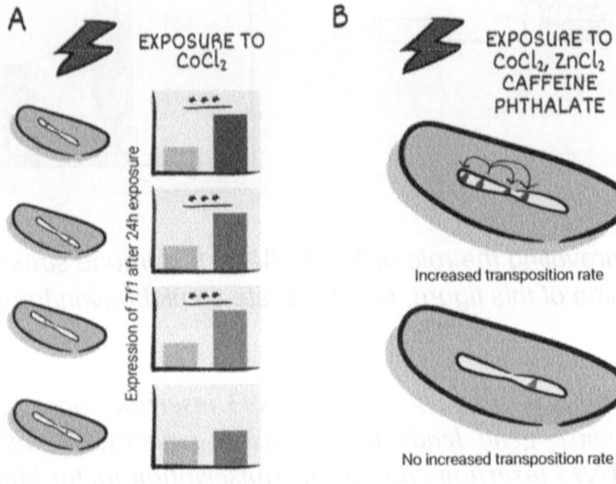

Figure 5.6. Tf1 *insertions induced by temperature and oxidative stress in* Schizosaccharomyces pombe *are also induced by other stresses. (A) The four strains from Feng et al. (2013) containing one* Tf1 *insertion each located in a particular gene promoter were subjected to $CoCl_2$ heavy metal stress. Measuring the expression of* Tf1 *after 24-h exposure to $CoCl_2$ revealed high levels of expression of* Tf1 *in three out of the four strains. (B) The four strains were then exposed to four other stresses – $CoCl_2$, $ZnCl_2$, caffeine and phthalate – and in most cases (11/16), they found an increased transposition rate (top); while for five cases, the stress did not trigger the transposition of the* Tf1 *insertion (bottom). For a color version of this figure, see www.iste.co.uk/huavan/transposable.zip*

To test if this increased transposition of *Tf1* after exposure to different stresses plays a role in the response to environmental shock, Esnault et al. performed a competition assay. The competition assay consisted of exposing different yeast cells, each with a different *Tf1* insertion, to different concentrations of $CoCl_2$ for several generations (Figure 5.7(A)). Afterwards, they checked if particular *Tf1* insertions were enriched in the surviving cell culture populations, suggesting that those insertions contributed to the survival, growth and reproduction of the yeast cells (Figure 5.7(A)). Indeed, they discovered an enrichment of some particular insertions of *Tf1* in the different cell cultures, suggesting that this TE is implicated in the regulation of a specific group of genes. When they analyzed the function of these genes, they discovered an enrichment for genes associated with the target of rapamycin (TOR) pathway: a total of 36 genes. The TOR pathway has a central role in cell growth and metabolism in all eukaryotic cells, and could be implicated in the metabolism of $CoCl_2$.

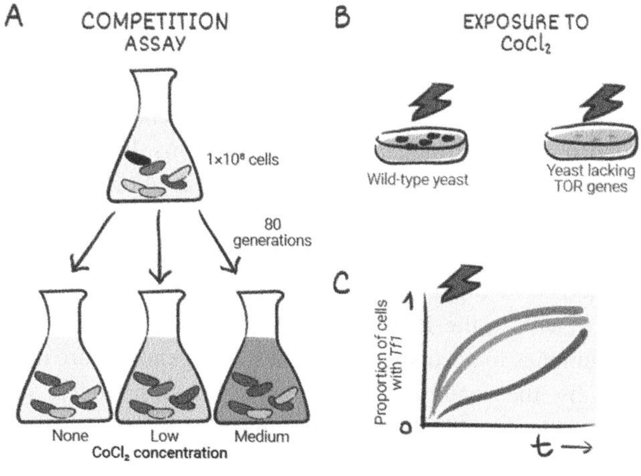

Figure 5.7. Tf1 *insertions play a role in the response to environmental shock. (A) Competition assay. A total of 1 × 10^8 cells each containing a different* Tf1 *insertion were exposed to three $CoCl_2$ levels: none, low and medium for 80 generations. Strains with particular* Tf1 *insertions were enriched in response to stress. (B) Yeast cells lacking TOR factors and wild-type (with TOR factors) cells were exposed to high levels of $CoCl_2$. Significant growth was only detected in yeast with TOR factors. (C) Relative proportion of cells with three particular* Tf1 *insertions, compared to normal cells (without the insertion) after exposure to $CoCl_2$. Cells containing the three* Tf1 *insertions outcompeted the wild-type cells lacking the insertions. For a color version of this figure, see www.iste.co.uk/huavan/transposable.zip*

The authors wanted to further demonstrate that the TOR pathway is implicated in the detoxification of $CoCl_2$. For that, they compared the survival of yeast cells lacking some of the genes related to the TOR pathway, that were identified in the competition assays, against a normal (wild-type) yeast cell. They found that yeast cells lacking some of these key TOR-related genes had a greater sensitivity to $CoCl_2$, while wild-type yeast could grow in exposure to this heavy-metal (Figure 5.7(B)). Finally, to demonstrate that specific *Tf1* insertions were sufficient to confer resistance to $CoCl_2$, they performed a competition assay comparing the survival of yeast cells with – and without – three *Tf1* insertions in the promoter of three genes: two genes participating in the regulation of the TOR pathway and one gene in the uptake of heavy metals. In cell cultures containing $CoCl_2$, cells with the insertions outcompeted the normal cells without insertions (Figure 5.7(C)), demonstrating that *Tf1* insertions play a role in stress response.

5.5. Conclusion

In this chapter, we have summarized several studies that showed how TEs from different families and through different molecular mechanisms contribute to the response of organisms to environmental stress conditions. From inducing increased gene expression and generating gene duplications that allow fruit flies to survive insecticides, to increasing expression of a disease gene in plants and affecting the expression of genes in a stress-response gene regulatory network in yeast, TEs are relevant for the ability of organisms to cope with stress. Since their discovery by Dr. Barbara McClintock in the 1940s, the evidence for an important role of TEs in stress response continues to accumulate. The idea that genomes are dynamic entities, initially rejected by the majority of the scientific community, is now widely accepted, based on the wealth of data providing evidence for the existence and relevance of TE insertions.

5.6. Acknowledgments

We thank Carmen Arboledas (CES Ramón y Cajal, Granada, Spain) for proofreading the chapter. Research at the González Lab is funded by the European Research Council (ERC) under the European Union's Horizon 2020 research and innovation program (H2020-ERC-2014-CoG-647900) and by the PID2020-115874GB-I00/AEI/ 10.13039/501100011033.

5.7. References

Anthony, N., Unruh, T., Ganser, D., Ffrench-Constant, R.H. (1998). Duplication of the Rdl GABA receptor subunit gene in an insecticide-resistant aphid, myzus persicae. *Mol. Gen. Genet.*, 260(2–3), 165–175.

Baduel, P. and Quadrana, L. (2021). Jumpstarting evolution: How transposition can facilitate adaptation to rapid environmental changes. *Curr. Opin. Plant Biol.*, 61, 102043.

Casacuberta, E. and González, J. (2013). The impact of transposable elements in environmental adaptation. *Mol. Ecol.*, 22(6), 1503–1517.

Catania, F., Kauer, M.O., Daborn, P.J., Yen, J.L., Ffrench-Constant, R.H., Schlötterer, C. (2004). World-wide survey of an Accord insertion and its association with DDT resistance in *Drosophila melanogaster*. *Mol. Ecol.*, 13(8), 2491–2504.

Chatterjee, A.G., Esnault, C., Guo, Y., Hung, S., McQueen, P.G., Levin, H.L. (2014). Serial number tagging reveals a prominent sequence preference of retrotransposon integration. *Nucleic Acids Res.*, 42(13), 8449–8460.

Chung, H., Bogwitz, M.R., McCart, C., Andrianopoulos, A., Ffrench-Constant, R.H., Batterham, P., Daborn, P.J. (2007). Cis-regulatory elements in the accord retrotransposon result in tissue-specific expression of the *Drosophila melanogaster* insecticide resistance gene Cyp6g1. *Genetics*, 175(3), 1071–1077.

Chuong, E.B., Elde, N.C., Feschotte, C. (2016). Regulatory evolution of innate immunity through co-option of endogenous retroviruses. *Science*, 351(6277), 1083–1087.

Daborn, P.J., Yen, J.L., Bogwitz, M.R., Le Goff, G., Feil, E., Jeffers, S., Tijet, N., Perry, T., Heckel, D., Batterham, P. et al. (2002). A single P450 allele associated with insecticide resistance in Drosophila. *Science*, 297(5590), 2253–2256.

Del Villar, S.G. and Jones, A.K. (2018). Cloning and functional characterisation of the duplicated RDL subunits from the pea aphid, acyrthosiphon pisum. *Int. J. Mol. Sci.*, 19(8), 2235.

Dubin, M.J., Mittelsten Scheid, O., Becker, C. (2018). Transposons: A blessing curse. *Curr. Opin. Plant Biol.*, 42, 23–29.

Esnault, C. and Levin, H.L. (2015). The long terminal repeat retrotransposons Tf1 and Tf2 of *Schizosaccharomyces pombe*. *Microbiol. Spectr.*, 3(4), MDNA3-0040-2014.

Esnault, C., Lee, M., Ham, C., Levin, H.L. (2019). Transposable element insertions in fission yeast drive adaptation to environmental stress. *Genome Res.*, 29(1), 85–95.

Feng, G., Leem, Y.E., Levin, H.L. (2013). Transposon integration enhances expression of stress response genes. *Nucleic Acids Res.*, 41(2), 775–789.

Ffrench-Constant, R.H., Rocheleau, T.A., Steichen, J.C., Chalmers, A.E. (1993a). A point mutation in a Drosophila GABA receptor confers insecticide resistance. *Nature*, 363(6428), 449–451.

Ffrench-Constant, R.H., Steichen, J.C., Ode, P.J. (1993b). Cyclodiene insecticide resistance in *Drosophila melanogaster* (Meigen) is associated with a temperature-sensitive phenotype. *Pestic. Biochem. Physiol.*, 46(1), 73–77.

Ffrench-Constant, R.H., Daborn, P.J., Feyereisen, R. (2005). Resistance and the jumping gene. *BioEssays*, 28(1), 6–8.

González, J. and Petrov, D. A. (2012). Evolution of genome content: Population dynamics of transposable elements in flies and humans. *Methods Mol. Biol.*, 855, 361–383.

Guio, L., Vieira, C., González, J. (2018). Stress affects the epigenetic marks added by natural transposable element insertions in *Drosophila melanogaster*. *Sci. Rep.*, 8(1), 1–10.

Hazzouri, K.M., Sudalaimuthuasari, N., Kundu, B., Nelson, D., Al-Deeb, M.A., Le Mansour, A., Spencer, J.J., Desplan, C., Amiri, K.M.A. (2020). The genome of pest *Rhynchophorus ferrugineus* reveals gene families important at the plant-beetle interface. *Commun. Biol.*, 3(1), 1–14.

Hua-Van, A., Le Rouzic, A., Boutin, T.S., Filée, J., Capy, P. (2011). The struggle for life of the genome's selfish architects. *Biol. Direct*, 6(1), 19.

Hunter, R.G. (2020). Stress, adaptation, and the deep genome: Why transposons matter. *Integr. Comp. Biol.*, 60(6), 1495–1505.

Kazazian, H.H., Wong, C., Youssoufian, H., Scott, A.F., Phillips, D.G., Antonarakis, S.E. (1988). Haemophilia A resulting from de novo insertion of L1 sequences represents a novel mechanism for mutation in man. *Nature*, 332(6160), 164–166.

Keller, E.F. (1983). *A Feeling for the Organism: The Life and Work of Barbara McClintock*. W.H. Freeman, San Francisco.

Le Goff, G. and Hilliou, F. (2016). Resistance evolution in *Drosophila*: The case of CYP6G1. *Pest Manag. Sci.*, 73(3), 493–499.

Lu, K., Song, Y., Zeng, R. (2021). The role of cytochrome P450-mediated detoxification in insect adaptation to xenobiotics. *Curr. Opin. Insect Sci.*, 43, 103–107.

McClintock, B. (1984). The significance of responses of the genome to challenge. *Science*, 226(4676), 792–801.

Payer, L.M. and Burns, K.H. (2019). Transposable elements in human genetic disease. *Nat. Rev. Genet.*, 20(12), 760–772.

Rech, G.E., Bogaerts-Márquez, M., Barrón, M.G., Merenciano, M., Villanueva-Cañas, J.L., Horváth, V., Fiston-Lavier, A.-S., Luyten, I., Venkataram, S., Quesneville, H. et al. (2019). Stress response, behavior, and development are shaped by transposable element-induced mutations in Drosophila. *PLOS Genet.*, 15(2), e1007900.

Remnant, E.J., Good, R.T., Schmidt, J.M., Lumb, C., Robin, C., Daborn, P.J., Batterham, P. (2013). Gene duplication in the major insecticide target site, Rdl, in *Drosophila melanogaster*. *Proc. Natl. Acad. Sci. USA*, 110(36), 14706–14710.

Rostant, W.G., Wedell, N., Hosken, D.J. (2012). Transposable elements and insecticide resistance. *Adv. Genet.*, 78, 169–201.

Saleh, A., Macia, A., Muotri, A.R. (2019). Transposable elements, inflammation, and neurological disease. *Front. Neurol.*, 10, 894.

Schrader, L. and Schmitz, J. (2019). The impact of transposable elements in adaptive evolution. *Mol. Ecol.*, 28(6), 1537–1549.

Sheng, C.W., Jia, Z.Q., Ozoe, Y., Huang, Q.T., Han, Z.J., Zhao, C.Q. (2018). Molecular cloning, spatiotemporal and functional expression of GABA receptor subunits RDL1 and RDL2 of the rice stem borer Chilo suppressalis. *Insect Biochem. Mol. Biol.*, 94, 18–27.

Thompson, M., Steichen, J.C., Ffrench-Constant, R.H. (1993). Conservation of cyclodiene insecticide resistance-associated mutations in insects. *Insect Mol. Biol.*, 2(3), 149–154.

Ullastres, A., Merenciano, M., González, J. (2021). Regulatory regions in natural transposable element insertions drive interindividual differences in response to immune challenges in *Drosophila. Genome Biol.*, 22(1), 1–30.

Wells, J.N. and Feschotte, C. (2020). A field guide to eukaryotic transposable elements. *Annu. Rev. Genet.*, 54, 539–561.

Wicker, T., Sabot, F., Hua-Van, A., Bennetzen, J.L., Capy, P., Chalhoub, B., Flavell, A., Leroy, P., Morgante, M., Panaud, O. et al. (2007). A unified classification system for eukaryotic transposable elements. *Nat. Rev. Genet.*, 8(12), 973–982.

Wilson, T.G. (1993). Transposable elements as initiators of insecticide resistance. *J. Econ. Entomol.*, 86(3), 645–651.

Zervudacki, J., Yu, A., Amesefe, D., Wang, J., Drouaud, J., Navarro, L., Deleris, A. (2018). Transcriptional control and exploitation of an immune-responsive family of plant retrotransposons. *EMBO J.*, 37(14), e98482.

6

Domestication (Exaptation) of Transposable Elements

Christopher ELLISON
Department of Genetics, Human Genetics Institute of New Jersey, Rutgers University, Piscataway, USA

6.1. Introduction

The work of Charles Darwin and Gregor Mendel laid the foundation for the field of evolutionary genetics. Darwin's theory of evolution by natural selection proposed that adaptation occurs because natural variation among individuals in a population allows some individuals to produce more offspring than others. Darwin, however, did not offer a clear explanation for the mechanisms that produce such variation, whereas Mendel's work provided such an explanation: mutations in genes are the source of heritable variation that natural selection acts upon. In other words, Mendelian inheritance provides the basis for Darwinian evolution.

We often think of mutations as a single nucleotide change, such as a change from cytosine to thymine, or an insertion or deletion of a small number of nucleotides. However, mutations can be much more complex, for example, inversions and translocations can change the locations of hundreds of genes along the chromosome. Insertions of transposable elements (TEs), either via copy-and-paste, or cut-and-paste activities, qualify as another form of complex mutation. Transposons may be many thousands of base pairs in size and carry multiple protein coding genes and non-coding sequences that can act as potent drivers of gene expression. Insertions of these elements at new locations in the genome are heritable and have been associated with a wide variety of phenotypic traits (Casacuberta and Gonzalez 2013).

Transposable Elements and Genome Evolution,
coordinated by Aurélie HUA-VAN and Pierre CAPY. © ISTE Ltd 2024.

Furthermore, the rate of transposon insertions can be much higher than the mutation rate, which means that novel traits derived from transposon activity can evolve at a faster rate compared to those that evolve from other types of mutations (Adrion et al. 2017). There is a growing body of work suggesting that transposons play an important role in the evolutionary process by providing genetic variation for natural selection to act upon.

There are two major ways by which TEs can affect an organism's phenotype. A TE can be disruptive: it can insert into the coding sequence of a gene, thereby inactivating it, or it can be constructive: a TE can provide the raw material for a genome feature, such as a gene, noncoding RNA, or promoter element, that can then be co-opted by the host genome for a new purpose. This co-option process is also known as exaptation or molecular domestication. In this chapter, we shall cover three major aspects of transposon domestication: host genes that are derived entirely from TEs or from TE-gene fusions (i.e. exonization); the ability of transposons to rewire gene regulatory networks; and the role of transposons in the formation of structural components of the genome (Figure 6.1).

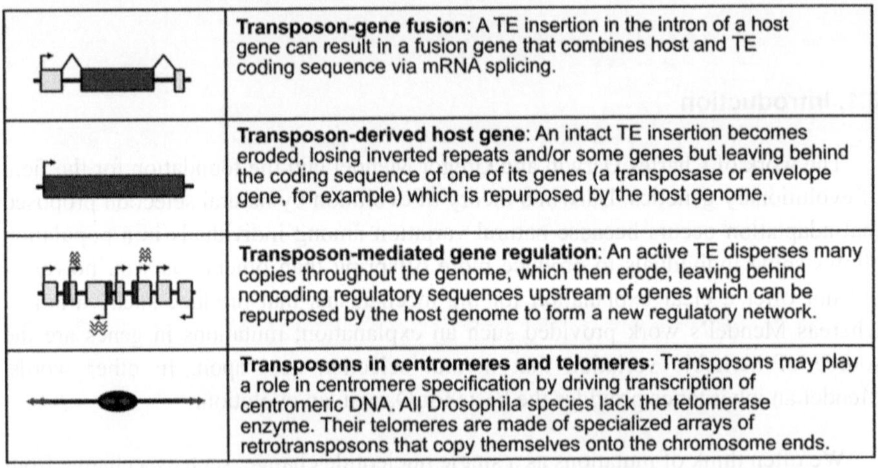

Figure 6.1. *The various way a transposon can be domesticated. Transposon-gene fusion, also called exonisation, is a variation of transposon-derived host genes (the coding sequence is domesticated), unlike TE-mediated gene regulation in which TE sequences (DNA) are used in cis. In centromere and telomeres, both DNA and TE protein sequences are involved. For a color version of this figure, see www.iste.co.uk/huavan/transposable.zip*

6.2. Host genes derived from transposons

There are now many examples of host genes that originated from a TE across a wide variety of species. In this section, we shall highlight a handful of examples while examining several themes that many of these gene acquisition events have in common.

6.2.1. *Repeated domestication of transposases*

Transposons carry a variety of genes that aid in their replication. Most DNA transposons only need a single gene to replicate, which is known as transposase. The transposase protein functions to mobilize the TE and is responsible for its cut-and-paste activity. Transposases have DNA binding activity and specifically recognize sequence elements present at the ends of the TEs that they mobilize (Haren et al. 1999). A large number of host genes that have been reported to have arisen from transposons were derived from transposases (Feschotte and Pritham 2007). Below, several of the best-characterized examples of transposase domestication from both animals and plants are described.

Innate immunity is an ancient, nonspecific immune response that is conserved across most multicellular organisms, which triggers an inflammatory response upon detection of a pathogen. However, jawed vertebrates (known as gnathostomes) also have an adaptive immune system (Cooper and Alder 2006). Adaptive immune systems depend upon a genetic mechanism to produce an astonishingly large repertoire of antigen-receptor genes, each of which is expressed in a different cell. These cells are known as T cells and B cells. Early in their development, they rearrange four gene segments (known as variable, diversity and joining segments (V(D)J)) to create the antigen binding regions of the T-cell and B-cell receptor genes (Figure 6.2). The proteins encoded by the *RAG1* and *RAG2* genes bind to specific sequences adjacent to the V(D)J segments and catalyze the recombination events that allow for the production of a large diversity of antigen-receptor genes (Cooper and Alder 2006). The *RAG1/RAG2* genes likely arose from an ancient member of the *Transib* superfamily of DNA transposons (Kapitonov and Jurka 2005). In the ancestor of jawed vertebrates, this transposon likely inserted into the exon of a receptor gene. Subsequent insertion of the same TE into additional exons likely allowed for "shuffling" of exons between insertions, a primitive version of the current V(D)J recombination system (Cooper and Alder 2006).

This adaptive immune system is a vertebrate-specific evolutionary innovation. As vertebrates diversified and adapted to novel environmental niches, they also likely encountered novel pathogens. An adaptive immune system would confer a strong selective advantage in such a situation and it is astounding that a domesticated transposase played a critical role in the evolution of this trait (Figure 6.2).

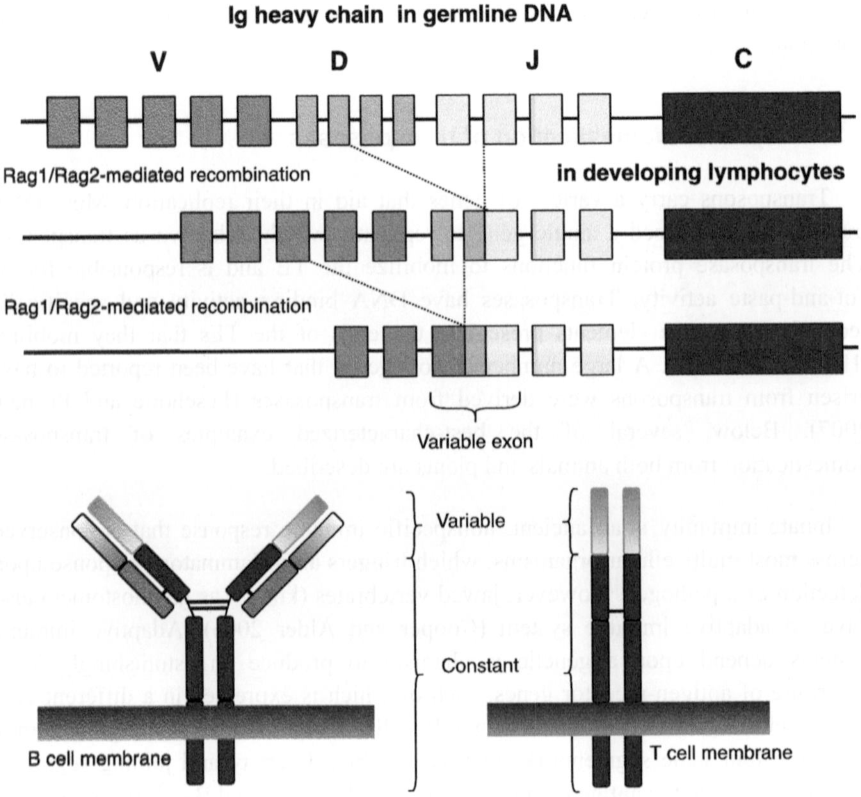

Figure 6.2. *The V(D)J recombination process allows recognition of a wide variety of antigens in the adaptive immune response in jawed vertebrates. T and B cells patrol the body and, if their membrane-bound receptor binds a foreign antigen, undergo proliferation and produce antibodies that target the antigen. The TE-derived Rag1 and Rag2 proteins catalyse V(D)J recombination, first between a D and J segment and then between a V segment and the newly formed DJ fusion. This process creates novel protein domains that form the variable region of the immunoglobulin (IG) heavy chain, allowing antibodies to recognize a wide repertoire of antigens. For a color version of this figure, see www.iste.co.uk/huavan/transposable.zip*

Another fascinating example of an evolutionary innovation caused by transposase domestication comes from plants. Plants optimize their growth by monitoring and responding to changing light conditions. Plants are able to use specialized pigments known as phytochromes to detect light from both the red and far red spectrums. The relative abundance of these two forms of radiation influences plant growth strategies. For example, plant leaves reflect most far red light, meaning that a plant growing in the shade of a larger plant will be exposed to a greater proportion of far red light (Franklin 2008). Shade intolerant plants in this situation commonly exhibit a "shade-avoidance response" by allocating energy resources towards elongation rather than leaf production in an attempt to escape the shade of their neighbors (Franklin 2008). The ancient phytochrome phyB, which is shared across angiosperms and gymnosperms, controls shade avoidance and inhibits seed germination in deep shade. However, an angiosperm-specific phytochrome known as phyA acts to promote germination and helps plants establish in deep shade. The evolution of this gene in angiosperms was likely a key innovation that allowed early flowering plants to colonize forests dominated by ferns and gymnosperms (Mathews 2005). The phyA signaling pathway is controlled by two transcription factors known as FHY3 and FAR1, both of which are domesticated transposases, similar to the maize *Mutator* DNA transposon (Lin et al. 2007). Interestingly, at least five other genes involved in the response to light in plants, collectively known as FAR1-Related Sequences (*FRS*), have been independently acquired from transposases (Joly-Lopez et al. 2016).

The recurrent domestication of transposases and the functions of the novel host genes are both related to the unique properties of the transposase protein. A recent survey of 596 tetrapod species identified 106 instances of host genes that were acquired from the transposases of DNA transposons (Cosby et al. 2021). These gene acquisitions involved fusion events that combine the DNA binding activity of the transposase with domains from other host genes to form a chimeric protein via alternative splicing (Cosby et al. 2021). Most transposases bind to sequence elements located within the TE that carries them. One amazing consequence of transposase domestication is that the newly acquired host gene will already be capable of binding to locations in the genome containing full or even partial fragments of the parent TE. If any of these sequences happen to be near a gene, a new regulatory pathway could easily evolve (Feschotte and Pritham 2007). This phenomenon, combined with the propensity for chimeric proteins composed of host gene and transposase domains, provides a rich source of evolutionary novelty.

6.2.2. *Repeated domestication of envelope genes*

Retrotransposons and endogenous retroviruses (ERVs), which replicate via an RNA intermediate, usually carry the *gag* and/or *env* genes, which encode proteins that encapsulate the RNA intermediate for transport within and between cells. Host genes that were acquired from the *gag/env* genes of retrotransposons have been co-opted to play a role in a variety of processes including the evolution of the mammalian placenta and brain development. One fascinating example of ERV domestication is that of the *syncytin* gene family. The ERV envelope proteins play a role in membrane fusion to help the virus gain entry into the host cell. Mammalian *syncytin* proteins play a critical role in the function of the placenta by creating a layer of fused cells that helps with nutrient exchange along the fetal/maternal boundary. *Syncytin* proteins have been recurrently domesticated from different ERV families at least 10 times in various groups of mammals (reviewed in Naville et al. (2016)) and also appear to have played a role in the evolution of a simplified placenta in live-bearing Mabuya lizards (Cornelis et al. 2017). The transition from egg-bearing (oviparity) to live-bearing (viviparity) represents a key innovation that likely played an important role in the evolutionary success of the mammalian lineage. It is fascinating to consider the fact that an ERV provided the raw materials that this important trait evolved from.

Another interesting case of retroelement domestication involves that of the neuronal protein known as Arc. *Arc* originated in the common ancestor of mammals, birds, reptiles and amphibians (Pastuzyn et al. 2018). Animals from all of these groups exhibit synaptic plasticity, which is the ability of the brain to process and store information by forming new structures. The *Arc* gene is known to play a role in synaptic plasticity and mutations in *Arc* have been linked to neurodevelopmental disorders (Bramham et al. 2010). Recent work has shed light on the origin and function of *Arc* (Ashley et al. 2018; Pastuzyn et al. 2018). This gene was acquired from the *gag* gene of a retrotransposon in the tetrapod ancestor (Pastuzyn et al. 2018). Interestingly, *Arc* appears to have retained a function similar to that of *gag*: it forms capsid-like particles that envelope RNA (Ashley et al. 2018; Pastuzyn et al. 2018). These particles are then used to transport *Arc* mRNA between neurons via extracellular vesicles (Ashley et al. 2018; Pastuzyn et al. 2018). There is still much to understand about the function of the protein but this unique form of intercellular transport is likely related to its important role in synaptic plasticity. An *Arc*-like gene has also been characterized in Drosophila and found to play a similar role (Ashley et al. 2018; Pastuzyn et al. 2018). Strikingly, Pastuzyn et al. have shown that the Drosophila Arc-like protein, which is found in all schizophoran flies, is not directly related to the tetrapod Arc and was instead independently acquired from an insect retrotransposon.

6.3. TEs can disperse noncoding regulatory sequences across the genome

The ability of transposons to affect gene expression was first proposed by Barbara McClintock, who envisioned transposons as "controlling elements" whose coordinated transposition turned genes on and off in a cell-type specific manner (Comfort 2001). We now know that active TE insertions tend to suppress gene expression, either by inactivating the genes themselves, or via spreading of heterochromatin. However, the ability of TE-derived sequences to activate gene expression is experiencing a resurgence of interest (Chuong et al. 2017). Transposons carry a variety of regulatory sequences that are necessary for successful replication. Retrotransposons and autonomous DNA transposons can contain promoters, intron/exon splice sites and poly-adenylation signal sequences. Furthermore, transposons must replicate specifically in the germline so that their new copies will be transmitted to the progeny. Many TEs therefore carry binding sites for host transcription factors that will promote their expression in germline or germline-adjacent tissues. A growing body of work has shown how these TE-derived regulatory sequences can be repurposed by the host genome to regulate the expression of host genes (Chuong et al. 2017). The replicative activity of TEs allows them to disperse their regulatory sequences across the genome. The sequences can then be repurposed, not just on a gene-by-gene basis but for the regulation of entire biological pathways. This phenomenon has come to be known as *rewiring gene regulatory networks*.

One of the most striking examples of this rewiring phenomenon is related to genes involved in the innate immune response in mammals (Chuong et al. 2016). In general, innate immunity involves the formation of physical and chemical barriers to foreign organisms within the body of the host. Although this basic response is evolutionarily conserved, there have been many species-specific and lineage-specific changes in the biological pathways underlying the innate immune response, likely to adapt to differences in infectious agents faced by different species. In mammals, inflammation is a major aspect of the innate immune response, which is mediated by signaling molecules known as interferons. Two transcription factors, known as IRF and STAT, initiate the inflammation response by binding to regulatory elements known as enhancers. In general, enhancers are noncoding regulatory sequences that control the spatial and temporal expression patterns of genes. In this case, the IRF and STAT-bound enhancers regulate interferon stimulated genes. Together, these transcription factors and the genes they regulate form an interferon-inducible gene regulatory network. The genes that are part of this regulatory network differ between different mammalian species, likely in response to different pathogens. Chuong et al. have shown that specific ERVs that carry IRF and/or STAT binding sites have "wired" new genes into this gene network by depositing these binding sites at new locations in their host genomes. It is eerily apropos that regulation of the innate immune response in mammals evolved from an ancient retroviral infection!

TE-derived regulatory sequences have been involved in a variety of evolutionary innovations in mammals, from pregnancy to embryogenesis (Dunn-Fletcher et al. 2018; Fuentes et al. 2018), but there are also examples of the role that TEs play in human-specific adaptations. One question that has fascinated biologists for decades is: what are that genetic changes that make us uniquely human compared to our primate ancestors? Many human-specific traits are rooted in the morphology of the human face and skull. For example, an upright posture, enlarged brain and complex speech patterns all required modifications of the human craniofacial complex. Prescott et al. (2015) profiled gene expression in the cranial neural crest cells of humans and chimpanzees and found that divergence in gene expression is associated with sequence divergence in enhancers. They were able to identify species-specific enhancers that regulate the expression of genes involved in craniofacial morphology. Strikingly, they found that almost half of these species-specific enhancers were acquired from TEs, mostly ERVs and LINE L1 elements. Their results suggest that transposon-derived regulatory sequences have played a major role in the evolution of human-specific phenotypic innovations.

Dosage compensation is a chromosome-specific form of gene regulation that is shared by many metazoan species with heteromorphic sex chromosomes. In species with XY sex chromosome systems, females have two copies of the X chromosome, whereas males only have a single copy. Two different mechanisms have evolved in order to equalize the X versus autosome gene expression levels between the sexes. In most mammalian species, one of the female X chromosomes in inactivated and transcriptionally silenced. In flies and many other insect species, the single X chromosome in males is hypertranscribed. Both mechanisms require chromosome-specific regulation of gene expression. Research in Drosophila has shown that the hypertranscription of the evolutionarily young X chromosome of *Drosophila miranda* evolved due to the activity of a helitron TE (Ellison and Bachtrog 2013). The *Drosophila miranda* neo-X chromosome is highly enriched for a non-autonomous helitron known as ISX. This element carries a sequence motif that is recognized by the Drosophila dosage compensation protein complex (DCC). Ellison et al. discovered that an ancestral element, known as ISY, acquired a 10 bp deletion which led to the formation of the DCC motif and which is the only sequence difference between the ISY and ISX elements. ISX dispersed this motif across the genome; however, autosomal insertions were likely deleterious and thus were eliminated from the population, whereas X-linked insertions were beneficial and spread through the population, with the end result being that the ISX element is highly enriched on the neo-X chromosome. This study highlights the fact that, due to their high insertion rate and because they carry ready-made regulatory elements, TEs can rapidly generate novel regulatory networks. The neo-X chromosome formed very recently in evolutionary time, yet was able to quickly become dosage compensated. It would take orders of magnitude longer for the DCC binding sites on the neo-X to evolve via small indels and point mutations. Interestingly, a different transposon may have played a role in the evolution of dosage compensation on the middle-aged X chromosome that is shared between *Drosophila miranda* and *Drosophila pseudoobscura* and there is evidence that the independently evolved neo-X chromosome of another Drosophila species, *Drosophila robusta*, also became dosage compensated due to the activity of a completely different TE (Ellison and Bachtrog 2013; Ellison and Bachtrog 2019).

6.4. TEs form structural components of the genome

Transposons can be co-opted by their host genome to serve as novel genes and to regulate the expression of host genes. These examples of TE domestication have been associated with a variety of phenotypic innovations. However, there is also evidence that TEs play a role in the formation of new species by causing reproductive isolation (Serrato-Capuchina and Matute 2018). A classic example of this phenomenon is that of hybrid dysgenesis (see Chapter 4), where normally silenced TEs are reactivated in hybrid progeny, causing sterility (Bregliano et al. 1980). However, TEs can also cause changes in chromosome structure, which can cause reproductive incompatibilities. One major way by which TEs affect chromosome structure is via the formation of chromosomal rearrangements: dispersed copies of the same TE can cause aberrant recombination between TE copies in different locations on the chromosome. However, there is also evidence that TEs can affect chromosome structure via a role in the specification of centromeres and telomeres. Differences in chromosomal rearrangements, centromere positioning and telomere regulation all have the potential to cause reproductive isolation between species and thus serve as three different mechanisms by which TEs can influence speciation.

6.4.1. *Transposons function in centromere specification*

Both centromeres and telomeres are fundamental components of genome organization across all eukaryotes, yet the genetic basis of these structures varies widely even among fairly closely related species. All centromeres are marked by specific histone H3 variants known as CENP-A in animals and CenH3 in plants.

It is currently unclear whether specific DNA sequences influence centromere positioning. Most centromeres are enriched for satellite DNA, however satellites are not sufficient to promote centromere formation and newly formed so-called neocentromeres have been found to lack satellite DNA (Nakano et al. 2003; Alonso et al. 2007; Klein and O'Neill 2018). Transcription of noncoding RNAs has previously been shown to be important for centromere formation and specific non-LTR retrotransposons have been identified within the core centromere sequence of a variety of organisms, from plants to animals (Chen et al. 2015; Chang et al. 2019). These two observations raise the possibility that TEs provide a source of promoters to catalyze transcription within the centromere (Klein and O'Neill 2018). Surprisingly, not only is there a potential role for TEs in determining centromere location, but CENP-B-like centromere binding proteins have independently evolved from domesticated TEs at least twice: once in mammals and once in fungi (Casola et al. 2008).

6.4.2. *Transposons act as telomeres in Drosophila*

The "end-replication" problem inherent to DNA replication was first described by James Watson in 1972. This problem arises during the replication of the "lagging strand" of DNA which requires constant priming and stops at the last primer, which is usually 70–100 bp from the end of the chromosome. Thus, chromosome ends are shortened by 70–100 bp every replication cycle. Canonical telomeres, such as those present in most mammals, are composed of short DNA repeats that are extended by the telomerase enzyme to prevent end-shortening due to incomplete replication. Other species have more complex telomeres composed of short DNA repeats interspersed with specific TEs (Louis 2002). Most Drosophila species have an extreme form of telomeres where telomerase has been lost and the chromosome ends are composed of head to tail arrays of specialized non-LTR retrotransposons (Biessmann et al. 1992; Levis et al. 1993). In most forms of TE domestication, the TE sequence is eroded, leaving behind a single gene or regulatory sequence that is no longer mobile. The telomeric retrotransposons of Drosophila represent a unique form of transposon domestication: these are fully mobile elements whose activity is constrained by the host genome such that they only replicate onto the ends of the chromosome. Domesticated elements that remain active are prone to evolve antagonistically with their host. Indeed, there is ample evidence that these telomeric transposons are simultaneously serving a critical role for their host genome while also evolving under conflict with their host genome (Lee et al. 2017; Saint-Léandre et al. 2019; Ellison et al. 2020). While the replacement of telomerase by active transposons may be unique to Drosophila, there is evidence that telomerase itself shares a common ancestor with all retrotransposons (Gladyshev and Arkhipova 2007).

6.5. Summary

The importance of TE-derived genes and regulatory sequences to the structure and function of host genomes is becoming more and more clear. There are now enough specific examples of TE domestication for the possibility of beginning to draw more general conclusions about this process. For example, it is clear that TE-derived genes do not gain wildly new functions when they are co-opted by their host genome. Instead, their function for the host is often similar to their function for the TE: transposases often become DNA binding proteins that help regulate host genes while envelope proteins help to package host macromolecules. Interestingly, there also seems to be specific biological pathways that are enriched for domesticated TEs, including immunity, antiviral activity and genome defense (Jangam et al. 2017). It is fascinating to consider that many factors used by host genomes to defend against infection were themselves acquired from infectious elements such as ERVs and TEs.

Another major question related to this topic is: how common are TE domestication events? Do the examples described above represent rare peculiarities of evolution or is TE domestication a major driving force of evolutionary innovation? The latter seems to be true, at least for some specific biological pathways. The most convincing evidence for this proposition is the degree of evolutionary convergence that has occurred for specific traits. Almost all of the examples described above show some amount of convergent evolution, where the same trait has evolved independently due to independent cases of TE domestication in different species. This convergence suggests that TE domestication is indeed an important driver of evolution and there are likely to be many more examples of the role of TE exaptation in evolutionary innovation that have yet to be uncovered.

6.6. References

Adrion, J.R., Song, M.J., Schrider, D.R., Hahn, M.W., Schaack, S. (2017). Genome-wide estimates of transposable element insertion and deletion rates in *Drosophila melanogaster*. *Genome Biol Evol*, 9(5), 1329–1340.

Alonso, A., Fritz, B., Hasson, D., Abrusan, G., Cheung, F., Yoda, K., Radlwimmer, B., Ladurner, A.G., Warburton, P.E. (2007). Co-localization of CENP-C and CENP-H to discontinuous domains of CENP-A chromatin at human neocentromeres. *Genome Biol*, 8(7), R148.

Ashley, J., Cordy, B., Lucia, D., Fradkin, L.G., Budnik, V., Thomson, T. (2018). Retrovirus-like Gag protein Arc1 binds RNA and traffics across synaptic boutons. *Cell*, 172(1–2), 262–274.e211.

Biessmann, H., Valgeirsdottir, K., Lofsky, A., Chin, C., Ginther, B., Levis, R.W., Pardue, M.L. (1992). HeT-A, a transposable element specifically involved in "healing" broken chromosome ends in *Drosophila melanogaster*. *Mol Cell Biol*, 12(9), 3910–3918.

Bramham, C.R., Alme, M.N., Bittins, M., Kuipers, S.D., Nair, R.R., Pai, B., Panja, D., Schubert, M., Soule, J., Tiron, A. et al. (2010). The Arc of synaptic memory. *Exp Brain Res*, 200(2), 125–140.

Bregliano, J.C., Picard, G., Bucheton, A., Pelisson, A., Lavige, J.M., L'Heritier, P. (1980). Hybrid dysgenesis in *Drosophila melanogaster*. *Science*, 207(4431), 606–611.

Casacuberta, E. and Gonzalez, J. (2013). The impact of transposable elements in environmental adaptation. *Mol Ecol*, 22(6), 1503–1517.

Casola, C., Hucks, D., Feschotte, C. (2008). Convergent domestication of pogo-like transposases into centromere-binding proteins in fission yeast and mammals. *Mol Biol Evol*, 25(1), 29–41.

Chang, C.H., Chavan, A., Palladino, J., Wei, X., Martins, N.M.C., Santinello, B., Mellone, B.G. (2019). Islands of retroelements are major components of *Drosophila* centromeres. *PLoS Biol*, 17(5), e3000241.

Chen, C.C., Bowers, S., Lipinszki, Z., Palladino, J., Trusiak, S., Bettini, E., Rosin, L., Przewloka, M.R., Glover, D.M, O'Neill, R.J. et al. (2015). Establishment of centromeric chromatin by the CENP-A assembly factor CAL1 requires FACT-mediated transcription. *Dev Cell*, 34(1), 73–84.

Chuong, E.B., Elde, N.C., Feschotte, C. (2016). Regulatory evolution of innate immunity through co-option of endogenous retroviruses. *Science*, 351(6277), 1083–1087.

Chuong, E.B., Elde, N.C., Feschotte, C. (2017). Regulatory activities of transposable elements: From conflicts to benefits. *Nat Rev Genet*, 18(2), 71–86.

Comfort, N.C. (2001). From controlling elements to transposons: Barbara McClintock and the Nobel Prize. *Trends Biochem Sci*, 26(7), 454–457.

Cooper, M.D. and Alder, M.N. (2006). The evolution of adaptive immune systems. *Cell*, 124(4), 815–822.

Cornelis, G., Funk, M., Vernochet, C., Leal, F., Tarazona, O.A., Meurice, G., Heidmann, O., Dupressoir, A., Miralles, A., Ramirez-Pinilla, M.P. et al. (2017). An endogenous retroviral envelope syncytin and its cognate receptor identified in the viviparous placental Mabuya lizard. *Proc Natl Acad Sci USA*, 114(51), E10991–E11000.

Cosby, R.L., Judd, J., Zhang, R., Zhong, A., Garry, N., Pritham, E.J., Feschotte, C. (2021). Recurrent evolution of vertebrate transcription factors by transposase capture. *Science*, 371(6531), eabc6405.

Dunn-Fletcher, C.E., Muglia L.M., Pavlicev, M., Wolf, G., Sun M.A., Hu, Y.C., Huffman, E., Tumukuntala, S., Thiele, K., Mukherjee A. et al. (2018). Anthropoid primate-specific retroviral element THE1B controls expression of CRH in placenta and alters gestation length. *PLoS Biol*, 16(9), e2006337.

Ellison, C.E. and Bachtrog, D. (2013). Dosage compensation via transposable element mediated rewiring of a regulatory network. *Science*, 342(6160), 846–850.

Ellison, C. and Bachtrog, D. (2019). Contingency in the convergent evolution of a regulatory network: Dosage compensation in *Drosophila*. *PLoS Biol*, 17(2), e3000094.

Ellison, C.E., Kagda, M.S., Cao, W. (2020). Telomeric TART elements target the piRNA machinery in *Drosophila*. *PLoS Biol*, 18(12), e3000689.

Feschotte, C. and Pritham, E.J. (2007). DNA transposons and the evolution of eukaryotic genomes. *Annu Rev Genet*, 41, 331–368.

Franklin, K.A. (2008). Shade avoidance. *New Phytol*, 179(4), 930–944.

Fuentes, D.R., Swigut, T., Wysocka, J. (2018). Systematic perturbation of retroviral LTRs reveals widespread long-range effects on human gene regulation. *Elife*, 7, e35989.

Gladyshev, E.A. and Arkhipova, I.R. (2007). Telomere-associated endonuclease-deficient Penelope-like retroelements in diverse eukaryotes. *Proc Natl Acad Sci USA*, 104(22), 9352–9357.

Haren, L., Ton-Hoang, B., Chandler, M. (1999). Integrating DNA: Transposases and retroviral integrases. *Annu Rev Microbiol*, 53, 245–281.

Jangam, D., Feschotte, C., Betran, E. (2017). Transposable element domestication as an adaptation to evolutionary conflicts. *Trends Genet*, 33(11), 817–831.

Joly-Lopez, Z., Hoen, D.R., Blanchette, M., Bureau, T.E. (2016). Phylogenetic and genomic analyses resolve the origin of important plant genes derived from transposable elements. *Mol Biol Evol*, 33(8), 1937–1956.

Kapitonov, V. V. and Jurka, J. (2005). RAG1 core and V(D)J recombination signal sequences were derived from *Transib* transposons. *PLoS Biol*, 3(6), e181.

Klein, S.J. and O'Neill, R.J. (2018). Transposable elements: Genome innovation, chromosome diversity, and centromere conflict. *Chromosome Res*, 26(1–2), 5–23.

Lee, Y.C., Leek, C., Levine, M.T. (2017). Recurrent innovation at genes required for telomere integrity in *Drosophila*. *Mol Biol Evol*, 34(2), 467–482.

Levis, R.W., Ganesan, R., Houtchens, K., Tolar, L.A., Sheen, F.M. (1993). Transposons in place of telomeric repeats at a *Drosophila* telomere. *Cell*, 75(6), 1083–1093.

Lin, R., Ding, L., Casola, C., Ripoll, D.R., Feschotte, C., Wang, H. (2007). Transposase-derived transcription factors regulate light signaling in *Arabidopsis*. *Science*, 318(5854), 1302–1305.

Louis, E.J. (2002). Are *Drosophila* telomeres an exception or the rule? *Genome Biol*, 3(10), REVIEWS0007.

Mathews, S. (2005). Phytochrome evolution in green and nongreen plants. *J Hered*, 96(3), 197–204.

Nakano, M., Okamoto, Y., Ohzeki, J., Masumoto, H. (2003). Epigenetic assembly of centromeric chromatin at ectopic alpha-satellite sites on human chromosomes. *J Cell Sci*, 116(Pt 19), 4021–4034.

Naville, M., Warren, I.A., Haftek-Terreau, Z., Chalopin, D., Brunet, F., Levin, P., Galiana, D., Volff, J.N. (2016). Not so bad after all: Retroviruses and long terminal repeat retrotransposons as a source of new genes in vertebrates. *Clin Microbiol Infect*, 22(4), 312–323.

Pastuzyn, E.D., Day, C.E., Kearns, R.B., Kyrke-Smith, M., Taibi, A.V., McCormick, J., Shepherd, J.D. (2018). The neuronal gene arc encodes a repurposed retrotransposon Gag protein that mediates intercellular RNA transfer. *Cell*, 172(1–2), 275–288, e218.

Prescott, S.L., Srinivasan, R., Marchetto, M.C., Grishina, I., Narvaiza, I., Selleri, L., Gage, F.H., Swigut, T., Wysocka, J. (2015). Enhancer divergence and cis-regulatory evolution in the human and chimp neural crest. *Cell*, 163(1), 68–83.

Saint-Leandre, B., Nguyen, S.C., Levine, M.T. (2019). Diversification and collapse of a telomere elongation mechanism. *Genome Res*, 29(6), 920–931.

Serrato-Capuchina, A. and Matute, D.R. (2018). The role of transposable elements in speciation. *Genes (Basel)*, 9(5), 254.

7

Horizontal Transfers and Transposable Elements

Emmanuelle LERAT
Laboratoire Biométrie et Biologie Évolutive (LBBE), CNRS, Université Claude Bernard Lyon 1, Villeurbanne, France

7.1. Introduction

Horizontal transfer (HT) is defined as the passage of genetic material between individuals of different species, as opposed to vertical transfer, which involves the transmission of genetic information from parents to offspring. The DNA transmitted by HT is integrated into the genome of the recipient species, where it can then be distributed vertically to the offspring. HTs were first described in prokaryotes at the very beginning of the 20th century, with Fred Griffith's 1928 experiments on the virulence of streptococci, followed by Joshua Lederg's 1947 experiments on *Escherichia coli*, revealing the molecular mechanism behind these transfers (see below). With the advent of the first complete genome sequencing in the late 1990s, several studies have demonstrated the preponderance of gene HTs between bacterial species. For example, the analysis of one of the first sequenced bacteria, *E. coli*, showed that almost 20% of the genes have been acquired by HT (Lawrence and Ochman 1998). The large number of HTs discovered in bacteria raised the question of how to reconstruct the phylogeny of species (Doolittle 1999), before methodologies were proposed to circumvent the problem (Lerat et al. 2003). Comparison of microorganisms from different domains of life has shown that HTs can occur even between bacteria and archaea (Aravind et al. 1998). For a long time,

Transposable Elements and Genome Evolution,
coordinated by Aurélie HUA-VAN and Pierre CAPY. © ISTE Ltd 2024.

HTs were thought to be possible only in prokaryotes, due to their unicellular and asexual nature, making it easier to integrate new genes into their genomes and transmit them through populations. However, HTs have also been detected in both unicellular (Etten and Bhattacharya 2020) and multicellular eukaryotes (Keeling and Palmer 2008). For example, studies carried out on phytoplankton (or plant plankton), which comprise all the cyanobacteria and microalgae found in water, revealed that a significant proportion of the genes enabling them to colonize new habitats originated in prokaryotes (Qiu et al. 2013; Schönknecht et al. 2013). Thus, gene HTs have been highlighted as potential drivers in the evolution of genomes across all domains of life.

As transposable elements (TEs) are genomic sequences, they can be transmitted horizontally in the same way as genes. The first evidence of HT of transposable elements (HTT) concerned a DNA transposon, the P element. In the early 1990s, a comparison of the sequence of this element found in two distantly related Drosophila species, *Drosophila willistoni* and *Drosophila melanogaster*, showed for the first time that HTs do not only involve genes (Daniels et al. 1990). Around the same time, the HT of another DNA transposon, the mariner element, was shown to have occurred between flies of the genera *Drosophila* and *Zaprionus*, the two groups having been separated for around 50 million years (Maruyama and Hartl 1991). Initially considered a rather marginal phenomenon, HTTs were eventually proposed to be far more numerous than what had been observed. In particular, HTs are now accepted as part of the lifecycle of a TE (Le Rouzic and Capy 2005; Schaack et al. 2010).

For a long time, the maintenance of TEs within genomes was considered a paradox. In fact, TEs do not, on the whole, directly benefit their host (Orgel and Crick 1980; Hickey 1982). Consequently, one may expect TEs to be gradually eliminated from genomes under the combined effect of several factors such as their rate of excision, transposition and mutation, the negative selection against deleterious insertions and other host-specific population genetic and regulatory parameters (Badge and Brookfield 1997; Brookfield and Badge 1997; Le Rouzic and Deceliere 2005; Le Rouzic et al. 2007). However, TEs can represent significant proportions in genomes, demonstrating their ability to maintain themselves for a very long time. The discovery that TEs can be transmitted horizontally offers them a new opportunity to persist by colonizing new genomes. The lifecycle of a TE (Figure 7.1) may begin, for example, with the HT of an active TE family into a new host genome.

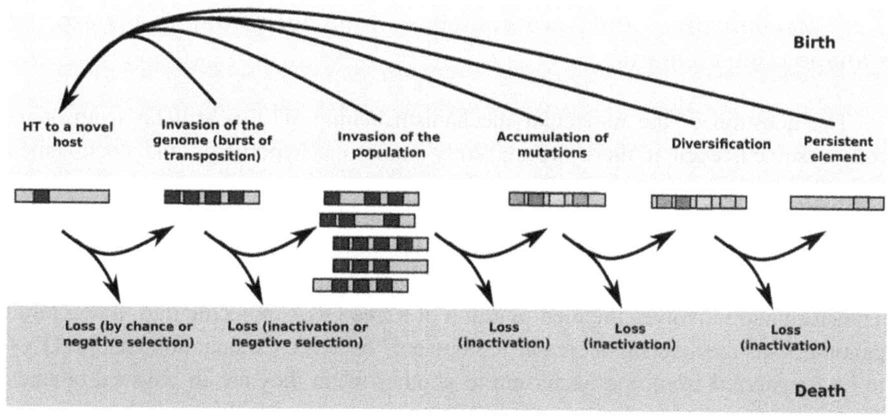

Figure 7.1. *Lifecycle of a TE family. Gray bars represent a genome, small rectangles represent TE copies and arrows represent transitions between different stages. Adapted from Schaack et al. (2010)*

If the host's regulatory system is ineffective, this will allow the new family to multiply and invade the genome (characterized by a burst of transposition). This invasion must take place through increases in the number of copies in the genome and in the frequency of these copies in the host species population. As the regulation of this new TE family sets up, copies can become progressively inactivated, with the accumulation of disabling mutations in their sequences, or even be lost, leading to the disappearance of the family. However, it remains possible for a family to persist by controlling its copy number or by acquiring preferred insertion sites in genome-safe areas. HT would be another way of escaping extinction, if it occurs before the family is inactivated.

The fact that HTs are part of the TE lifecycle implies that all classes of TEs can potentially be affected. Although all classes of TEs can be subject to HTs, it has been noted that there is a bias toward DNA transposons, which account for the vast majority of cases reported to date (Gilbert and Feschotte 2018), although in some species, such as Drosophila, LTR retrotransposons are particularly prone to HTs (Loreto et al. 2008). The explanation put forward for the prevalence of DNA transposons would be that they possess intrinsic characteristics that enable them to colonize genomes and therefore have a greater chance of sustaining themselves after transfer, such as simpler genetic organization and lower dependence on host factors necessary for their transposition.

7.2. Mechanisms and prerequisites for horizontal transfers of transposable elements

The question of the molecular mechanism behind HTT is still far from being fully resolved, even if there are currently numerous hypotheses and documented cases that point the way forward.

Work carried out on prokaryotes where HTs were first highlighted has pointed to three mechanisms: transformation, conjugation and transduction (Villa et al. 2019). Transformation involves the incorporation of a free DNA molecule into the bacterial genome. This mechanism occurs in "competent" bacteria. Conjugation enables DNA to be transferred from one bacterium to another when they are in physical contact. Transduction uses a viral vector, the bacteriophage, which integrates into the bacterial genome. The success of these mechanisms is facilitated by the fact that prokaryotes are unicellular, without a nucleus and reproduce asexually. In eukaryotes, other mechanisms have been proposed, but there is currently no concrete proof of their action(s). There are, however, a large number of observations that point to the means and prerequisites necessary for the phenomenon to occur.

7.2.1. *Direct transmission*

A number of mechanisms intrinsic to TEs may enable them to be transferred horizontally. For example, several LTR retrotransposons have the capacity to form viral particles. Indeed, some of these elements possess an open reading frame (ORF) encoding a particular protein called ENV (Vicient et al. 2001; Laten et al. 2003). This gene, which is found in the sequences of retroviruses in which it enables entry into the host cell, confers an infectious character on them (Coffin et al. 1997). Comparative analyses of the ENV protein between several TEs and viruses have shown it to be homologous (Lerat and Capy 1999). Experiments with Drosophila LTR retrotransposons have shown that they can form viral particles and thus infect cells (Kim et al. 1994; Leblanc et al. 2000).

TEs could also be transmitted directly between hosts via the formation of episomes (Kempken and Kück 1998). An episome is an extrachromosomal DNA molecule that can become circularized and likely results from the deletion of copies inserted into the genome under the action of non-homologous recombination (Mourier 2016). Episomes that possess all the information required for transposition have been observed for different classes of TEs, both in in vitro and in vivo systems (Flavell and Ish-Horowicz 1983; Sundaresan and Freeling 1987; Hirochika and Otsuki 1995; O'Brochta et al. 2009). It is therefore possible that TEs in this circularized form can be transmitted directly, particularly if they are found in extracellular vesicles.

Extracellular vesicles are small membrane sacs that can be released from all cell types of all prokaryotic and eukaryotic organisms (van Niel et al. 2018). Two types can be distinguished: exosomes, which are produced by exocytosis, and larger microvesicles, which are released from maternal cells via budding of a plasma membrane. These vesicles were initially seen as a means for cells to get rid of harmful or unnecessary molecules (Johnstone et al. 1987). However, they are now recognized as vectors of biological material between cells. In particular, they have the capacity to carry DNA and messenger RNA (Cai et al. 2016). Analysis of microvesicles produced by cancer cells has shown that they contain RNA from retrotransposons corresponding to endogenous retroviruses and L1- and Alu-like elements (Balaj et al. 2011). This transfer appears to enable integration into the genome and expression of TEs, as experimentally determined (Fischer et al. 2016; Kawamura et al. 2019). Furthermore, it has been proposed that retroviruses use these extracellular vesicles as a mode of infection (Gould et al. 2003).

Another direct mechanism leading to the transmission of genetic material between different species is the phenomenon of introgression, which enables the production of fertile hybrids through various crosses, including backcrosses with one of the parental species. This phenomenon has been observed between closely related species in both animals and plants. Indeed, if the genomic region subjected to introgression contains TEs, these can be introduced into a new species by these means. This mechanism has been proposed to explain the numerous cases of HTT observed in species close to *D. melanogaster* (Lerat et al. 2011; Modolo et al. 2014).

7.2.2. *Transmission by a viral vector*

Direct transmission of TEs, however, is difficult and rare. Another possibility involves an intermediary as a vector. For this role, viruses are specifically indicated (Gilbert and Cordaux 2017). In particular, some viruses are known to have the ability to infect multiple species, making them interesting potential candidates as DNA vectors between different species (Geoghegan et al. 2017).

It has been known for several years that certain viruses can incorporate genes into their genome, in particular oncogenes, and be responsible for the development of certain cancers (Vogt 2012). This is the case, for example, with Rous sarcoma virus, in which an oncogene responsible for the development of cancerous tumors in chickens was identified for the first time (Martin 1970). Other examples have shown that the presence of genes in viruses can be commonplace, like the presence of human genes in herpes viruses and poxviruses (Holzerlandt et al. 2002; Hughes and Friedman 2005). In insects, too, the presence of host DNA insertions within virus genomes has been described (Fraser et al. 1983).

TE sequences have also been found to be integrated into the genome of viruses. For example, genome analysis of a baculovirus, the nuclear polyhedrosis virus that normally infects the lepidopteran *Autographa californica* (alfalfa autograph) following repeated infections of cultured cells of the lepidopteran *Trichoplusia ni* (cabbage looper), showed the presence of a copia-like retrotransposon (the TED element) from infected cells (Miller and Miller 1982). In the same virus, inserts derived from TEs from the genome of the lepidopteran *Spodoptera frugiperda* (fall armyworm) were found (Carstens 1987). Similarly, a complete and potentially active DNA transposon was detected in another baculovirus after infection of larvae of the lepidopteran *Cryptophlebia leucotreta* (false codling moth) from the genome of this butterfly (Jehle et al. 1995). More recently, the sequence of a non-autonomous MITE-like element homologous to a *Piggybac* DNA transposon was identified in the genome of an iridovirus, isolated from larvae of simulids, diptera of the genus *Simulium* (Piégu et al. 2013). In the genome of another iridovirus, "*invertebrate iridescent virus* 6", sequences corresponding to a SINE-like element found in leafhopper genomes have been identified (Yang et al. 2019). Consequently, insect viruses appear to be able to carry TEs horizontally between different species. TEs also have the capacity to colonize the genomes of giant viruses. Indeed, it has been shown for *Pandoravirus salinus*, a virus infecting amoebae of the genus *Acanthamoeba*, that around 30 copies of MITE-like elements are present in its genome (Sun et al. 2015). Vertebrate viruses can also be capable of transferring TEs. As a result, poxviruses have been proposed as vectors for the HT of retrotransposons between reptiles and mammals (Piskurek and Okada 2007). Genome analysis of the taterapox virus infecting rodents revealed the presence of a SINE-type element only described in lizards and snakes.

Another virus-related phenomenon could also enable HTTs. Some viruses encapsidate their own RNA, but can also encapsidate RNA from their host. Analysis of the content of virus particles from the Flock House virus (FHV) and omega virus (NwV) infecting insects showed that between 1 and 5% of the content corresponded to several types of DNA transposons and retrotransposons (Routh et al. 2012). Another example showed that in virus particles from the cucumber necrosis virus, the highest proportion of host RNA corresponded to retrotransposons (Ghoshal et al. 2015).

The various examples given above illustrate specific cases and do not allow us to appreciate the proportion that such a mechanism may represent in nature. For this reason, more global and systematic approaches have been undertaken. For example, analysis of populations of the baculovirus *A. californica* multiple nucleopolyhedrovirus (AcMNPV) obtained after infection of *T. ni* caterpillars and sequenced at a very great depth (100,000 ×) has shown that two DNA transposons can mobilize in this virus at a non-negligible frequency (one insertion for around 8,500 viral genomes) (Gilbert et al. 2014). These two DNA transposons are described as having recently been transferred horizontally between insect species that can be infected by this virus. This study has therefore demonstrated the potential role of viruses as vectors of HTT between animal species. Following this work, a more systematic search for integrated host sequences in the genomes of some 20 baculovirus populations retrieved after infection of caterpillars of two moth species showed that almost 5% of the viruses contained at least one host sequence (Gilbert et al. 2016). Among these sequences, several TEs could be identified as having been horizontally transferred. Very deep sequencing of the same virus isolated from *Spodoptera exigua* larvae (the beet moth) revealed numerous structural variations (deletions, inversions, duplications and insertions), and in particular, the presence of numerous TE insertions (Loiseau et al. 2020). In the same study, structural variants were also detected in iridoviruses and a herpes virus. These latest studies provide further evidence for the role of viruses in HTT.

7.2.3. *Transmission via host–pathogen interactions*

Viruses have the ability to be transmitted directly between hosts. Another type of interaction, involving host–pathogen relationships, can enable the transmission of genetic material. In certain cases, a pathogen may target several host species, thus serving as a vector for horizontal transmission. For example, parasitoid wasps lay their eggs inside the larvae of various insects, including Lepidoptera. It has been shown that the genome of these wasps contains the sequence of a bracovirus in provirus form, which is expressed in the ovaries and enables the formation of DNA-containing particles. These particles are then injected into the host at the time of oviposition. Analysis of the sequence of one of these bracoviruses led to the identification of a TE likely to have originated in the wasp (Dupuy et al. 2011). More recently, the presence of bracovirus DNA sequences has been detected in the genomes of several Lepidoptera (Gasmi et al. 2015). Consequently, analysis of the inserted sequences showed that they contained genes from wasps, thus highlighting a gene HT between animals via a viral vector for which the donor, recipient and vector are clearly identified. Interestingly, the transferred genes, which are expressed in Lepidoptera, could enable the latter to adapt to counter wasp parasitism. Host–pathogen interactions therefore appear to represent an ideal scenario for horizontal transmission.

The transfer of several families of DNA transposons has been demonstrated between invertebrates and vertebrates (Gilbert et al. 2010). For example, analysis of sequences from *Rhodnius prolixus*, a hematophagous insect parasite of various animals (including humans, in which it is the vector of Chagas disease), has revealed the presence of four TE families that are found in the genomes of different species as far apart as the opossum and the squirrel monkey. One of these TE families has even been identified in a snail, *Lymnaea stagnalis*, known to be the vector of several vertebrate fluke species. Similarly, phylogenetic sequence analysis of a non-LTR retrotransposon, *BovB*, sporadically found in several vertebrate species, has suggested that it may have been transmitted by ticks, *Bothriocroton hydrosauri* and *Amblyomma limbatum,* known to infest both mammals and reptiles (Walsh et al. 2013). Furthermore, it has been shown that a canid parasitic nematode, *Ancylostoma caninum*, contains in its genome an active DNA transposon family with very close proximity to an element found in humans, indicating that this element has probably been transferred from the host to the parasite (Laha et al. 2007).

Another potential vector for HTTs is the endosymbiotic *Wolbachia* bacterium. This bacterium, which lives inside the cells of its hosts, can itself be horizontally transferred between hosts (Raychoudhury et al. 2009). In addition, it can also transmit its DNA to its hosts. For example, sequences of the bacterium, ranging from a few hundred base pairs to virtually its entire genome, have been found to be inserted in the genomes of four insects and four nematodes (Dunning Hotopp et al. 2007). In one described case of gene transfer between *Wolbachia* and the *Aedes aegypti* mosquito, these genes were shown to be expressed, demonstrating that such transfers are not simply dead-end routes, but can have a functional impact (Klasson et al. 2009). So, although no examples of TE transfers have been observed via this bacterium, it could nonetheless be a serious potential vector.

Finally, microsporidia could also be potential vectors. These organisms form a group of intracellular parasites comprising over 1,500 species, phylogenetically related to a group of fungi known as Cryptomycetes. These organisms are potentially capable of infecting the entire animal kingdom, in particular insects and humans. A comparative analysis of the TE content of several microsporidia species has highlighted several cases of HTTs from insects to microsporidia (Parisot et al. 2014).

7.2.4. *Factors for successful horizontal transfer of transposable elements*

HTTs must be associated with a number of factors to ensure its success, that is, the transferred TE can be maintained at least in one population and then in the species (Silva et al. 2004). First, for the transfer to take place, certain opportunities must present themselves to allow the TE to circulate between organisms, and in particular its incorporation into the germline of the recipient species in the case of animals. This underlines the need for vectors or the possibility of cross-fertilization between species. Different species can also be linked via host–pathogen relationships, either through the sharing of common pathogens or directly through the interaction between pathogen and host. In all cases, it is essential that the donor and recipient species, in addition to being contemporary, are sympatric, that is, living in the same geographical area, enabling them to meet regularly.

In addition to the opportunities that can influence the frequency of HTTs, once the TE has been integrated into its new host genome, it will have to increase its copy number in the individual in which the transfer took place, as well as spreading throughout the population. As a result, some host properties can become limiting, such as the effective population size (Ne), selection against deleterious insertions, the presence or absence of particular host factors necessary for transposition or enabling it to be regulated, in particular via epigenetic mechanisms. On the other hand, genetic drift, positive selection, domestication and exaptation may be favorable to the fixation of the new element. Certain characteristics of the TEs themselves may also be limiting or facilitating factors, such as their promoter regions. For example, in vitro studies of the activity of the promoter regions of two DNA transposons *Bari1* and *Bari2* have shown that they are capable of transcribing a reporter gene in cultured cells from widely separated organisms (Drosophila, human, yeast and bacteria) (Palazzo et al. 2017).

7.3. Bioinformatics methods for detecting horizontal transfer of transposable elements

The detection of HTs in general can be difficult, since many factors come into play which can blur the identification of the event, such as sequence degradation or stochastic loss. Nevertheless, a number of methods have been developed to enable such identifications based on several pieces of evidence (Figure 7.2.):

– The presence of phylogenetic incongruences between the species tree and the TE tree suspected to have undergone an HT. Indeed, if the transfer has taken place between two very distant species, the phylogenetic tree of the TE will show the two species in question close together in the tree, contrary to their position in the species tree.

– Uneven distribution of the TE within several species. In this case, the TE is not homogeneously present in a set of closely related species.

– A very high degree of sequence identity between TEs of different species, incompatible with the divergence time of these species. At the time of transfer, the TE sequence arriving in a recipient species is identical to that present in the donor species, whereas if homologous gene sequences are compared between donor and recipient species, they will denote the evolutionary history of both species, including a particular sequence divergence.

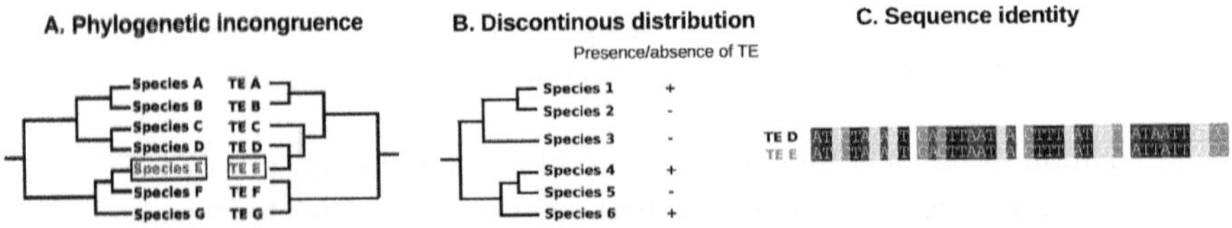

Figure 7.2. *Evidence of HTT that can be observed and searched for*

The first methods to be developed were mainly aimed at detecting gene HTs. Two main types of methods were developed: surrogate methods, which infer HTs at the level of genomic regions if their nucleotide composition (GC content, codon usage, di- and tri-nucleotide frequencies) differs from the average composition of the genome, and comparative analysis methods, which are mainly based on phylogenetic analyses (de Carvalho and Loreto 2012). However, it has been shown that approaches based on codon usage and base composition are not sufficient in themselves to reliably attest HTs (Lerat et al. 2000, 2002; Koski et al. 2001). Generally, the most robust way to conclude to an HT is to combine several pieces of evidence within a statistical framework that allows the validity of hypotheses to be tested (Loreto et al. 2008; Wallau et al. 2012). More recently, another methodology has been proposed based on the construction of networks, taking into account ecological factors linking different species (Venner et al. 2017).

There are currently only two bioinformatics methods dedicated specifically to HTT detection, *HTdetect* (Modolo et al. 2014) and *VHICA* (Wallau et al. 2016).

7.3.1. *HTdetect*

The principle of *HTdetect* (see: https://github.com/l-modolo/htdetect) is to compare the sequences of two entire genomes in order to detect, without any a priori concerning the type of sequences (genes, intergenic sequences or TEs), regions that are more identical than expected between the two species under consideration. To achieve this, genomes are aligned and a threshold identity value is defined, corresponding to the mode of the distribution of the percentage identity calculated along the alignment. The genomes are then compared using BLASTN (Altschul et al. 1990) and the hits for which the calculated percentage identity is significantly greater than or equal to the threshold value are selected to represent "potentially transferred sequences" (Figure 7.3(A)).

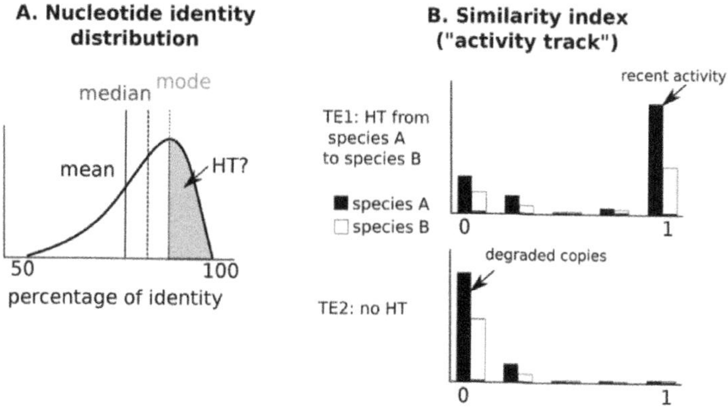

Figure 7.3. *Steps for determining horizontally transferred sequences by comparing complete genomes*

The next step is to validate or not validate these sequences. In the case of TEs annotated in these regions, the validation procedure involves taking into account the dynamics of TEs in the genomes under consideration. The starting hypothesis is that when a TE arrives in a genome via a HT, it escapes the host's defense mechanisms for a certain period of time and can replicate rapidly in this new host (Anxolabéhère et al. 1988; Le Rouzic and Capy 2005; Schaack et al. 2010). We can therefore expect to observe a large number of similar copies for this TE in one of the two genomes following transposition bursts (Lerat et al. 2011). Thus, a similarity index (called an "activity track") is calculated for each copy in a given genome by comparing it to the most similar pair of copies detected between the two genomes. This index varies between 0 (for old and degraded copies) and 1 (for copies with a low level of divergence and therefore recently active) (Figure 7.3(B)). This index can also be used to infer the direction of transfer. This approach has the advantage of not having any a priori about the horizontally transferred sequences and allows us to focus on TE types with no coding capacity, such as SINEs. The main disadvantage of this method is that species comparisons have to be made pairwisely, which limits its scope of application.

7.3.2. *VHICA*

The VHICA method (Vertical and Horizontal Transfer Consistency Analysis, see: https://rdrr.io/cran/vhica/) is based on a combined approach that takes into account both codon usage bias and sequence divergence. The comparison is made between several species. In this way, by comparing the synonymous substitution rate (dS) and the codon usage bias of vertically transmitted genes, on the one hand, and TEs, on the other hand, a correlation is calculated between these values for each pair of species and for each type of sequence. An expected linear regression line is estimated from the correlation values obtained for the genes (Figure 7.4(A)).

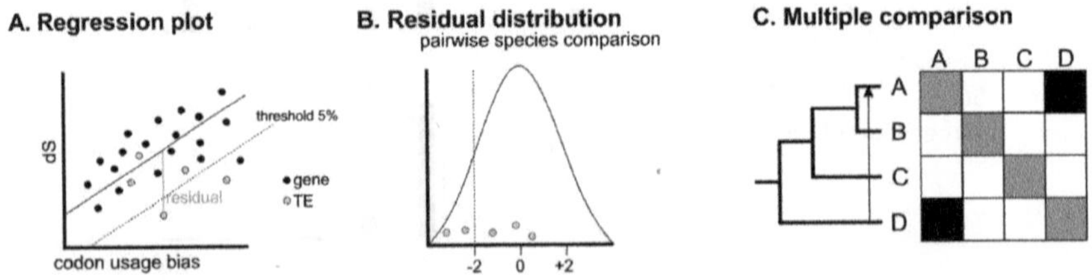

Figure 7.4. *Steps in identifying HTTs using the VHICA method*

The residuals for each gene and each TE are calculated from the regression line and compared with the residual distribution of reference genes (Figure 7.4(B)). A HT is inferred between two species when there is a statistically significant deviation from the residual distribution of vertically transmitted genes (Figure 7.4(C)). The method can be used if certain biological and statistical conditions have been met, such as the reference genes used being vertically transmitted, the genes and TEs being subject to the same rate of evolution, the relationship between codon usage bias and synonymous substitution rate being linear, or the distribution of residues following a normal distribution. The main disadvantages of the method are that only TEs with coding capacity can be considered, and that assumptions must be verified for each dataset analyzed.

7.4. Documented examples of horizontal transfers of transposable elements

There are now a large number of well-documented cases of HTTs. In this section, we will mention just a few examples for which the phenomenon is unambiguous, and which are considered to be the first of their kind to be described.

Historically, the first case described involved the *P* element, a DNA transposon identified in the model species *D. melanogaster*. This element was first shown to be responsible for a particular phenomenon known as hybrid dysgenesis (Kidwell 1983). This phenomenon results from the mobilization of a TE in a cross between a female lacking the element and a carrier male, and leads to sterility in the offspring, whereas the reciprocal crossing produces normal offspring. More detailed studies of this element have shown that its distribution in Drosophila species is highly fragmentary, with a near-absence in species close to *D. melanogaster* and, on the contrary, high prevalence in a more distant group of species including *D. willistoni*, a species endemic to the American continent (Daniels et al. 1990). In addition, comparison of the sequences of this element between these species showed that they were identical with the exception of one nucleotide, which was not compatible with the divergence between the species dating back more than 50 million years. Thus, it was possible to demonstrate that the transfer probably took place in the 1950s and that, since then, the *P* element has invaded all *D. melanogaster* populations, since all populations captured in the wild since this period possess this element, unlike populations captured before this date and maintained in the laboratory. This element has also recently been found in the genome of *Drosophila simulans,* a species that has been separated from *D. melanogaster* for 4 million years, where it arrived by HT from the latter (Kofler et al. 2015). Analysis of the genomes of several natural populations suggests that the invasion of the element into this species must have taken place between 2006 and 2014. This exceptional example of HTT is the most emblematic case described to date, particularly because it could be detected almost at the moment of its arrival. However, the mechanism behind the transfer remains unknown to this day.

Depending on the age of the event, horizontally transferred TEs can be identified simultaneously in different species. For example, when analyzing the DNA transposons present in the genome sequence of *Otolemur garnettii*, a small primate living in Africa, an unknown DNA transposon family called *SPACE INVADER* (*SPIN*) was identified (Pace et al. 2008). A search of the GenBank database for homologous sequences to establish the distribution of this element in different species led to the identification of numerous highly similar sequences in five mammalian species (Malagasy hedgehog, little brown bat, mouse, rat and opossum) and two tetrapods (green anole and xenopus). The authors propose that *SPIN* has been the subject of several HTs, the oldest dating back around 40 million years and the most recent around 15 million years. In particular, they suggest that different tetrapod species were infected at close evolutionary times by this element, whose source species would be the same in each case, without being able to determine its taxonomic identity.

In plants, the number of HTTs seems paradoxically very low, despite the ease with which these organisms hybridize across species, which could facilitate the transfer of genetic material, and the absence of specific germ lines. For a long time, no HTs were documented, either at gene or TE level. As a result, the first example described dates back to 2006, when Diao et al. (2006) sequenced *Mu*-like elements (*MULE*), a DNA transposon, in eight grass species of the *Setaria* genus in order to compare them with *MULE* elements from more distantly related plant species such as rice and maize. Their results showed that a high degree of sequence similarity was observed between some elements of two closely related *Setaria* species (bird's millet, *Setaria italica* and giant setary, *Setaria faberi*) and an element found in rice, *Oryza sativa* spp *japonica* and spp *indica.* This strong similarity is incompatible with vertical transmission, since *Setaria* and rice have diverged for around 50 million years. In addition, by searching for these sequences using Southern blots hybridization, it was possible to show that this element does not have a continuous distribution among the grass species analyzed. In particular, the element is present in several bamboo species, but absent in the grass subfamilies *Chloridoideae* and *Pooideae*. The authors therefore proposed that the HT event occurred several million years ago between the ancestors of the *Bambusoideae* and *Panicoideae*.

As mentioned above, in some rare cases, it has been shown that HTTs can also occur between different life domains. For example, when annotating the TEs in the genome of *Arachis duranensis,* one of the wild relatives of the cultivated peanut, a new non-LTR retrotransposon of the RTE type was identified (Gao et al. 2018). Initially, elements of this type were described in nematodes and are mainly found in animals, but rarely in plants. A GenBank database search with the element found in *A. duranensis* showed that this type of element is present in 81 plant species, including 30 flowering plants. The authors then used these different sequences to search for homologs in animal species, highlighting sequences present in arthropods with percentages identity incompatible with the divergence time between plants and arthropods, which is around 1.6 billion years. The authors' hypothesis is that TH occurred in the ancestor of all flowering plants from an ancestral arthropod species. This example is only the second case of HTT between plants and animals, with the example of another non-LTR retrotransposon transferred more recently from arthropods around 340 million years ago (Lin et al. 2016).

The examples presented in this and previous sections are very likely to represent only the tip of the iceberg. Indeed, there are probably many more cases of HTTs in insects (Herédia et al. 2004; Lerat et al. 2011; Sormacheva et al. 2012; Bargues and Lerat 2017; Peccoud et al. 2017), vertebrates (Hamada et al. 1997; Kordis and Gubensek 1999; Volff et al. 2000), plants (Roulin et al. 2008) or between animals from very distant domains (Dupeyron et al. 2014; Zhang et al. 2014; Suh et al. 2016). Some global studies have provided estimates of the number of possible HTT cases for different types of species. For example, a systematic analysis of 195 insect species determined that at least 2,248 cases of HTT were likely to have occurred over the last 10 million years (Peccoud et al. 2017). Considering only vertebrates, systematic analysis of 307 genomes inferred a minimum of 975 cases of HTT between species that diverged more than 120 million years ago (Zhang et al. 2020).

Similarly, by studying 40 plant genomes, it was possible to demonstrate that 65% of them showed at least one case of HTT, and that these transfers could have occurred between very distantly related species. The results suggest, extrapolating for the 13,551 monocot and dicot genera, that around 2 million HTTs may have taken place (Baidouri et al. 2014).

As a result, HTT events are particularly frequent. The question that arises is whether these events can have a particular impact on the evolution of species. These questions will be addressed in the next section.

7.5. The impact of horizontal transfers of transposable elements

Given that TEs play an important role in the evolution of their host through various effects such as chromosomal rearrangements on a greater or lesser scale, changes in the regulation of neighboring genes or the contribution of new exonic sequences, it is possible that some cases of HTTs may have a significant functional impact on their new host.

7.5.1. *Adaptation and creation of gene novelties*

In *D. melanogaster*, one dozen potentially adaptive TEs have been described, including a copy of a *Bari* DNA transposon inserted between two genes involved in the degradation of juvenile hormone, which plays a role in regulating insect development, life-history traits and fitness (González et al. 2009). The presence of this copy has the effect of decreasing the level of expression of both genes, with a consequent increase in the level of juvenile hormone. This leads to a drop in viability and an increase in the development time of flies carrying this insertion. However, more recent work has shown that the presence of this insertion also results in resistance to oxidative stress via over-expression of neighboring genes (Guio et al. 2014). This explains why an insertion that seems at first sight to have a negative effect on its host's fitness is nevertheless adaptive under certain conditions. Interestingly, the *Bari* element has previously been described as having been horizontally transferred between *D. melanogaster, Drosophila simulans* and *Drosophila yakuba* (Modolo et al. 2014; Wallau et al. 2016).

A helitron family, a particular type of DNA transposon, has been identified in several bat species belonging to the Vespertilionidae family, while being absent from species of the Miniopteridae family, a sister group, but also from other mammalian species (Thomas et al. 2011). Given its discontinuous phylogenetic distribution, this TE was probably horizontally transferred around 16–25 million years ago (Thomas et al. 2011). A detailed analysis of this TE family in the *Myotis lucifugus* bat genome has shown that this TE has been a source of genomic innovation due to its ability to contain different host sequences such as gene fragments or regulatory regions (Thomas et al. 2014).

Another case allowing adaptation of recipient species was recently described in wheat pathogenic fungi (McDonald et al. 2019). In this study, the HT of a gene responsible for wheat pathogenicity, ToxA, was demonstrated between three fungal species *Parastagonospora nodorum*, *Pyrenophora tritici-repentis*, and *Bipolaris sorokiniana* separated by several million years (McDonald et al. 2018). Interestingly, HTs also involve adjacent regions of this gene spanning several tens of thousands of base pairs. Characterization of these regions has shown that they are enclosed between two terminal inverted repeats (TIRs) that typically flank DNA transposons. In this example, the insertion of a host gene inside a DNA transposon enabled it to be transmitted horizontally between different species of fungal pathogens sharing similar environments and hosts.

Horizontally transferred TEs can generate genetic novelty at the origin of adaptive changes. By tracing the TEs responsible for known phenotypic changes in the literature, Gilbert and Feschotte (2018) were therefore able to highlight that many cases are probably linked to HTs. For example, the LTR retrotransposon *Tcs2* is responsible for altering the expression of the Ruby gene responsible for blood orange staining. This TE shows a higher than expected sequence identity with a sequence found in asparagus, despite the fact that these two plants have been separated for 148 million years. It is therefore likely that this element was horizontally transferred.

7.5.2. *The appearance of new species*

TEs have been proposed as being able to enable the reproductive isolation of different taxa at different levels (pre-mating, pre- or post-zygote formation) (for review, see Serrato-Capuchina and Matute (2018)). If this is the case, newly arrived TEs in a genome by HT could have the capacity to lead to speciation events.

There are a number of cases where the link between TEs and speciation has been proposed. For example, investigation of the genome of the Atlantic salmon, *Salmosalar*, has shown that several TE families have undergone transposition bursts that coincide with the separation time of species in the Salmoninae group (de Boer et al. 2007). Some of these TEs also show HT characteristics, probably via parasites. The authors therefore proposed that recurrent HTTs may have led to massive rearrangements of the genomes of these fishes at different times during speciation.

Similarly, in a plant-pathogenic fungus, *Leptosphaeria maculans*, it was shown that its genome was invaded by TEs presumably acquired by HT, and that their activity enabled the generation of chromosomal inversions leading to the reproductive isolation of this species, but also allowed this pathogen to adapt to its host plants (Grandaubert et al. 2014).

7.6. Conclusion

Contrary to long-held beliefs, HT of TEs seems to be particularly common in all living organisms. Although many questions remain unanswered concerning the exact mechanisms by which they are transferred, this remains an important stage in their lifecycle, enabling them to maintain themselves in genomes and to be at the origin of genetic novelties.

7.7. References

Altschul, S.F., Gish, W., Miller, W., Myers, E.W., Lipman, D.J. (1990). Basic local alignment search tool. *Journal of Molecular Biology*, 215(3), 403–410.

Anxolabéhère, D., Kidwell, M.G., Periquet, G. (1988). Molecular characteristics of diverse populations are consistent with the hypothesis of a recent invasion of *Drosophila melanogaster* by mobile P elements. *Molecular Biology and Evolution*, 5(3), 252–269.

Aravind, L., Tatusov, R.L., Wolf, Y.I., Walker, D.R., Koonin, E.V. (1998). Evidence for massive gene exchange between archaeal and bacterial hyperthermophiles. *Trends in Genetics*, 14(11), 442–444.

Badge, R.M. and Brookfield, J.F. (1997). The role of host factors in the population dynamics of selfish transposable elements. *Journal of Theoretical Biology*, 187(2), 261–271.

Baidouri, M.E., Carpentier, M.C., Cooke, R., Gao, D., Lasserre, E., Llauro, C., Mirouze, M., Picault, N., Jackson, S.A., Panaud, O. (2014). Widespread and frequent horizontal transfers of transposable elements in plants. *Genome Research*, 24(5), 831–838.

Balaj, L., Lessard, R., Dai, L., Cho, Y.-J., Pomeroy, S.L., Breakefield, X.O., Skog, J. (2011). Tumour microvesicles contain retrotransposon elements and amplified oncogene sequences. *Nature Communications*, 2, 180.

Bargues, N. and Lerat, E. (2017). Evolutionary history of LTR-retrotransposons among 20 Drosophila species. *Mobile DNA*, 8(1), 7.

de Boer, J.G., Yazawa, R., Davidson, W.S., Koop, B.F. (2007). Bursts and horizontal evolution of DNA transposons in the speciation of pseudotetraploid salmonids. *BMC Genomics*, 8, 422.

Brookfield, J.F. and Badge, R.M. (1997). Population genetics models of transposable elements. *Genetica*, 100(1–3), 281–294.

Cai, J., Wu, G., Jose, P.A., Zeng, C. (2016). Functional transferred DNA within extracellular vesicles. *Experimental Cell Research*, 349(1), 179–183.

Carstens, E.B. (1987). Identification and nucleotide sequence of the regions of *Autographa californica* nuclear polyhedrosis virus genome carrying insertion elements derived from *Spodoptera frugiperda*. *Virology*, 161(1), 8–17.

de Carvalho, M.O. and Loreto, E.L.S. (2012). Methods for detection of horizontal transfer of transposable elements in complete genomes. *Genetics and Molecular Biology*, 35(4 Suppl.), 1078–1084.

Coffin, J.M., Hughes, S.H., Varmus, H.E. (1997). The interactions of retroviruses and their hosts. In *Retroviruses,* Coffin, J.M., Hughes, S.H., Varmus, H.E. (eds). Cold Spring Harbor Laboratory Press, Cold Spring Harbor, NY.

Daniels, S.B., Peterson, K.R., Strausbaugh, L.D., Kidwell, M.G., Chovnick, A. (1990). Evidence for horizontal transmission of the P transposable element between Drosophila species. *Genetics*, 124(2), 339–355.

Diao, X., Freeling, M., Lisch, D. (2006). Horizontal transfer of a plant transposon. *PLoS Biology*, 4(1), e5.

Doolittle, W.F. (1999). Phylogenetic classification and the universal tree. *Science*, 284(5423), 2124–2129.

Dunning Hotopp, J.C., Clark, M.E., Oliveira, D.C.S.G., Foster, J.M., Fischer, P., Muñoz Torres, M.C., Giebel, J.D., Kumar, N., Ishmael, N., Wang, S. et al. (2007). Widespread lateral gene transfer from intracellular bacteria to multicellular eukaryotes. *Science*, 317(5845), 1753–1756.

Dupeyron, M., Leclercq, S., Cerveau, N., Bouchon, D., Gilbert, C. (2014). Horizontal transfer of transposons between and within crustaceans and insects. *Mobile DNA*, 5(1), 4.

Dupuy, C., Periquet, G., Serbielle, C., Bézier, A., Louis, F., Drezen, J.-M. (2011). Transfer of a chromosomal Maverick to endogenous bracovirus in a parasitoid wasp. *Genetica*, 139(4), 489–496.

Etten, J.V. and Bhattacharya, D. (2020). Horizontal gene transfer in eukaryotes: Not if, but how much? *Trends in Genetics*, 36(12), 915–925.

Fischer, S., Cornils, K., Speiseder, T., Badbaran, A., Reimer, R., Indenbirken, D., Grundhoff, A., Brunswig-Spickenheier, B., Alawi, M., Lange, C. (2016). Indication of horizontal DNA gene transfer by extracellular vesicles. *PloS One*, 11(9), e0163665.

Flavell, A.J. and Ish-Horowicz, D. (1983). The origin of extrachromosomal circular copia elements. *Cell*, 34(2), 415–419.

Fraser, M.J., Smith, G.E., Summers, M.D. (1983). Acquisition of host cell DNA sequences by baculoviruses: Relationship between host DNA insertions and FP mutants of *Autographa californica* and *Galleria mellonella* nuclear polyhedrosis viruses. *Journal of Virology*, 47(2), 287–300.

Gao, D., Chu, Y., Xia, H., Xu, C., Heyduk, K., Abernathy, B., Ozias-Akins, P., Leebens-Mack, J.H., Jackson, S.A. (2018). Horizontal transfer of non-LTR retrotransposons from arthropods to flowering plants. *Molecular Biology and Evolution*, 35(2), 354–364.

Gasmi, L., Boulain, H., Gauthier, J., Hua-Van, A., Musset, K., Jakubowska, A.K., Aury, J.-M., Volkoff, A.-N., Huguet, E., Herrero, S. et al. (2015). Recurrent domestication by lepidoptera of genes from their parasites mediated by bracoviruses. *PLoS Genetics*, 11(9), e1005470.

Geoghegan, J.L., Duchêne, S., Holmes, E.C. (2017). Comparative analysis estimates the relative frequencies of co-divergence and cross-species transmission within viral families. *PLoS Pathogens*, 13(2), e1006215.

Ghoshal, K., Theilmann, J., Reade, R., Maghodia, A., Rochon, D. (2015). Encapsidation of host RNAs by cucumber necrosis virus coat protein during both agroinfiltration and infection. *Journal of Virology*, 89(21), 10748–10761.

Gilbert, C. and Cordaux, R. (2017). Viruses as vectors of horizontal transfer of genetic material in eukaryotes. *Current Opinion in Virology*, 25,16–22.

Gilbert, C. and Feschotte, C. (2018). Horizontal acquisition of transposable elements and viral sequences: Patterns and consequences. *Current Opinion in Genetics & Development*, 49, 15–24.

Gilbert, C., Schaack, S., Pace, J.K., Brindley, P.J., Feschotte, C. (2010). A role for host-parasite interactions in the horizontal transfer of transposons across phyla. *Nature*, 464(7293), 1347–1350.

Gilbert, C., Chateigner, A., Ernenwein, L., Barbe, V., Bézier, A., Herniou, E.A., Cordaux, R. (2014). Population genomics supports baculoviruses as vectors of horizontal transfer of insect transposons. *Nature Communications*, 5, 3348.

Gilbert, C., Peccoud, J., Chateigner, A., Moumen, B., Cordaux, R., Herniou, E.A. (2016). Continuous influx of genetic material from host to virus populations. *PLoS Genetics*, 12(2), e1005838.

González, J., Macpherson, J.M., Petrov, D.A. (2009). A recent adaptive transposable element insertion near highly conserved developmental loci in *Drosophila melanogaster*. *Molecular Biology and Evolution*, 26(9), 1949–1961.

Gould, S.J., Booth, A.M., Hildreth, J.E.K. (2003). The Trojan exosome hypothesis. *Proceedings of the National Academy of Sciences of the United States of America*, 100(19), 10592–10597.

Grandaubert, J., Lowe, R.G.T., Soyer, J.L., Schoch, C.L., Van de Wouw, A.P., Fudal, I., Robbertse, B., Lapalu, N., Links, M.G., Ollivier, B. et al. (2014). Transposable element-assisted evolution and adaptation to host plant within the *Leptosphaeria maculans-Leptosphaeria biglobosa* species complex of fungal pathogens. *BMC Genomics*, 15, 891.

Guio, L., Barrón, M.G., González, J. (2014). The transposable element *Bari-Jheh* mediates oxidative stress response in Drosophila. *Molecular Ecology*, 23(8), 2020–2030.

Hamada, M., Kido, Y., Himberg, M., Reist, J.D., Ying, C., Hasegawa, M., Okada, N. (1997). A newly isolated family of short interspersed repetitive elements (SINEs) in coregonid fishes (whitefish) with sequences that are almost identical to those of the SmaI family of repeats: Possible evidence for the horizontal transfer of SINEs. *Genetics*, 146(1), 355–367.

Herédia, F., Loreto, E.L.S., Valente, V.L.S. (2004). Complex evolution of gypsy in Drosophilid species. *Molecular Biology and Evolution*, 21(10), 1831–1842.

Hickey, D.A. (1982). Selfish DNA: A sexually-transmitted nuclear parasite. *Genetics*, 101(3–4), 519–531.

Hirochika, H. and Otsuki, H. (1995). Extrachromosomal circular forms of the tobacco retrotransposon Tto1. *Gene*, 165(2), 229–232.

Holzerlandt, R., Orengo, C., Kellam, P., Albà, M.M. (2002). Identification of new herpesvirus gene homologs in the human genome. *Genome Research*, 12(11), 1739–1748.

Hughes, A.L. and Friedman, R. (2005). Poxvirus genome evolution by gene gain and loss. *Molecular Phylogenetics and Evolution*, 35(1), 186–195.

Jehle, J.A., Fritsch, E., Nickel, A., Huber, J., Backhaus, H. (1995). TCl4.7: A novel lepidopteran transposon found in *Cydia pomonella* granulosis virus. *Virology*, 207(2), 369–379.

Johnstone, R.M., Adam, M., Hammond, J.R., Orr, L., Turbide, C. (1987). Vesicle formation during reticulocyte maturation. Association of plasma membrane activities with released vesicles (exosomes). *The Journal of Biological Chemistry*, 262(19), 9412–9420.

Kawamura, Y., Sanchez Calle, A., Yamamoto, Y., Sato, T.-A., Ochiya, T. (2019). Extracellular vesicles mediate the horizontal transfer of an active LINE-1 retrotransposon. *Journal of Extracellular Vesicles*, 8(1), 1643214.

Keeling, P.J. and Palmer, J.D. (2008). Horizontal gene transfer in eukaryotic evolution. *Nature Reviews. Genetics*, 9(8), 605–618.

Kempken, F. and Kück, U. (1998). Evidence for circular transposition derivatives from the fungal hAT-transposon restless. *Current Genetics*, 34(3), 200–203.

Kidwell, M.G. (1983). Evolution of hybrid dysgenesis determinants in *Drosophila melanogaster*. *Proceedings of the National Academy of Sciences of the United States of America*, 80(6), 1655–1659.

Kim, A., Terzian, C., Santamaria, P., Pélisson, A., Purd'homme, N., Bucheton, A. (1994). Retroviruses in invertebrates: The gypsy retrotransposon is apparently an infectious retrovirus of *Drosophila melanogaster*. *Proceedings of the National Academy of Sciences of the United States of America*, 91(4), 1285–1289.

Klasson, L., Kambris, Z., Cook, P.E., Walker, T., Sinkins, S.P. (2009). Horizontal gene transfer between *Wolbachia* and the mosquito *Aedes aegypti*. *BMC Genomics*, 10, 33.

Kofler, R., Hill, T., Nolte, V., Betancourt, A.J., Schlötterer, C. (2015). The recent invasion of natural *Drosophila simulans* populations by the P-element. *Proceedings of the National Academy of Sciences of the United States of America*, 112(21), 6659–6663.

Kordis, D. and Gubensek, F. (1999). Horizontal transfer of non-LTR retrotransposons in vertebrates. *Genetica*, 107(1–3), 121–128.

Koski, L.B., Morton, R.A., Golding, G.B. (2001). Codon bias and base composition are poor indicators of horizontally transferred genes. *Molecular Biology and Evolution*, 18(3), 404–412.

Laha, T., Loukas, A., Wattanasatitarpa, S., Somprakhon, J., Kewgrai, N., Sithithaworn, P., Kaewkes, S., Mitreva, M., Brindley, P.J. (2007). The bandit, a new DNA transposon from a hookworm-possible horizontal genetic transfer between host and parasite. *PLoS Neglected Tropical Diseases*, 1(1), e35.

Laten, H.M., Havecker, E.R., Farmer, L.M., Voytas, D.F. (2003). SIRE1, an endogenous retrovirus family from Glycine max, is highly homogeneous and evolutionarily young. *Molecular Biology and Evolution*, 20(8), 1222–1230.

Lawrence, J.G. and Ochman, H. (1998). Molecular archaeology of the *Escherichia coli* genome. *Proceedings of the National Academy of Sciences*, 95(16), 9413–9417.

Le Rouzic, A. and Capy, P. (2005). The first steps of transposable elements invasion: Parasitic strategy vs. genetic drift. *Genetics*, 169(2), 1033–1043.

Le Rouzic, A. and Deceliere, G. (2005). Models of the population genetics of transposable elements. *Genetical Research*, 85(3), 171–181.

Le Rouzic, A., Boutin, T.S., Capy, P. (2007). Long-term evolution of transposable elements. *Proceedings of the National Academy of Sciences of the United States of America*, 104(49), 19375–19380.

Leblanc, P., Desset, S., Giorgi, F., Taddei, A.R., Fausto, A.M., Mazzini, M., Dastugue, B., Vaury, C. (2000). Life cycle of an endogenous retrovirus, ZAM, in *Drosophila melanogaster*. *Journal of Virology*, 74(22), 10658–10669.

Lerat, E. and Capy, P. (1999). Retrotransposons and retroviruses: Analysis of the envelope gene. *Molecular Biology and Evolution*, 16(9), 1198–1207.

Lerat, E., Biémont, C., Capy, P. (2000). Codon usage and the origin of P elements. *Molecular Biology and Evolution*, 17(3), 467–468

Lerat, E., Capy, P., Biémont, C. (2002). Codon usage by transposable elements and their host genes in five species. *Journal of Molecular Evolution*, 54(5), 625–637.

Lerat, E., Daubin, V., Moran, N.A. (2003). From gene trees to organismal phylogeny in prokaryotes: The case of the γ-Proteobacteria. *PLoS Biology*, 1(1), 1–9.

Lerat, E., Burlet, N., Biémont, C., Vieira, C. (2011). Comparative analysis of transposable elements in the melanogaster subgroup sequenced genomes. *Gene*, 473(2), 100–109.

Lin, X., Faridi, N., Casola, C. (2016). An ancient Transkingdom horizontal transfer of Penelope-like retroelements from arthropods to conifers. *Genome Biology and Evolution*, 8(4), 1252–1266.

Loiseau, V., Herniou, E.A., Moreau, Y., Lévêque, N., Meignin, C., Daeffler, L., Federici, B., Cordaux, R., Gilbert, C. (2020). Wide spectrum and high frequency of genomic structural variation, including transposable elements, in large double-stranded DNA viruses. *Virus Evolution*, 6(1), vez060.

Loreto, E.L.S., Carareto, C.M.A. Capy, P. (2008). Revisiting horizontal transfer of transposable elements in *Drosophila*. *Heredity*, 100(6), 545–554.

Martin, G.S. (1970). Rous sarcoma virus: A function required for the maintenance of the transformed state. *Nature*, 227(5262), 1021–1023.

Maruyama, K. and Hartl, D.L. (1991). Evidence for interspecific transfer of the transposable element mariner between *Drosophila* and *Zaprionus*. *Journal of Molecular Evolution*, 33(6), 514–524.

McDonald, M.C., Ahren, D., Simpfendorfer, S., Milgate, A., Solomon, P.S. (2018). The discovery of the virulence gene ToxA in the wheat and barley pathogen *Bipolaris sorokiniana*. *Molecular Plant Pathology*, 19(2), 432–439.

McDonald, M.C., Taranto, A.P., Hill, E., Schwessinger, B., Liu, Z., Simpfendorfer, S., Milgate, A., Solomon, P.S. (2019). Transposon-mediated horizontal transfer of the host-specific virulence protein ToxA between three fungal wheat pathogens. *mBio*, 10(5), e01515–e01519.

Miller, D.W. and Miller, L.K. (1982). A virus mutant with an insertion of a copia-like transposable element. *Nature*, 299(5883), 562–564.

Modolo, L., Picard, F., Lerat, E. (2014). A new genome-wide method to track horizontally transferred sequences: Application to *Drosophila*. *Genome Biology and Evolution*, 6(2), 416–432.

Mourier, T. (2016). Transposable elements and circular DNAs. *Mobile Genetic Elements*, 6(6), e1240748.

van Niel, G., D'Angelo, G., Raposo, G. (2018). Shedding light on the cell biology of extracellular vesicles. *Nature Reviews Molecular Cell Biology*, 19(4), 213–228.

O'Brochta, D.A., Stosic, C.D., Pilitt, K., Subramanian, R.A., Hice, R.H., Atkinson, P.W. (2009). Transpositionally active episomal hAT elements. *BMC Molecular Biology*, 10, 108.

Orgel, L.E. and Crick, F.H. (1980). Selfish DNA: The ultimate parasite. *Nature*, 284(5757), 604–607.

Pace, J.K., Gilbert, C., Clark, M.S., Feschotte, C. (2008). Repeated horizontal transfer of a DNA transposon in mammals and other tetrapods. *Proceedings of the National Academy of Sciences of the United States of America*, 105(44), 17023–17028.

Palazzo, A., Caizzi, R., Viggiano, L., Marsano, R.M. (2017). Does the promoter constitute a barrier in the horizontal transposon transfer process? Insight from Bari transposons. *Genome Biology and Evolution*, 9(6), 1637–1645.

Parisot, N., Pelin, A., Gasc, C., Polonais, V., Belkorchia, A., Panek, J., El Alaoui, H., Biron, D.G., Brasset, E., Vaury, C. et al. (2014). Microsporidian genomes harbor a diverse array of transposable elements that demonstrate an ancestry of horizontal exchange with metazoans. *Genome Biology and Evolution*, 6(9), 2289–2300.

Peccoud, J., Loiseau, V., Cordaux, R., Gilbert, C. (2017). Massive horizontal transfer of transposable elements in insects. *Proceedings of the National Academy of Sciences of the United States of America*, 114(18), 4721–4726.

Piégu, B., Guizard, S., Spears, T., Cruaud, C., Couloux, A., Bideshi, D.K., Federici, B.A., Bigot, Y. (2013). Complete genome sequence of invertebrate iridescent virus 22 isolated from a blackfly larva. *The Journal of General Virology*, 94(Pt 9), 2112–2116.

Piskurek, O. and Okada, N. (2007). Poxviruses as possible vectors for horizontal transfer of retroposons from reptiles to mammals. *Proceedings of the National Academy of Sciences of the United States of America*, 104(29), 12046–12051.

Qiu, H., Price, D.C., Weber, A.P.M., Reeb, V., Yang, E.C., Lee, J.M., Kim, S.Y., Yoon, H.S., Bhattacharya, D. (2013). Adaptation through horizontal gene transfer in the cryptoendolithic red alga *Galdieria phlegrea*. *Current Biology*, 23(19), R865–R866.

Raychoudhury, R., Baldo, L., Oliveira, D.C.S.G., Werren, J.H. (2009). Modes of acquisition of *Wolbachia*: Horizontal transfer, hybrid introgression, and codivergence in the *Nasonia* species complex. *Evolution; International Journal of Organic Evolution*, 63(1), 165–183.

Roulin, A., Piegu, B., Wing, R.A., Panaud, O. (2008). Evidence of multiple horizontal transfers of the long terminal repeat retrotransposon *RIRE1* within the genus *Oryza*. *The Plant Journal: For Cell and Molecular Biology*, 53(6), 950–959.

Routh, A., Domitrovic, T., Johnson, J.E. (2012). Host RNAs, including transposons, are encapsidated by a eukaryotic single-stranded RNA virus. *Proceedings of the National Academy of Sciences of the United States of America*, 109(6), 1907–1912.

Schaack, S., Gilbert, C., Feschotte, C. (2010). Promiscuous DNA: Horizontal transfer of transposable elements and why it matters for eukaryotic evolution. *Trends in Ecology and Evolution*, 25(9), 537–546.

Schönknecht, G., Chen, W.-H., Ternes, C.M., Barbier, G.G., Shrestha, R.P., Stanke, M., Bräutigam, A., Baker, B.J., Banfield, J.F., Garavito, R.M. et al. (2013). Gene transfer from bacteria and archaea facilitated evolution of an extremophilic eukaryote. *Science*, 339(6124), 1207–1210.

Serrato-Capuchina, A. and Matute, D.R. (2018). The role of transposable elements in speciation. *Genes*, 9(5), 254.

Silva, J.C., Loreto, E.L., Clark, J.B. (2004). Factors that affect the horizontal transfer of transposable elements. *Current Issues in Molecular Biology*, 6(1), 57–72.

Sormacheva, I., Smyshlyaev, G., Mayorov, V., Blinov, A., Novikov, A., Novikova, O. (2012). Vertical evolution and horizontal transfer of CR1 non-LTR retrotransposons and Tc1/mariner DNA transposons in Lepidoptera species. *Molecular Biology and Evolution*, 29(12), 3685–3702.

Suh, A., Witt, C.C., Menger, J., Sadanandan, K.R., Podsiadlowski, L., Gerth, M., Weigert, A., McGuire, J.A., Mudge, J., Edwards, S.V. et al. (2016). Ancient horizontal transfers of retrotransposons between birds and ancestors of human pathogenic nematodes. *Nature Communications*, 7, 11396.

Sun, C., Feschotte, C., Wu, Z., Mueller, R.L. (2015). DNA transposons have colonized the genome of the giant virus *Pandoravirus salinus*. *BMC Biology*, 13(1), 38.

Sundaresan, V. and Freeling, M. (1987). An extrachromosomal form of the Mu transposons of maize. *Proceedings of the National Academy of Sciences of the United States of America*, 84(14), 4924–4928.

Thomas, J., Sorourian, M., Ray, D., Baker, R.J., Pritham, E.J. (2011). The limited distribution of Helitrons to vesper bats supports horizontal transfer. *Gene*, 474(1–2), 52–58.

Thomas, J., Phillips, C.D., Baker, R.J., Pritham, E.J. (2014). Rolling-circle transposons catalyze genomic innovation in a mammalian lineage. *Genome Biology and Evolution*, 6(10), 2595–2610.

Venner, S., Miele, V., Terzian, C., Biémont, C., Daubin, V., Feschotte, C., Pontier, D. (2017). Ecological networks to unravel the routes to horizontal transposon transfers. *PLoS Biology*, 15(2), e2001536.

Vicient, C.M., Kalendar, R., Schulman, A.H. (2001). Envelope-class retrovirus-like elements are widespread, transcribed and spliced, and insertionally polymorphic in plants. *Genome Research*, 11(12), 2041–2049.

Villa, T.G., Feijoo-Siota, L., Sánchez-Pérez, A., Rama, J.L.R., Sieiro, C. (2019). Horizontal gene transfer in bacteria, an overview of the mechanisms involved. In *Horizontal Gene Transfer: Breaking Borders Between Living Kingdoms*, Villa, T.G., Viñas, M. (eds). Springer International Publishing, Cham.

Vogt, P.K. (2012). Retroviral oncogenes: A historical primer. *Nature Reviews Cancer*, 12(9), 639–648.

Volff, J.N., Körting, C., Schartl, M. (2000). Multiple lineages of the non-LTR retrotransposon *Rex1* with varying success in invading fish genomes. *Molecular Biology and Evolution*, 17(11), 1673–1684.

Wallau, G.L., Ortiz, M.F., Loreto, E.L.S. (2012). Horizontal transposon transfer in eukarya: Detection, bias, and perspectives. *Genome Biol. Evol.*, 4(8), 689–699.

Wallau, G.L., Capy, P., Loreto, E., Le Rouzic, A., Hua-Van, A. (2016). VHICA, a new method to discriminate between vertical and horizontal transposon transfer: Application to the *Mariner* family within *Drosophila*. *Molecular Biology and Evolution*, 33(4), 1094–109.

Walsh, A.M., Kortschak, R.D., Gardner, M.G., Bertozzi, T., Adelson, D.L. (2013). Widespread horizontal transfer of retrotransposons. *Proceedings of the National Academy of Sciences*, 110(3), 1012–1016.

Yang, Q., Zhang, Y., Andika, I.B., Liao, Z., Kondo, H., Lu, Y., Cheng, Y., Li, L., He, Y., He, Y. et al. (2019). Horizontal transfer of a retrotransposon from the rice planthopper to the genome of an insect DNA virus. *Journal of Virology*, 93(6), e01516–18.

Zhang, H.-H., Feschotte, C., Han, M.-J., Zhang, Z. (2014). Recurrent horizontal transfers of *Chapaev* transposons in diverse invertebrate and vertebrate animals. *Genome Biol. Evol.*, 6(6), 1375–1386.

Zhang, H.-H., Peccoud, J., Xu, M.-R.-X., Zhang, X.-G., Gilbert, C. (2020). Horizontal transfer and evolution of transposable elements in vertebrates. *Nature Com.*, 11(1), 1362.

8

Genome Invasion Dynamics

Arnaud LE ROUZIC and Aurélie HUA-VAN

Évolution, Génomes, Comportement et Écologie (EGCE), CNRS, IRD, Université Paris-Saclay, Gif-sur-Yvette, France

8.1. The lifecycle of transposable elements

Transposable elements (TEs) are defined by their capacity to duplicate within genomes and invade populations. They are repeated sequences, contrasting with "regular" genic and intergenic sequences usually present only once per haploid genome. TE copies originate from a common ancestral sequence by a series of replicative transposition events, forming lineages (usually called TE families) that gather dozens, hundreds, or sometimes thousands of elements.

The number of TE copies in genomes depends on the species, the TE family and a complex series of events in the population history. In most prokaryotes, TE diversity is reduced (one or a handful of TE families), and the number of copies is generally low (often less than 10 insertions). In animals and plants, genomes typically host hundreds or thousands of TE copies, with a wider diversity among TE families. This chapter mainly focuses on TE dynamics in eukaryotic sexual species; TE dynamics in asexuals will be briefly described in section 8.5.

The analysis of genome content suggests that most TE families do not persist in a long-term equilibrium state. Indeed, for many TE families, only relic (degraded) sequences can be found, indicating that the TE is no longer able to transpose. In contrast, active TEs, whose sequence is complete and functional, are often deprived

Transposable Elements and Genome Evolution,
coordinated by Aurélie HUA-VAN and Pierre CAPY. © ISTE Ltd 2024.

of ancestral relatives, as if they were very recent in the genome. This has led genome evolutionary biologists to propose a "TE lifecycle" model, in which TE families actively invade the genome before being inactivated, and finally degraded and lost (Kidwell and Lisch 2001). The total TE content in the genome is then maintained as the result of a turnover, with new TE families replacing the old ones.

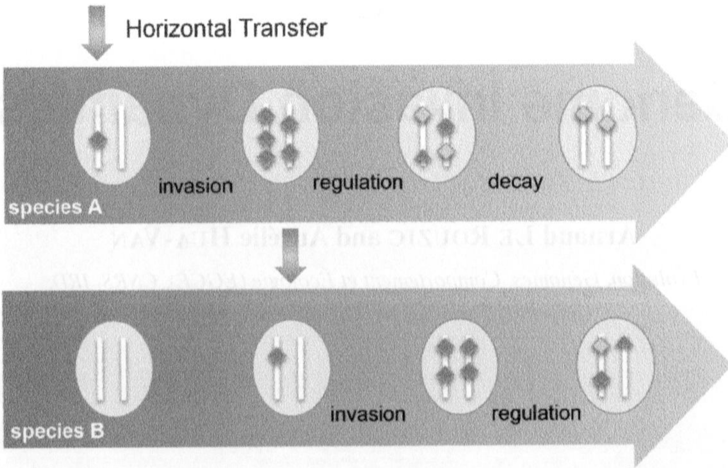

Figure 8.1. *Illustration of the lifecycle of a TE family. The host diploid genome is represented as two linear chromosomes; active TEs are dark diamonds. In most cases, the new element starts colonizing the genome after a horizontal transfer from another species. The TE amplifies and spreads, before being regulated. Inactive TE copies then accumulate mutations and deletions, and are slowly eliminated. The TE family can survive by undergoing a transfer to another species before decaying. The whole cycle takes thousands of generations. For a color version of this figure, see www.iste.co.uk/huavan/transposable.zip*

Genome invasion starts from a single copy in a single individual, being transmitted subsequently across generations, along with the rest of the genome. The origin of the new TE has to be determined on a case by case basis: mutation or recombination from existing TE copies, genome exchange (introgression) during rare hybridization events between distant populations or close species, etc. It has been established that new TE invasions often originate from horizontal transfers across species. Horizontal transfers are rare events for genes in eukaryotes, but have been proven to be frequent (at the evolutionary scale) for TEs (more details about horizontal transfers are provided in Chapter 7).

The dynamics of a TE family in a species can be presented as a series of stages (Figure 8.1). The first "invasion" stage after the arrival of a new TE copy relies on the intrinsic amplification ability of the TE, which duplicates in genomes and spreads in the population (more details in section 8.2). The increase in copy number is generally quickly limited by several mechanisms, including transposition regulation and natural selection against the accumulation of copies (section 8.3). Transposition then stops, and TE copies start to accumulate mutations (section 8.4). This lifecycle is specific to sexual organisms, in which TEs can easily spread – both in the genome and in the population. In asexual organisms, TE dynamics relies on different mechanisms, and it is the topic of section 8.5. Section 8.6 introduces bioinformatics methods to study TE dynamics from genomic data.

8.2. Transposable elements as parasites of sexual reproduction

8.2.1. *Amplification*

There is a wide diversity of TEs and transposition mechanisms (Chapters 1 and 2). Traditionally, TEs are classified into two classes: Class I TEs are retrotransposons, in which transposition relies on a RNA intermediate, and Class II TEs (transposons), in which DNA (at least one strand) physically changes location in the genome (no RNA intermediate).

In spite of deeply different transposition mechanisms, a wide diversity of TE families has managed to achieve a substantial evolutionary success. If Class I families are usually more successful because they are intrinsically replicative, some Class II TEs can also reach impressive copy number. TE family distribution varies across the tree of life, with Class I TEs more prevalent in large mammal genomes, in yeast, or in maize, while Class II TEs dominate in bacteria, in nematode worms, or in rice. Many organisms (including insects) display a more or less balanced TE content (Feschotte and Pritham 2007). Differences in copy number (between TE families or between species), observed in assembled genomes, can be explained by variation in transposition rates and variation of the duration of the active transposition stage. Differences can also be observed between populations of the same species. It is probable that environment and demography also play a role.

8.2.2. *The selfish DNA model*

TEs are very unusual genomic entities; understanding their properties and evolutionary history was particularly challenging for evolutionary and molecular biologists. After their discovery, considerable progress was made on the molecular mechanisms of transposition and on the TE distribution in the genome of model species, including bacteria, yeasts, plants (maize) and animals (drosophila). However, the evolutionary relevance of TEs was only poorly understood, especially in eukaryotes, as there was a dramatic mismatch between the large number of known TE copies and their elusive physiological role. A convincing evolutionary scenario, known as the "selfish DNA" model, arose in the early 1980s, based on the combination of two new ideas in evolutionary biology: the relevance of the gene-centric point of view in evolution (Dawkins 1976) and the idea that evolution is not necessarily driven by adaptation (Gould and Lewontin 1979).

The selfish DNA theory (Orgel and Crick 1980; Doolittle and Sapienza 1980) posits that the success of TEs is due to their capacity to transpose and replicate in genomes, rather than a putative function that would benefit their host. Sexual reproduction and recombination allows for the distribution of TE copies in all gametes. In such a context, the invasion of genomes (increase in TE copy number per individual) and the invasion of the population (increase in the frequency of individuals carrying TEs) are two consequences of the same transposition process and cannot be decoupled.

Figure 8.2 illustrates how both transposition and sexual reproduction are a prerequisite for a successful TE invasion. When the TE does not transpose, the total number of copies in the population is constant, and the element remains absent in most individuals. In absence of sexual reproduction, transposition increases the genomic copy number in the clonal lineage carrying TEs, but the element does not spread. Successful invasion under the selfish DNA model can only occur when both transposition and sexual reproduction are combined. The dynamics of TEs in asexual lineages is detailed in section 8.5.

The selfish DNA theory has been subsequently reinforced by formal mathematical models, which have demonstrated its compatibility with the traditional population genetic framework (Charlesworth and Charlesworth 1983). These models have shown that TE invasion is possible, even if TE insertions are deleterious. The theory predicts that a high transposition rate in the early stages of the invasion is critical, while transposition rates could decrease later on without consequences for the TE success (Le Rouzic and Capy 2005), a prediction that convincingly fits TE invasion observations.

8.3. Limiting the spread

A constant transposition rate per copy would lead to an exponential increase in the copy number, which cannot be sustained for a long time without filling the genome with TEs, disrupting genes and driving the population to extinction. Even if large-genome species harbor thousands of TEs, transposition activity rarely seems out of control; and although TE-driven population extinctions cannot be excluded, TE amplification is generally constrained by various mechanisms.

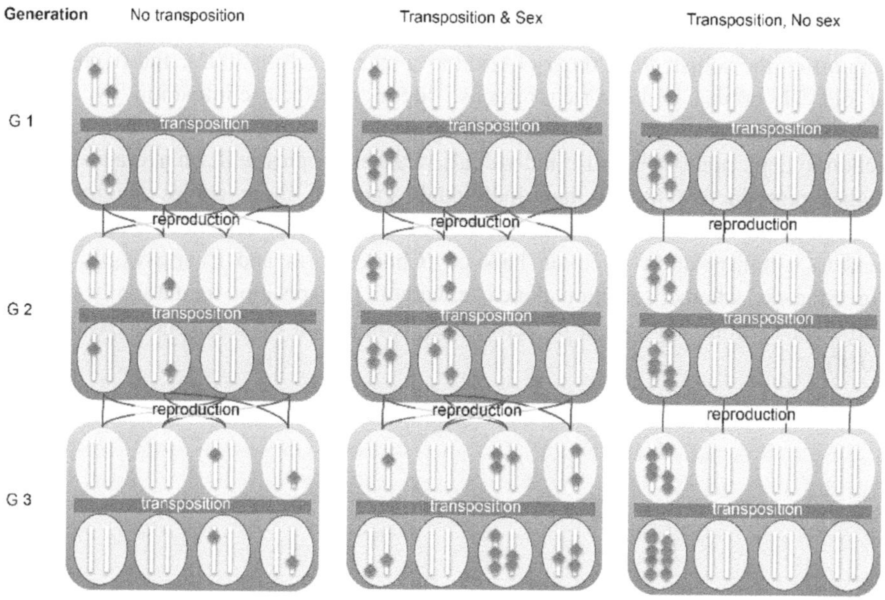

Figure 8.2. *Illustration of the selfish DNA theory applied to TEs. In absence of transposition (left panel), TEs segregate in the population but are rapidly diluted. Transposition (central panel) increases the probability to be transmitted to all offspring, ensuring the success of the invasion. Asexual reproduction (right panel) does not allow TEs to spread in the population, in spite of an efficient genome invasion. For a color version of this figure, see www.iste.co.uk/huavan/transposable.zip*

8.3.1. *Natural selection*

When inserting and multiplying in the genome, TEs are susceptible to disturbing cell functions. Active TEs can be deleterious in many ways (Nuzhdin 2000): additional copies increase the genome size, and thus the total amount of DNA to replicate, which can have a substantial cost (longer replication time and more energy), especially in small-genome organisms; TEs can disrupt genes and modify gene regulation, splicing or protein sequences themselves; active TEs use part of the costly cell resources to produce their transposition machinery; the transposition process can leave double-strand DNA breaks that prevents DNA to replicate; finally, repeated sequences along the genome may disturb chromosome pairing during meiosis and generate various chromosomal alterations – including inversions, segmental deletions, duplications and translocations – due to ectopic recombinations.

Although the exact effect of TE copies is hard to quantify (and may depend on the biology of the organism), indirect evidence is accumulating towards the existence of a general cost of TEs in genomes, including (i) the lower number of TEs in asexuals, supporting the selfish DNA theory (Arkhipova and Meselson 2005), and (ii) the lack of old TE copies in gene-rich regions, suggesting their elimination by natural selection (Barrón et al. 2014). Selection against TE insertions tend to decrease the genomic copy number because individuals with less copies will, on average, have a higher reproductive fitness than individuals carrying more copies (Figure 8.3). Transposition can, at least in theory, compensate natural selection, which would lead to a transposition–selection balance (Charlesworth and Charlesworth 1983). Nevertheless, whether selection alone (i.e. without transposition regulation) would be sufficient to limit the spread of active TEs remains an open question (Arkhipova 2018).

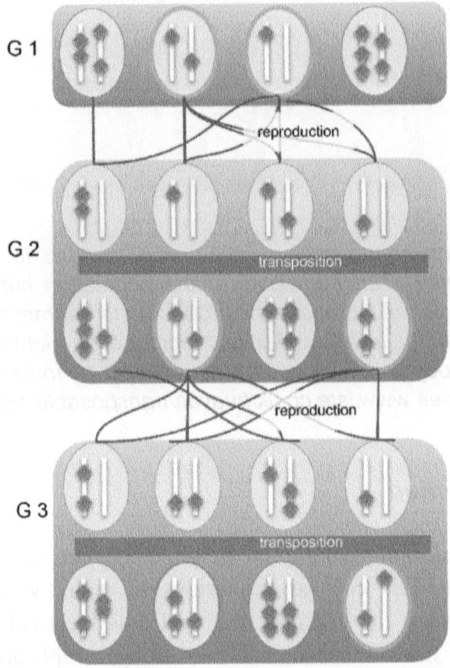

Figure 8.3. *Illustration of how selection and transposition can lead to a dynamic equilibrium in TE copy number. Reproductive fitness is higher (thick contour lines) in low TE copy number individuals. Due to selection, each reproduction episode is thus associated with a drop in TE copy number. New TE copies are then generated by transposition. When both phenomena compensate, the average copy number is maintained in spite of a deleterious effect of TE insertions. For a color version of this figure, see www.iste.co.uk/huavan/transposable.zip*

Active transposition generates specific mutations in genomes (mainly insertions and deletions). Random genomic mutations are generally neutral or detrimental, but rare genetic novelties can also be co-opted by natural selection. The observation that TEs are globally deleterious does not preclude that some specific TE insertions might happen to be advantageous. Indeed, there are multiple examples of individual TE copies that have been positively selected, either because of a reuse of the TE proteins for another cell function or for the influence of a TE insertion on the regulation of neighboring genes (Miller et al. 1997; Schrader and Schmitz 2019). However, due to recombination in sexually reproducing species, selection acts independently on TE copies and rare positively selected TE copies are unlikely to affect the invasion dynamics of the whole family.

8.3.2. *Transposition regulation*

In multicellular organisms, most cells (somatic cells) do not transmit their genome to the next generation, and the mutagenic effect of transposition is expected to be deleterious for both TEs and the host. For instance, active transposition can disrupt key oncogenes and participate to tumorogenesis (Burns 2017), which harms both the host and TEs, as the TE transmission to the next generation is compromised. Transposition activity is thus frequently controlled by regulation mechanisms, as detailed in Chapter 5. Transposition regulation in somatic cells can be secured by host functions, but some regulation mechanisms are also brought by the TEs themselves, and are therefore referred to as "self-regulation".

In contrast, there is a conflict between TEs and hosts in the germline, in which transposition events are heritable. While the host would benefit from limiting the number of potentially harmful mutations transmitted to the offspring, only transpositionally active TEs can spread in the population, even at a cost for the host.

In terms of genome invasion dynamics, predictions differ, depending on the nature of regulation. The first theoretical models assumed a decrease in the transposition rate when the copy number is high, leading to an equilibrium state (Charlesworth and Charlesworth 1983), but most studies failed in giving support for such a "physiological" regulation. Instead, TE families seem to be in a permanent disequilibrium. The current view concerning TE regulation favors the genomic regulation pathway, in which regulation is triggered by a mutational event (change in the DNA). This category encompasses the recently discovered pi-RNA regulation mechanism (Brennecke et al. 2007; Kelleher et al. 2020), which is a general genomic protection against mobile sequences: the insertion of a TE copy into a specific region of the genome (pi-cluster) triggers the production of specific anti-TE small RNAs, which are able to silence the whole family. This regulatory insertion is heritable and can be transmitted to the offspring, ensuring specific TE regulation across generations (Kofler 2019).

8.4. Long-term evolution

While natural selection and regulation can limit the spread of active TEs, long-surviving TE sequences accumulate mutations and become degraded, which can also modify the activity level of TEs. TE sequences can be altered by a wide spectrum of mutations. As any genomic sequence, TEs can be affected by DNA replication errors or genomic large deletions that can reduce the copy number. In addition, specific mutational pressures can alter TE sequences: (i) an increase in the C → T mutation rates, because TE sequences are frequently methylated; (ii) internal or 3' deletions due to incomplete or defective transposition, and (iii) for LTR retroelements, internal deletions occurring because of ectopic recombinations between the LTRs (leaving a so-called solo-LTR). As a result, TE sequences tend to degrade, generation after generation, due to the accumulation of alterations. However, those altered TE copies can still have an influence on the dynamic of the whole family.

8.4.1. *Interactions between TE copies*

Mutations in the TE sequence can affect their functional properties by modifying the transposition machinery (change in mRNA expression or change in the protein sequence), or by altering the capacity of the transposition machinery to recognize the TE and promote transposition. Mutated TEs can then become either immobile (unable to transpose but able to produce transposition-related proteins) or non-autonomous (able to transpose, provided some transposition machinery produced by other copies from the same genome). The evolutionary consequences are drastically different for both categories, as immobile copies are doomed and cannot maintain over evolutionary time, while non-autonomous copies can form evolutionary-successful sub-families. For instance, miniature inverted-repeat transposable elements (MITEs) are widespread Class II non-autonomous TEs, and short interspaced nuclear elements (SINEs) are Class I non-autonomous sequences. Such non-autonomous TEs can be efficient enough to outcompete their autonomous counterparts, and even drive the whole family to extinction (Robillard et al. 2016).

The complexity of interactions between TEs and the rest of the genome has often been compared to the interactions between species in an ecosystem (Brookfield 2005; Le Rouzic et al. 2007; Venner et al. 2009). Such interactions can remind mutualism (such as among various genes in the genome), parasitism (TEs vs. genes), up to hyper-parasitism (autonomous vs. non-autonomous TEs). This "genome ecology" analogy is further developed in Chapter 9.

8.4.2. *Selection on TE sequences*

In most genomes, a large majority of TE-related sequences are relics, that is, they have lost both transposition and protein production functions. This observation raises complex (and yet unanswered) questions about the strength of conservative selection on TE sequences. In "regular" genes, mutations that degrade protein-coding sequences tend to be eliminated by natural selection, as the transmission of the non-functional allele across generations is jeopardized by the decrease in reproductive fitness. Such a mechanism is ineffective for "selfish" TEs, as they are not subject to natural selection (a defective TE does not prevent the host from reproducing).

TE sequences are rather subject to intra-genomic selection, due to competition among TEs: TE copies displaying the highest transposition rate will be over-represented in the next generation. However, when TEs from the same family are able to cross-mobilize, conservative selection on protein-coding TE sequences is expected to be low because all copies from the same family share the same transposition machinery. The contribution of each copy to the pool of transposition-related proteins is thus unlikely to affect the competitive ability of individual TEs. This situation is a typical example of a "tragedy of the commons", where the benefits of individual contributions is shared among the populations, including the bad contributors (such as non-autonomous TEs). The absence of conservative selection has been confirmed by empirical observations, especially for Class II element sequences (Zhang et al. 2020). In contrast, conservative selection can be detected in TEs whose copies tend to promote their own transposition more frequently, such as some Class I (LINE or LTR) elements (Wei et al. 2001).

The absence of conservative selection on the sequence of TEs should lead to their loss from the tree of life, which is obviously not the case. Indeed, even if selection on sequence is weak for most of the TE lifecycle, there exists a considerable amount of selection on TEs when they are horizontally transferred across species. Horizontal transfers usually encompass short DNA sequences, and only one copy is expected to start the TE invasion, so that only fully functional TEs can jump across species. Such a drastic horizontal transfer selective filter is supported by empirical data (Zhang et al. 2020), and is likely to explain – to a great extent – how functional TEs maintain in spite of the low level of intra-species conservative selection.

8.5. The intriguing case of asexuals

The selfish DNA model provides a reference framework to explain TE amplification in large eukaryotic genomes: the "selfish" invasion of the genome relies on sexual reproduction to spread TE copies in the whole population. In addition, recombination improves the efficiency of natural selection, by generating new genotypes, favoring the elimination of the most deleterious TE insertions.

But what happens in species with no sexual reproduction, and then no recombination? When a TE copy is inserted in the genome of a unique individual, typically after a horizontal transfer event, it may amplify and invade the genome of the clonal lineage. In absence of recombination, we expect an accumulation of insertions, since they cannot be removed. However, the lack of elimination of deleterious insertions will also tend to decrease the fitness of the clonal lineage carrying TEs. This, combined with its inability to invade the genome of the other clones of the population, make the evolutionary success of the TE-carrying clone quite unlikely, if TE copies are on average neutral, and virtually impossible if TEs are even slightly deleterious (Figure 8.2). Worse, it drives the whole lineage to extinction. Hence, the prediction of the theory is far from straightforward (Flores-Ferrer et al. 2021). It is usually considered that asexual species should be deprived of TEs (Lockton and Gaut 2010; Castagnone-Sereno and Danchin 2014), especially in small populations (Arkhipova and Meselson 2005; Dolgin and Charlesworth 2006; Boutin et al. 2012).

The question is worth asking, because reduced or absent recombination is widespread. This is, for instance, the case in organisms such as bacteria or archaea, and in some eukaryotes who have lost the capacity to reproduce sexually (including plant or fungal lineages, as well as a few animals, such as rotifera). It is also true for organisms displaying various reproduction modes (featuring self-fertilization and parthenogenetic reproduction), in which recombination is indeed limited. Overall, theoretical predictions about the lack of TEs in asexuals have not been confirmed by empirical data. Prokaryotes have less TEs than most eukaryotes, both in terms of TE copies and TE diversity; however, they still have some. In eukaryotes, comparison between asexually and sexually related species has constantly provided mixed results. Experimental evolution on species able to reproduce sexually or asexually has also provided contradictory results (Schaack et al. 2010; Bast et al. 2019; Chen and Zhang 2021).

This mismatch between theory and genomic data suggests that some assumptions of the population genetic models are violated. For instance, even asexual lineages regularly exchange DNA; such transfers could theoretically replace sex and allow for selfish spread. TEs are frequently transferred between eukaryotes, and prokaryotes are also known to exchange a lot, through conjugation, transduction or transformation. Genetic load could also be reduced by limiting transposition (prokaryotic TEs are usually tightly regulated) or using specific insertion sites. Furthermore, population size and then demography may also influence the fate of TEs in genomes. An interesting alternative hypothesis lies in the possibility of a non-selfish spread, assuming that some TE insertions are beneficial and invade the population through positive selection. In this scenario, all TE copies do not need to be beneficial, as the lack of recombination can drive slightly deleterious insertions along with the beneficial ones. In such a scenario, long-term persistence of active TE copies requires a constant flux of beneficial insertions, which is only possible when the population is never well adapted – for instance, in unstable or fluctuating environments (Wright and Schoen 1999; McGraw and Brookfield 2006).

8.6. Transposable element genomics

A wide range of theoretical work was performed in the past to understand and predict the dynamics of TEs. In real organisms, TE dynamics can be approached by several proxies, including the transposition rate, age and length of the transposition bursts, or the insertion frequency at the various TE-containing loci. A powerful approach to study TE evolution within – and among – species relies on combining recent technological advances in molecular biology (new generation sequencing) and computer sciences (bioinformatics). Several methods can be used, depending of the characteristics of the species, including the observation of the evolution of the TEs over several generations, the analysis of assembled genomes, or the comparative analysis of genomic sequences obtained from several populations.

8.6.1. *Transposition rates obtained in the lab*

Transposition rates are usually calculated as the number of replications per TE copy and per generation from at least two consecutive generations. In eukaryotes such as drosophila, detecting new copies was first achieved using old techniques such as Southern blot or in situ hybridization, and then quantitative PCR. Nowadays, whole genome sequencing offers an easy and cost-effective solution. Our knowledge about transposition rates mainly comes from estimation of Class I elements, in *Drosophila melanogaster*, performed on mutation accumulation lines, lab stocks, or in experimental evolution. Reported transposition rates range between 10^{-4} to 10^{-3} transposition events per copy and per generation for active TEs settled in populations (Maside et al. 2000; Adrion et al. 2017), to much higher rates measured in controlled experiments during transpositions bursts, from 0.1 to 0.4 transposition events per copy and per generation (Robillard et al. 2016). Overall, the rate of replicative transposition is not widely different across TE families, and the success of the genome invasion does not seem to depend on the transposition mechanisms. Transposition rates may depend on the genetic backgrounds and environmental conditions (such as temperatures), but remain in the same range between Class I and Class II TEs. Transpositions rate estimates are also available for a few other species, such as human and yeast, although differences in generation time and number of genomic copies makes meaningful comparison especially tricky.

8.6.2. *Comparative genomics of assembled genomes*

Any new assembled genome needs to be annotated for TEs before being annotated for genes. From the TE annotations, sequences can be extracted. Sequences belonging to the same family share a common ancestor, which they derived by replicative transposition while accumulating mutations. TE copies from the same family are generally similar enough to be fully aligned and compared. The distribution of the divergence among copies is widely used to infer the age of the family and the timing of TE amplification.

The divergence among TE sequences reflects the time elapsed since this common ancestor. Phylogenetic reconstruction methods can be applied on the TE sequence alignment in order to reconstruct the evolutionary history of TE families by inferring how TE sequences derive from each other. Each branching event in the resulting phylogenetic tree corresponds to a transposition event, which generated two copies that accumulated mutations independently. These transposition events can be dated (in genetic divergence, or in million years if a molecular clock is applied). Their distribution gives a direct indication of the dynamics along time (Figure 8.4). It can be noted, however, that the quality of genome assembly has a large impact on the ability to exhaustively detect element insertions.

Existing data suggests that, in spite of different copy numbers among TE families, the host species has a considerable influence on the TE lifecycle. In Drosophila, for instance, only new, horizontally transferred TE families seem to be able to invade genomes. The invasion is very rapid (transposition "bursts"), after which transposition stops and the element decays (Wallau et al. 2014). A similar pattern was identified in bats (Ray et al. 2008). In primates, and especially in the human lineage, TEs also undergo cycles of transposition bursts followed by quiet stages, but very few horizontal transfers have been documented; transposition bursts rather start from a TE already present in the genome and which got reactivated (the reactivation mechanism being unclear (Khan et al. 2006)). In contrast, some TE families have been shown to be active at a low (but constant) rate for a long time in the fungus *Fusarium oxysporum* (Le Rouzic et al. 2013), which does not preclude ongoing bursts of old elements in specific strains.

Comparative genomics is a research program aiming at studying similarities and differences in genome composition and evolutionary dynamics among species. Comparing TE sequence divergence makes it possible to reconstruct the evolutionary history of TE families, the species in which they proliferated and the species from which they were eliminated. The detection of horizontal transfers, for instance, is based on comparative approaches (Wallau et al. 2012).

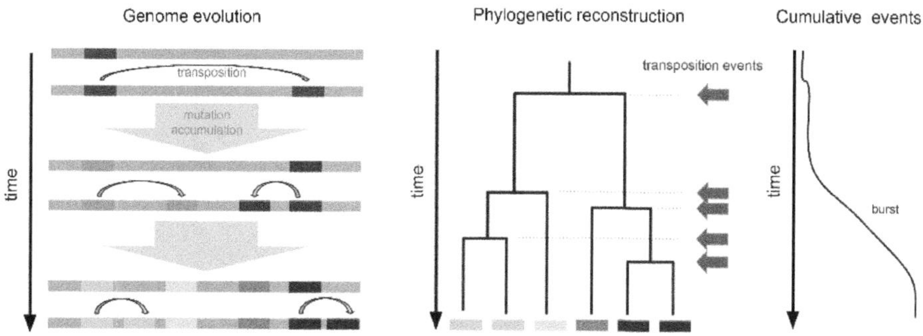

Figure 8.4. *Illustration of the reconstruction of transposition dynamics from genome sequence data. Left: Immediately after transposition, TE template and copies are identical (same color). In the course of evolutionary time, copies accumulate mutations independently, and their sequence diverges. Right: from the alignment of homologous TE copies, a phylogeny can be reconstructed. Branching in the phylogenetic tree correspond to transposition events (horizontal arrows). Cumulative curve of transposition events allows us to visualize the length and intensity of transposition bursts. For a color version of this figure, see www.iste.co.uk/huavan/transposable.zip*

8.6.3. *Population genomics*

Finally, the availability of a genome assembly makes it possible to easily compare individuals or populations of the same species. For this, short reads sequencing (such as Illumina) is usually enough to detect new insertion sites and calculate insertion frequency. The pool-seq method (DNA is extracted from several individuals of the same population) is widely used for tiny organisms such as drosophila and has proven very efficient. With enough individuals sampled, the frequency of a TE insertion at any given locus can be estimated. Active TE families are usually characterized by numerous insertion sites with low frequency, while for old, inactive families, detectable insertions sites are present at high frequency or fixed (i.e. in all individuals/populations). Several bioinformatics tools can perform population comparisons. Access to the insertion frequencies offers the opportunity to precisely study the dynamics of TEs in relation to various external parameters, such as recent invasions, geographical or bioclimatic factors (Kapun et al. 2020; Mérel et al. 2021). It also allows us to detect abnormally high frequency in some particular loci, supposed to correspond to adaptive insertions – that is, insertions bringing an advantage to the host (Chapter 4).

8.7. Conclusion

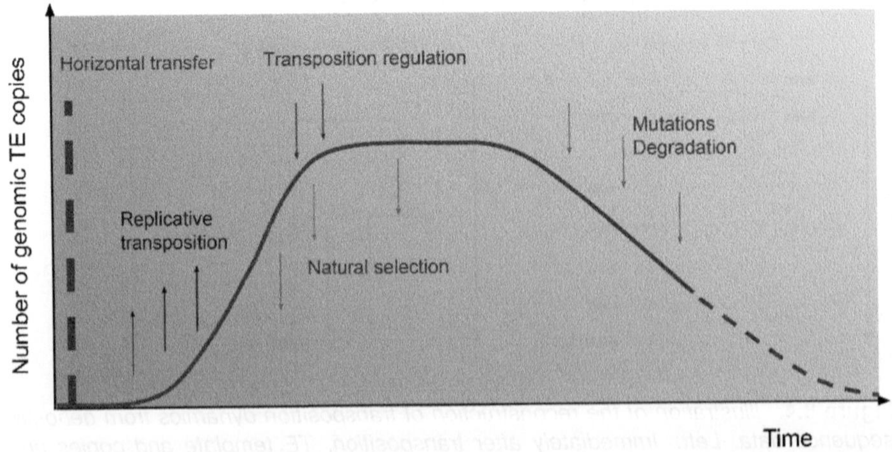

Figure 8.5. *Copy number dynamics during the TE lifecycle. When a new TE arrives in a genome (by horizontal transfer in most cases), it starts invading the genome of the species through replicative transposition. Regulation mechanisms limit the spread, and together with natural selection against deleterious TE insertions, stabilizes the total copy number. Transposition stops and TE sequences are slowly degraded by accumulating mutations and deletions. For a color version of this figure, see www.iste.co.uk/huavan/transposable.zip*

The growing availability of genome sequences has confirmed the robustness of the TE lifecycle model. TE families do not persist forever in genomes, and the TE content of species is dynamic, old families being replaced by new ones in a general turnover (Figure 8.5). Replicative transposition ensures an exponential increase in copy number at the onset of the TE invasion; the copy number then stabilizes relatively rapidly under the combined effect of transposition regulation and natural selection. At this point, transposition stops, and TE sequences slowly accumulate mutations, until they are too degraded to be identified as such.

Although this general scheme seems to be quite universal, at least in eukaryotes, a substantial amount of issues remain to be explored. Why does the copy number differ across TE families and species? What is the effect of natural selection, and what are the mechanisms involved? What is the importance of interactions between TEs from the same families and from different families? Why are horizontal transfers of TEs so frequent? Why are there more TEs in asexual species than predicted by the theory? The development of new genomics tools and concepts will likely play a considerable role when addressing such questions.

8.8. References

Adrion, J.R., Song, M.J., Schrider, D.R., Hahn, M.W., Schaack, S. (2017). Genome-wide estimates of transposable element insertion and deletion rates in *Drosophila melanogaster*. *Genome Biology and Evolution*, 9(5), 1329–1340.

Arkhipova, I.R. (2018). Neutral theory, transposable elements, and eukaryotic genome evolution. *Molecular Biology and Evolution*, 35(6), 1332–1337.

Arkhipova, I.R. and Meselson, M. (2005). Deleterious transposable elements and the extinction of asexuals. *Bioessays*, 27(1), 76–85.

Barrón, M.G., Fiston-Lavier, A.-S., Petrov, D.A., González, J. (2014). Population genomics of transposable elements in *Drosophila*. *Annual Review of Genetics*, 48, 561–581.

Bast, J., Jaron, K.S., Schuseil, D., Roze, D., Schwander, T. (2019). Asexual reproduction reduces transposable element load in experimental yeast populations. *Elife*, 8, e48548.

Boutin, T.S., Le Rouzic, A., Capy, P. (2012). How does selfing affect the dynamics of selfish transposable elements? *Mobile DNA*, 3(1), 1–9.

Brennecke, J., Aravin, A.A., Stark, A., Dus, M., Kellis, M., Sachidanandam, R., Hannon, G.J. (2007). Discrete small RNA-generating loci as master regulators of transposon activity in *Drosophila*. *Cell*, 128(6), 1089–1103.

Brookfield, J.F. (2005). The ecology of the genome—mobile DNA elements and their hosts. *Nature Reviews Genetics*, 6(2), 128–136.

Burns, K.H. (2017). Transposable elements in cancer. *Nature Reviews Cancer*, 17(7), 415–424.

Castagnone-Sereno, P. and Danchin, E.G.J. (2014). Parasitic success without sex—the nematode experience. *J Evol Biol*, 27(7), 1323–1333.

Charlesworth, B. and Charlesworth, D. (1983). The population dynamics of transposable elements. *Genetics Research*, 42(1), 1–27.

Chen, P. and Zhang, J. (2021). Asexual experimental evolution of yeast does not curtail transposable elements. *Mol Biol Evol*, 38(7), 2831–2842.

Dawkins, R. (1976). *The Selfish Gene*. Oxford University Press, Oxford.

Dolgin, E.S. and Charlesworth, B. (2006). The fate of transposable elements in asexual populations. *Genetics*, 174(2), 817–827.

Doolittle, W.F. and Sapienza, C. (1980). Selfish genes, the phenotype paradigm and genome evolution. *Nature*, 284(5757), 601–603.

Feschotte, C. and Pritham, E.J. (2007). DNA transposons and the evolution of eukaryotic genomes. *Annu. Rev. Genet.*, 41, 331–368.

Flores-Ferrer, A., Nguyen, A., Glémin, S., Deragon, J.-M., Panaud, O., Gourbière, S. (2021). The ecology of the genome and the dynamics of the biological dark matter. *Journal of Theoretical Biology*, 518, 110641.

Gould, S.J. and Lewontin, R.C. (1979). The spandrels of San Marco and the Panglossian paradigm: A critique of the adaptationist programme. *Proceedings of the Royal Society of London. Series B. Biological Sciences*, 205(1161), 581–598.

Kapun, M., Barrón, M.G., Staubach, F., Obbard, D.J., Wiberg, R.A.W., Vieira, J., Goubert, C., Rota-Stabelli, O., Kankare, M., Bogaerts-Márquez, M. et al. (2020). Genomic analysis of European *Drosophila melanogaster* populations reveals longitudinal structure, continent-wide selection, and previously unknown DNA viruses. *Mol Biol Evol*, 37(9), 2661–2678.

Kelleher, E.S., Barbash, D.A., Blumenstiel, J.P. (2020). Taming the turmoil within: New insights on the containment of transposable elements. *Trends in Genetics*, 36(7), 474–489.

Khan, H., Smit, A., Boissinot, S. (2006). Molecular evolution and tempo of amplification of human line-1 retrotransposons since the origin of primates. *Genome Research*, 16(1), 78–87.

Kidwell, M.G. and Lisch, D.R. (2001). Perspective: Transposable elements, parasitic DNA, and genome evolution. *Evolution*, 55(1), 1–24.

Kofler, R. (2019). Dynamics of transposable element invasions with piRNA clusters. *Molecular Biology and Evolution*, 36(7), 1457–1472.

Le Rouzic, A. and Capy, P. (2005). The first steps of transposable elements invasion: Parasitic strategy vs. genetic drift. *Genetics*, 169(2), 1033–1043.

Le Rouzic, A., Dupas, S., Capy, P. (2007). Genome ecosystem and transposable elements species. *Gene*, 390(1–2), 214–220.

Le Rouzic, A., Payen, T., Hua-Van, A. (2013). Reconstructing the evolutionary history of transposable elements. *Genome Biology and Evolution*, 5(1), 77–86.

Lockton, S. and Gaut, B.S. (2010). The evolution of transposable elements in natural populations of self-fertilizing *Arabidopsis thaliana* and its outcrossing relative *Arabidopsis lyrata*. *BMC Evol Biol*, 10, 10.

Maside, X., Assimacopoulos, S., Charlesworth, B. (2000). Rates of movement of transposable elements on the second chromosome of *Drosophila melanogaster*. *Genetics Research*, 75(3), 275–284.

McGraw, J.E. and Brookfield, J.F. (2006). The interaction between mobile DNAs and their hosts in a fluctuating environment. *Journal of Theoretical Biology*, 243(1), 13–23.

Mérel, V., Gibert, P., Buch, I., Rodriguez Rada, V., Estoup, A., Gautier, M., Fablet, M., Boulesteix, M., Vieira, C. (2021). The worldwide invasion of *Drosophila suzukii* is accompanied by a large increase of transposable element load and a small number of putatively adaptive insertions. *Mol Biol Evol*, 38(10), 4252–4267.

Miller, W.J., McDonald, J.F., Pinsker, W. (1997). Molecular domestication of mobile elements. In *Evolution and Impact of Transposable Elements*. Capy, P. (ed.). Springer, Dordrecht.

Nuzhdin, S.V. (2000). Sure facts, speculations, and open questions about the evolution of transposable element copy number. In *Transposable Elements and Genome Evolution*. McDonald, J.F. (ed.). Springer, Dordrecht.

Orgel, L.E. and Crick, F.H. (1980). Selfish DNA: The ultimate parasite. *Nature*, 284(5757), 604–607.

Ray, D.A., Feschotte, C., Pagan, H.J., Smith, J.D., Pritham, E.J., Arensburger, P., Atkinson, P.W., Craig, N.L. (2008). Multiple waves of recent DNA transposon activity in the bat, *Myotis lucifugus*. *Genome Research*, 18(5), 717–728.

Robillard, É., Le Rouzic, A., Zhang, Z., Capy, P., Hua-Van, A. (2016). Experimental evolution reveals hyperparasitic interactions among transposable elements. *Proceedings of the National Academy of Sciences*, 113(51), 14763–14768.

Schaack, S., Pritham, E.J., Wolf, A., Lynch, M. (2010). DNA transposon dynamics in populations of *Daphnia pulex* with and without sex. *Proc Biol Sci*, 277(1692), 2381–2387.

Schrader, L. and Schmitz, J. (2019). The impact of transposable elements in adaptive evolution. *Molecular Ecology*, 28(6), 1537–1549.

Venner, S., Feschotte, C., Biémont, C. (2009). Dynamics of transposable elements: Towards a community ecology of the genome. *Trends in Genetics*, 25(7), 317–323.

Wallau, G.L., Ortiz, M.F., Loreto, E.L.S. (2012). Horizontal transposon transfer in eukarya: Detection, bias, and perspectives. *Genome Biology and Evolution*, 4(8), 801–811.

Wallau, G.L., Capy, P., Loreto, E., Hua-Van, A. (2014). Genomic landscape and evolutionary dynamics of mariner transposable elements within the *Drosophila* genus. *BMC Genomics*, 15(1), 1–19.

Wei, W., Gilbert, N., Ooi, S.L., Lawler, J.F., Ostertag, E.M., Kazazian, H.H., Boeke, J.D., Moran, J.V. (2001). Human L1 retrotransposition: Cispreference versus trans complementation. *Molecular and Cellular Biology*, 21(4), 1429–1439.

Wright, S.I. and Schoen, D.J. (1999). Transposon dynamics and the breeding system. *Genetica*, 107(1), 139–148.

Zhang, H.H., Peccoud, J., Xu, M.R.X., Zhang, X.G., Gilbert, C. (2020). Horizontal transfer and evolution of transposable elements in vertebrates. *Nature Communications*, 11(1), 1–10.

9

The Ecology of Transposable Elements

Pierre CAPY[1], Christian BIÉMONT[2] and Cristina VIEIRA[2]

[1] *Évolution, Génomes, Comportement et Écologie (EGCE), CNRS, Université Paris-Saclay, Gif-sur-Yvette, France*

[2] *Laboratoire Biométrie et Biologie Évolutive (LBBE), CNRS, Université Claude Bernard Lyon 1, Villeurbanne, France*

9.1. Introduction

Transposable elements (TEs) are not living entities, but like viruses, they are sequences that cannot be found outside cells and can only multiply within them. Consequently, while these cells represent a suitable environment, the evolution of TEs within a genome obeys constraints that may come from the elements themselves, genomes or the interaction between the two. This may lead to their proliferation, degeneration or to them being silenced without being destroyed.

Over the last few decades, the vision of the intracellular environment as an ecosystem in which TEs could evolve has developed. The various stages in the life of a TE within a cell, a population and a species, as well as recent observations and results leading to this conclusion, are presented in this chapter.

9.1.1. *Lifecycle of a transposable element*

First and foremost, and even if this may at first appear to be an abuse of language, it is possible to speak of the "lifecycle" of TEs within the genome of a

 © ISTE Ltd 2024.

species. The phases of this cycle are summarized in Figure 9.1 and detailed in Figure 9.2.

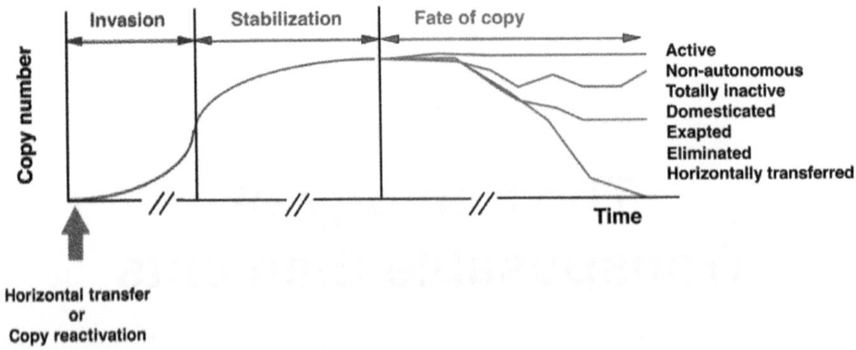

Figure 9.1. *Lifecycle of a family of transposable elements. See text and Figure 9.2 for details. For a color version of this figure, see www.iste.co.uk/huavan/transposable.zip*

First, a TE active copy can be introduced by horizontal transfer or a copy previously inactivated in the host genome can be reactivated. Initially, this copy is present in a single copy in a single individual. It can therefore be lost very quickly by simple genetic drift if the transposition rate (i.e. the replication rate of the copy) is low (Le Rouzic and Capy 2005). To pass this first hurdle, the transposition rate must be relatively high and/or be accompanied by active diffusion of the element in the population.

Then, two processes will take place through the colonization of the genome and invasion of the host population.

During the colonization process, the genome in which the TE arrives is either naive, that is, it has never been confronted with this element, or contains active or inactive copies of this family. If it is a naive genome, it is very likely that the host genome is not capable of regulating transposition, so the number of copies will increase rapidly. In cases where the genome has already been in the presence of the introduced TE family, and despite the possible absence of active copies, non-autonomous full copies or relic elements may persist in the genome. These traces of an old or recent invasion can very quickly regulate the transposition activity of the new copy and block its multiplication in the host genome, notably via epigenetic phenomena (see Chapter 4). This system of transposition regulation resembles what antibodies can do against a virus in vertebrates and can be compared to a kind of immunity against new invasions (Muller et al. 2013; Duc et al. 2019).

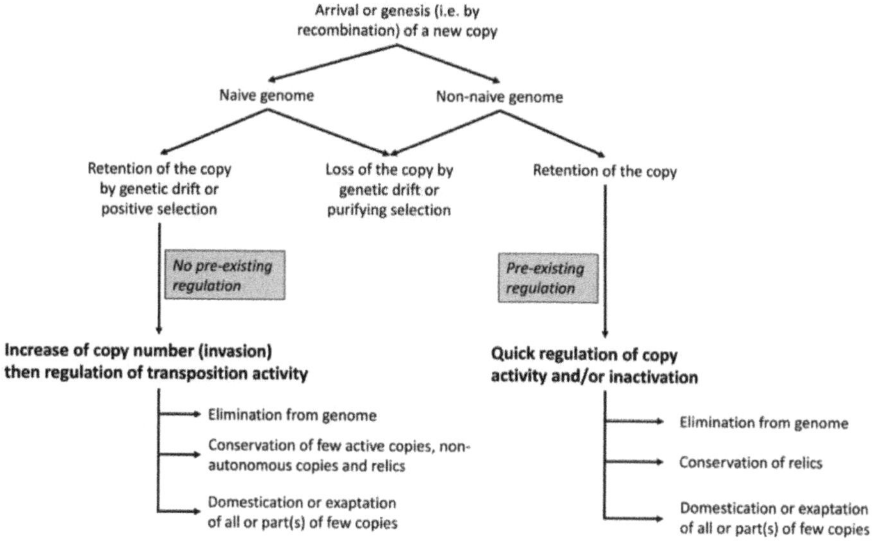

Figure 9.2. *Fate of a newly arrived copy in a naive genome (i.e. one that has never seen this transposable element, left), or in a non-naive genome, in which regulation pre-exists (right)*

With regard to the dissemination of TEs in a host population, its success will depend on the effect of the insertions on the phenotype. For example, it has been repeatedly shown in Drosophila that when TEs promote population adaptation, this can lead to an increase in the frequency of insertions in that population (Gonzalez et al. 2010; Barron et al. 2014; Villanueva-Cañas et al. 2017; Bourgeois and Boissinot 2019).

Finally, the regulation of transposition activity and the sedentarization of copies can lead to their degradation by mutation accumulation, to their elimination by recombination, or even to their domestication if one or more insertions are beneficial (Daborn et al. 2002; DeBarry et al. 2006; Wicker et al. 2018; Villanueva-Cañas et al. 2019).

9.1.2. *Variability in the number of copies*

The large number of genomes currently sequenced and analyzed clearly reveals major differences in both the quantity and diversity of TEs; genes sometimes accounting for only a few percent. Variability in the average number of copies within a genome is the result of several synergistic and/or sometimes antagonistic factors. The main forces at work here are summarized in Figure 9.3. Since most insertions are deleterious or neutral, natural selection and genetic drift will be major forces. Selection will tend to reduce the number of copies, except in the case of advantageous insertions, while genetic drift can either reduce or increase this number. The balance between these two forces is mainly modulated by the effective population size (*Ne* = number of breeding individuals). Thus, when *Ne* is large, natural selection takes precedence over genetic drift, and when *Ne* is small, genetic drift is the major evolutionary force and can lead to the fixation of slightly deleterious mutations (in this case insertions) (Wright 1931; Kimura and Crow 1963, 1964). Beyond these two mechanisms, transposition leads to an increase in copy number, while deletions following ectopic recombination or excisions not followed by re-insertion (for *Class II* elements) tend to decrease copy number.

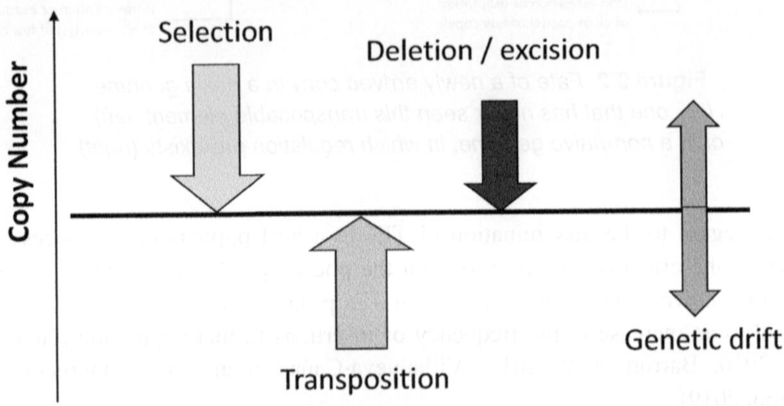

Figure 9.3. *Evolutionary forces affecting copy number within a family of transposable elements. See text for details. For a color version of this figure, see www.iste.co.uk/huavan/transposable.zip*

Over the last few decades, several hypotheses have been put forward in an attempt to understand the strong differences in the TE content of different species. In this context, it has been suggested that genome size may be correlated with the accumulation of TEs (Lynch and Conery 2003). Indeed, the smaller the effective population size, the larger the genome size. This is probably linked to the fact that, despite their deleterious effects, TEs can be maintained in genomes through genetic drift (see above). However, this correlation is much debated (Charlesworth and Barton 2004; Daubin 2004; Lynch 2004; Vinogradov 2004a, 2004b; Sessegolo et al. 2016), as many phenomena need to be taken into account when considering smaller scales than the entire tree of life, as in isopods of the *Asellidae* family (Lefébure et al. 2017) or ancestral and invasive populations of *Drosophila suzukii* (Mérel et al. 2021). Thus, although it has never really been studied, there is probably, under a neutral model, a threshold of TE accumulation beyond which the addition of new copies will have a very limited impact on the adaptive value of individuals, as they are likely to be inserted into regions where there are already TEs. This can also give rise to clusters of elements, which even if inserted into each other can be responsible for regulating the activity of all the copies of a TE family via small RNAs, as has been demonstrated in *Drosophila melanogaster* and *Drosophila simulans* (see Mérel et al. 2020; Saint-Léandre et al. 2020).

9.1.3. *What does the genome represent for a transposable element?*

TEs need the transcription and translation machinery of cells to multiply. As with viruses, this raises two questions: (i) their origin and (ii) what genomes represent for TEs. This is the subject of this chapter.

Given the ubiquitous distribution of TEs in the tree of life, it would seem that they are essential components of genomes. So, would a genome devoid of TEs be an "evolutionary dead end"? In this respect, it is worth pointing out that very rare species appear to have no TEs. It has been clearly shown that some have lost their TEs, while others have acquired TEs through horizontal transfer, as in bdelloid rotifers (Gladyshev et al. 2008). Moreover, in some species, such as the malaria vector *Plasmodium falciparum*, which initially appeared to be devoid of TEs (Gardner et al. 2002), it is now accepted that their genomes have had TEs; some have been domesticated and it is not impossible that some potentially active elements remain (Durand et al. 2006; Bohme et al. 2018).

While the negative effects of TE insertions are predominant in the short term, they cannot be over the long term. This is why TEs are considered the "toolbox" of genomes, insofar as their transposition is a source of genetic variability that can lead to rapid adaptation of species following a change in their environment. However, TEs cannot remain permanently active. In a genome, after the invasion phase, mutations (insertion-deletions, point mutations) lead to the emergence of copies, which may be dead (no longer mobilizable) or non-autonomous (non-active but mobilizable via active elements of the same family). Epigenetic regulation of TEs can also take place via several mechanisms: cytosine methylation; modification of chromatin structure; regulation by small RNAs (see Chapter 4). Under stress, changes to the TE epigenome can lead to reactivation of active copies previously silenced and/or altered expression of genes located near TE insertions.

9.2. Cellular, population and specific dynamics

9.2.1. *The dynamics of transposable elements vary from cell to cell*

In a multicellular organism, the regulation of TE activity is not the same in all tissues, particularly when the somatic and germline are individualized (for a review, see Haig 2016). For example, in 12 vertebrate species wide variations exist between the somatic and germline, with greater expression in the testes (Pasquesi et al. 2020). Similar results have been found in *D. melanogaster* and *D. simulans*, with geographical variations (Saint-Léandre et al. 2017, 2020). In both cases, differential regulation by small *piwi*RNAs (or *pi*RNAs) is observed. As TE activity can vary with the age of organisms and be the cause of many diseases, it is not surprising to observe that it is repressed at transcriptional and post-transcriptional levels in the somatic lineage (Pasquesi et al. 2020).

Beyond sex cells, disparities in activity may also exist between different somatic tissues. In several metazoans, TE activity does not appear to be subject to the same constraints in the central nervous system as in other somatic tissues. Thus, it has been repeatedly observed that TE activity is much higher in the brain, suggesting their contribution to neuronal plasticity but also their involvement in several diseases (Sfera et al. 2018; Ahmadi et al. 2020; Evans and Erwin 2020; Treiber and Waddell 2020). The *pi*RNA pathway proteins *Aubergine* and *Argonaute 3*, involved in TE *silencing* in the Drosophila female germline, are underexpressed in parts of the central nervous system (CNS) such as pedunculated body neurons in the cerebellum (Perrat et al. 2013). This downregulation leads to increased TE activity and may contribute to the maintenance of genomic heterogeneity and hence neuronal plasticity in this region. Other analyses also suggest a deleterious impact of this TE mobility in the brain insofar as it may contribute to age-related neuronal decline (Li et al. 2013). Furthermore, this higher TE activity in the brain appears to be linked to the emergence of several neurodegenerative diseases such as dry AMD geographic atrophy (Kaneko et al. 2011), X-fragile syndrome (Tan et al. 2012) and many other abnormalities (for recent reviews, see Ahmadi et al. 2020; Evans and Erwin 2020; Ochoa Thomas et al. 2020). That said, TE activity in the brain is still a hotly debated topic (Treiber and Waddell 2017).

9.2.2. *Interactions between groups of transposable elements within a genome*

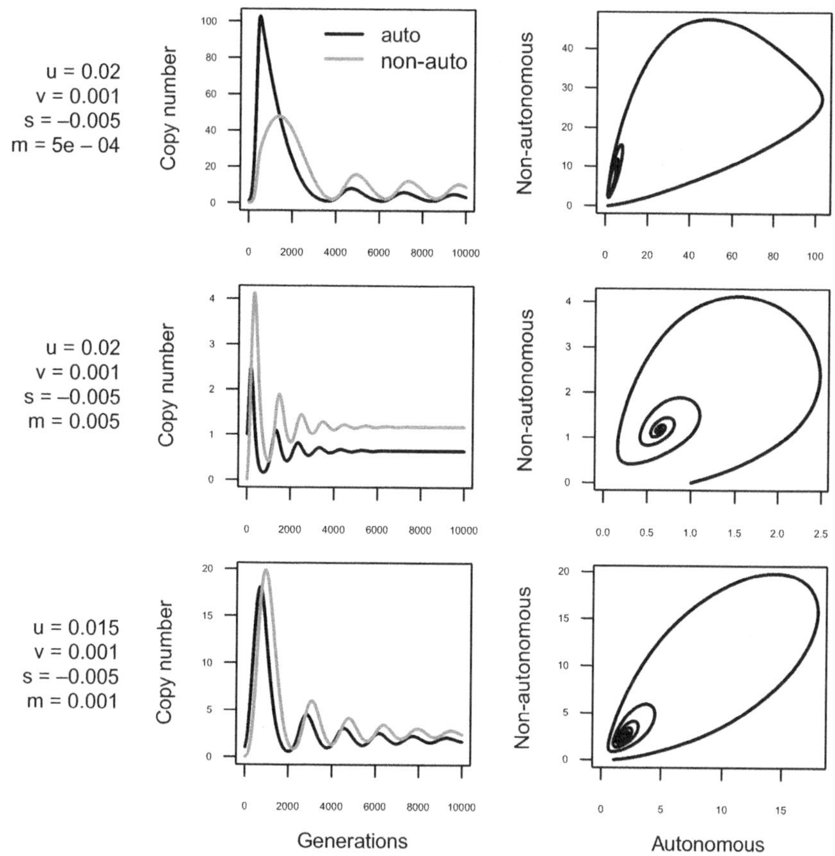

Figure 9.4. *Evolution of the number of autonomous and non-autonomous copies, according to the model described in Le Rouzic and Capy (2006). In this case, only autonomous copies provide the transposition machinery for their own mobility, as well as for that of non-autonomous copies. u = transposition rate; v = excision (loss) rate; s = impact of each copy on host fitness; m = probability of an autonomous copy to become non-autonomous*

As discussed repeatedly in this book, several types of TE copies from the same family with structural and functional differences can coexist within the same genome. In other words, not all copies present in a genome are necessarily active. If some copies are inactive and can never move again, this does not prevent them from having possible impacts on the expression profile of adjacent genes, or from being partially exapted to new functions. Other copies are non-autonomous but can be mobilized in *trans*. More precisely, a single non-autonomous copy in a genome cannot transpose, as it is unable to encode the factors required for its mobility. On the other hand, if there is another copy in the genome capable of providing it with these factors, it will be able to transpose.

In a genome, simulation of the dynamics of these two types of copies (autonomous and non-autonomous), taking into account their interactions, can reveal oscillations reminiscent of the Lotka–Volterra model in ecology (Lotka 1925; Volterra 1926; Le Rouzic and Capy 2006). Through this approach, they can appear depending on certain ranges of parameter values, such as the impact of copies on individual fitness, transposition, excision and mutation rates. Interactions between copies can be viewed as similar to those in the Lotka–Volterra competition model. In other words, the two types of copies "compete" for a single resource, namely the proteins required for their multiplication, such as transposase for *Class II* elements. Autonomous copies will decrease as these copies accumulate mutations that will eventually transform them from autonomous copies to non-autonomous copies (see Figure 9.4).

9.2.3. *Transposable element environment and activity*

The relationship between TEs and the environment has been suspected from the very beginning of their discovery. In her 1984 article, Barbara McClintock (1984) retraced the steps leading to the idea that TEs can be activated following a change in their environment, whether intra-cellular stresses or a wider environment, and whether the stresses are biotic or abiotic.

It is in this context that the relationships between environment and TE activity have been addressed by several teams, in both germline and somatic cells. The general idea behind this type of approach can be summarized as follows. Two phases can be considered: a short response phase, generally affecting the somatic lineage, and a long response phase affecting the germline. Indeed, when an organism is subjected to stress, there is initially a vital need to respond rapidly to this stress. To achieve this, defense genes may be activated, for example, by the release of transcriptional activators and/or epigenetic modifications. On the other hand, if the stress is maintained over several generations, more profound changes, particularly in the germ line, may occur (Capy et al. 2000). This can be illustrated by several studies, notably in *D. melanogaster* (Arnault et al. 1997; Garcia Guerreiro 2012; Rech et al. 2019), yeast (Esnault et al. 2019), fungal pathogens (Fouche et al. 2020), corn (Makarevitch et al. 2015) and tobacco (Mhiri et al. 1997; Grandbastien 1998).

In *N. tabacum* and tomato, *Tnt1* elements respond to biotic stresses, such as attack by pathogenic fungi, and abiotic stresses, such as mechanical injury (Grandbastien 1998). In the U3 region of the LTRs of this retrotransposon, it is possible to find several short motifs corresponding to transcription activators, which are also present upstream of plant defense genes (Grandbastien et al. 1997; Mhiri et al. 1997). From an evolutionary point of view, this raises the question of the origin of these motifs, that is, whether natural selection has retained the TEs that have recovered these motifs by inserting themselves close to the defense genes, or whether they are TE motifs that have been exapted by the defense genes. At present, it is difficult to answer this question without a detailed analysis of the dynamics of these motifs, especially as the two hypotheses are not completely exclusive.

The relationship between TEs and environmental conditions is often put forward to justify the long-term maintenance of TEs in genomes, insofar as they can be a source of genetic variability, the raw material necessary for all evolution (McDonald 1993; Capy et al. 1997; Rebollo et al. 2012). Over the last 10 years, there has been a sharp increase in the number of publications addressing this subject, particularly following detailed studies of the regulation of TE activity by epigenetic factors, whether cytosine methylations as in *Arabidopsis thaliana* (Bouyer et al. 2017), RNAs (*si*RNA, *pi*RNA) such as those described in *D. melanogaster* (Aravin et al. 2007; Brennecke et al. 2008; Cappucci et al. 2019; Mérel et al. 2020; Saint-Leandre et al. 2020), plants (McCue et al. 2012), or long noncoding RNAs (*Linc*RNA) again in plants (Wang et al. 2017) or chromatin states (Guio et al. 2014, 2018).

Today, even if not all TEs are sensitive to stresses, this TE/environment relationship is widely recognized, as evidenced by several recent reviews on the subject (Grandbastien 2015; Horváth et al. 2017; Lanciano and Mirouze 2018).

9.2.4. *Scales of structuring: from the copy to the ecosystem*

Natural populations are generally not continuous entities without barriers across their range. In many cases, they present a metapopulation structure, that is, they are made up of a set of sub-populations or demes, exchanging migrants. The diagram of a metapopulation is shown in Figure 9.5.

If migration rates between sub-populations are not too high, this type of structure is highly effective in maintaining genetic variability within a population or species. In fact, the evolutionary dynamics within each sub-population are independent of each other and can tend toward different genetic balances depending on the environmental specificities of each deme. Furthermore, if a sub-population dies out, the space left vacant can be recolonized by migrants from neighboring demes.

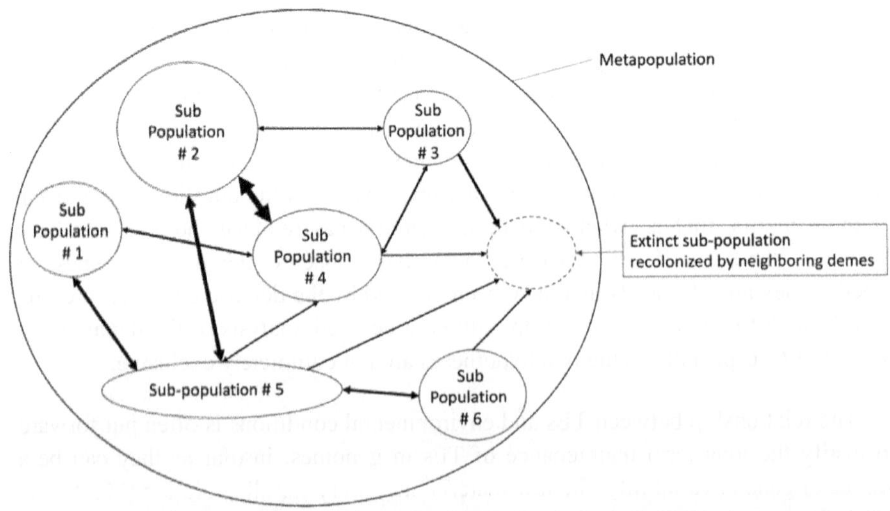

Figure 9.5. *Structuring of a metapopulation into interconnected subpopulations (demes). The thickness of the arrows indicates the intensity of migration between subpopulations. The empty niche left by an extinct subpopulation (dotted line) is generally recolonized by neighboring demes. For a color version of this figure, see www.iste.co.uk/huavan/transposable.zip*

In this context, from a population biology point of view, four interlocking levels can be defined:

– an individual belongs to a sub-population;

– a set of sub-populations belongs to a metapopulation;

– a set of metapopulations belongs to a species in an ecosystem;

– a set of species belongs to an ecosystem.

The transition to TEs simply requires the addition of a new level: that of the genome. If we wish to make a correspondence with population biology, we can assimilate, within a genome, each copy of a TE to an individual (see Box 9.1) and all of the copies of the same family in this genome to a sub-population. Each individual, in the ecological sense of the term, therefore represents a sub-population. TEs can multiply within an individual's genome and pass from one individual to another via genetic exchanges such as reproduction. These exchanges correspond to migrations in the metapopulation paradigm. Approaches of this type have been developed by several groups (Brookfield 2005a; Deceliere et al. 2005; Le Rouzic et al. 2007; Venner et al. 2007, 2009).

Ecology	**Genome**
An individual	A copy of TEs
A population	A set of copies of TEs from the same family
A community	A set of copies of different families of TEs
A species	A family or sub-family of TEs
Specific richness	The number of families of TEs
Relative abundance of species	Number of copies of each family of TEs
Birth rate	The transposition rate
Mortality rate	Deletion rate
An ecological niche	The set of parameters defining the interactions between a family of TEs and its genomic environment
Demographic stochasticity	Variation in copy number within families of TEs due to random variations in transposition and excision rates
The ecological drift	Random loss of TE copy diversity due to demographic stochasticity
Migration	Mixing of TE populations through intra-species crosses or horizontal transfers between different species
Speciation	Divergence of the DNA sequence of TEs from the reference sequence, leading to a new family of TEs
An ecosystem	A genome

Box 9.1. *Ecological community versus TE community analogies used in this chapter (Kidwell and Lisch 1997; Brookfield 2005a; Mauricio 2005; Le Rouzic et al. 2007; Venner et al. 2009)*

If the metapopulation model can be adapted to genomes between individuals, it can also be adapted to ecosystems. In this case, the paradigm's sub-populations are species, and the migrations are transfers of genetic material between species within an ecosystem. These horizontal transfers can take place in two main ways:

– On the one hand, when species are phylogenetically close, there may be transfers of TEs between these related species, if they still interbreed and give rise to fertile individuals. In this case, we are talking about introgression, of which several cases have been described between closely related species in the wild.

– On the other hand, when species are phylogenetically distant, the exchange of TEs between two species is generally the result of a horizontal transfer. This latter hypothesis can be advanced in the absence of any artifact masking a vertical transfer, as may be the case when the basic rules for constructing phylogenies are not respected (Capy et al. 1994). This being the case, while transfer mechanisms are well documented in prokaryotes (García-Aljaro et al. 2017), they are not known in eukaryotes, even though several recent publications illustrate this particularity of TEs being relatively easily transferred from one species to another without going through the sexual pathway (Loreto et al. 2008; Wallau et al. 2016; Gilbert and Feschotte 2018; de Melo and Wallau 2020; Gilbert et al. 2021). This point is presented in detail in Chapter 7.

This multi-layered structuring enables TEs not only to diffuse very rapidly into a species' natural populations and maintain themselves there, as will be explained later, but also to colonize species whose ranges overlap, or even share the same ecosystem or parasites. For example, the rapid spread of a recently arrived element in a naive species can be illustrated by the colonization of all *D. melanogaster* populations by the *P* element (Brégliano and Kidwell 1983; Kidwell 1983; Anxolabéhère et al. 1985; Good et al. 1989), which was transferred from *Drosophila willistoni* in the 1940s–1950s (Daniels et al. 1990). This invasion is not the only one in this species, and several transfers are documented for the elements *I*, *hobo*, and *tirant* (Bucheton et al. 1992; Bonnivard et al. 2000; Schwarz et al. 2021) or even for the *P* element in *D. simulans* (Kofler et al. 2015a) and more widely in many Drosophila species (Loreto et al. 2008; Wallau et al. 2016).

9.3. The "genome ecology" approach

The preceding chapters clearly show that TEs have had and continue to have a major influence on genetic diversity by inducing structural and functional modifications, the raw material for natural selection and genetic drift. They thus have a major impact on species adaptation, whether in terms of micro- or macro-evolution (Biémont and Vieira 2006; Biémont 2010). It is therefore essential to understand whether differences in the dynamics, structure and composition of TE communities between genomes and between TEs of the same genome reflect a form of organization or are the result of stochastic processes (transposition, excision, recombination, selection, drift, etc.). The existence of a large number of TE families and a large number of copies has led to draw an analogy with the vision of ecosystems in ecology. An ecosystem can be characterized as a network of interactions between different species in a given habitat. Each element in the network benefits from the ecosystem, but the ecosystem benefits from what each element, alone or in interaction, can contribute. Interactions are therefore very important, as they can have emergent properties that no single element could develop on its own. This may explain the resilience/robustness of certain ecosystems, or their fragility. For example, if an ecosystem element disappears, the ecosystem may collapse sooner or later, or adapt in various ways. This can range from replacing the missing element to reorganizing the network of interactions within the ecosystem. If we consider genomes as ecosystems, the cell is the habitat, and TEs are one of the components of the network of interactions within this cell. The genome ecosystem is made up of all of these components and their interactions. Box 9.1 summarizes the main analogies encountered in the literature and used to analyze TEs and their genomes as ecosystems.

While the modeling of TE dynamics within genomes has been the subject of numerous studies (see previous chapters), the approach based on our knowledge of an analogy between interactions between TEs within genomes and interactions between species within ecosystems is more recent. The first studies to introduce the term "genome ecology" (Kidwell and Lisch 1997) focused on two types of TEs, which could be competitive, cooperative or have a host/parasite relationship. Other ecology approaches could be applied to TEs and genomes (Song and Schaack 2018; Boxes 9.2 and 9.3) but analyses are recent and therefore still limited. In the following, we will confine ourselves to approaches based on ecological niche theory (Chase and Leibold 2003) and neutralist biodiversity theory (Hubell 2001), which have been the subject of much debate in ecology and for which we have the most significant experimental data.

9.3.1. *Ecological niche theory*

The ecological niche theory is based on the existence of interactions between species and with their environment (Chase and Leibold 2003). Species depend on resources, compete with each other and are subject to predators and/or parasites. Stable coexistence between competing species is possible if the niche is fragmented. Thus, the community of TEs presents a stable composition as long as they possess diverse characteristics enabling them to occupy different genomic niches and limit their competition, for example, for the metabolic components necessary for transposition, for the DNA sequences in which they can integrate or to resist the various host molecular mechanisms that control their expression

(RNAi, methylation, etc.). According to this theory, an increase in the diversity of genomic niches can lead to an increase in the number of TE families coexisting in the genome.

In genomes with a high proportion of TEs, particularly in plants, many copies of TEs are merely relics of old insertions, and their sequences are truncated, heavily modified and fragmentary. These relics interfere with the analysis of the evolutionary history of TEs and their interactions with the genome when all TE families are considered. Taking into account intact copies of a number of well-defined TE families, Stitzer et al. (2019) observed in maize a high diversity of evolutionary strategies (insertions far from genes, in areas with low recombination rates, in heterochromatic regions, in regions poor or rich in genes or rich in GC, etc.). Each of these families thus depends on the evolution of distinct ecological niches. These authors conclude that the different TE families coexist over time, compete for space and cooperate. These thoughts are illustrated by the work of Saylor et al. (2013) on the *Bos taurus* genome. Based on the analogy between the distribution of species along a transect and the distribution of TEs along chromosomes, these authors report the impact of gene density on the TE community.

9.3.2. *Neutralist theory of biodiversity*

The neutralist biodiversity theory presents a complementary, even alternative explanation to the ecological niche theory. It accounts for the establishment of a complex community of TEs based solely on random processes (see Linquist et al. 2013; Linquist et al. 2015; Arkhipova 2018). This theory is inspired by the neutralist vision of Kimura (1983) and the selfish DNA (Doolittle and Sapienza 1980; Orgel and Crick 1980) of population genetics. Thus, it is not the differences in the characteristics of TEs that explain the dynamics and structure of their community, but the existence of stochastic variations at the level of each TE copy. For example, according to Hubell (2001) a community of TEs (see Box 9.1) can be stable because of a balance between extinction and the emergence of new copies via immigration and speciation phenomena, or by horizontal transfers between species, as has been shown in the case of populations subdivided into interactions (metapopulations: Levins 1969; for TEs: see also Deceliere et al. 2005). It is important to note, however, that even though TEs represent a significant proportion of genomes, they share genomic space with other repeated elements such as mini- and micro-satellites, simple repeats, *t*RNAs, miRNAs, *sn*RNAs, *sc*RNAs, etc. and of course genes. It is the whole genome that is the ecosystem, not just the portion of the genome represented by TEs alone. Thus, Serra et al. (2013) have concluded that the neutralist theory accounts for most of the genomic data concerning the relative abundance and diversity of the set of repeated elements in a large number of eukaryotic genomes.

One of the criticisms to the TE distribution analyses is that their taxonomic relationship is too limited. By analyzing 42 nematode genomes spanning around 500 million years, Szitenberg et al. (2016) conclude that there is no effect of life history, GC content or variation in processes controlling the expression of TEs by RNA*i*. For these authors, stochastic processes appear to be the major determinants of TE diversity.

Linquist et al. (2013, 2015) have stressed the need to distinguish between short-term dynamics studies, which really concern community ecology approaches (which they call "transposon ecology"), and "evolutionary" studies using population genetics methods (which they call "genome ecology" (see Figure 9.6)). Genome ecology then specifically concerns the impact of external ecological factors, both cellular and external to the organism, on the genome. The ecology of TEs thus concerned how ecological factors expressed within the genome affect the abundance and distribution of TEs (see Box 9.2 for a presentation of some ecological theories that can be used on TEs and are not based on evolutionary processes). By analogy with population genetics studies that estimate the proportion of phenotypic variance explained by genetic and environmental factors, these authors assessed the proportion of variance in the abundance and distribution of TEs due to ecological and evolutionary factors on genomic data from a set of phylogenetically distant mammals and a group of closely related arthropods. Variation at the level of TE lineages (not families) is explained by ecological factors (genome size, GC-poor or GC-rich regions) in closely related taxa, while evolutionary factors (phylogenetic distance between genomes) explain a high proportion of variation at family level among the most distantly related taxa. Similarly, using a simulation in which TEs are deleterious, do not evolve over generations (no recombination, no coevolution) and are not subject to horizontal transfer, Kremer et al. (2020) observe that although a large proportion of TEs are eliminated from genomes and their host population may become extinct, in some cases, TEs not only persist but accumulate in the genome. However, such accumulation is only possible under certain conditions, notably low aggressiveness of the TEs. This is an idea put forward by Doolittle et al. (1984) and illustrated by Le Rouzic and Capy (2005). This invasion and the maintenance of a stable density are made possible by the fact that TEs are able to produce new inactive copies into which they can insert themselves and are then no longer subject to selection. These new copies represent a new habitat for TEs, resulting in an increase in genome size. This process is reminiscent of the observation in ecology that species can maintain themselves in an ecosystem if they generate their own habitat, a process known as "ecosystem engineering" (Jones et al.

1994; Hastings et al. 2007; Pearce 2011). On the other hand, interactions between active (autonomous) and inactive (non-autonomous) copies can also lead to super-parasitism: non-autonomous copies parasitizing autonomous copies (Robillard et al. 2016).

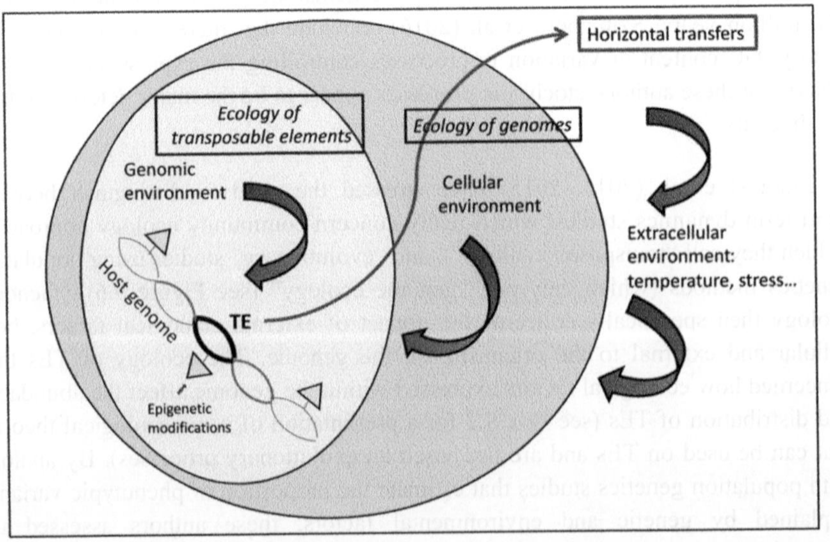

Figure 9.6. *All of the interactions undergone by TEs in their nuclear, cellular and external environments. The distribution of TEs along chromosomes depends on a variety of factors: transposition and excision rates, recombination rates, GC/AT density, chromatin opening or closing, presence of repeated sequences, etc. and their expression is regulated by histones, methylations, acetylations, small RNAs, etc. TEs are thus dependent on modifications of both their cellular environment and the environment of the host organism, which act via epigenetic modifications of the DNA molecule. Together, these interactions lead to the host organism's phenotype, behavior, susceptibility to certain diseases and even certain neuronal functions. For the complex interactions between TEs and their cellular and extracellular environments (Capy et al. 2000; Bourque et al. 2018; Saylor et al. 2019). For a color version of this figure, see www.iste.co.uk/huavan/transposable.zip*

Taken together, these experiments lead to the conclusion that "neutral" processes such as transposition, excision and dispersal etc. are sufficient to explain the diversity and abundance of species (and therefore of TEs) within their ecosystems, without appealing to species-specific characteristics. But a neutral global model does not necessarily imply that the processes behind it are all neutral, and we know of many examples involving selection against the harmful effects of certain TE insertions, or selection for advantageous mutations.

Functional redundancy: High biodiversity contributes to the persistence and stability of ecosystems.

Keystone species: A species that has a disproportionate effect on its environment in relation to its abundance. It can be, but does not have to be, a predator.

Lotka–Volterra model: Stable coexistence between two species is possible if intra-specific competition has a greater effect than inter-specific competition.

Island biogeography theory: Species diversity on an island is positively related to island size, but negatively related to the island's distance from the mainland.

Ecological niche theory: Stable coexistence between competing species is possible if the niche is fragmented.

Neutralist theory of biodiversity: Species diversity, distribution and relative abundance can be explained by the stochasticity of demographic processes (mortality rates, birth rates, immigration).

Box 9.2. *Example of ecological models and theories of interest to genomes and TEs. Adapted from Song and Schaack (2018)*

Arms race (Red Queen dynamics): Coevolution between pairs of antagonistic species (host-parasites).

Evolution of virulence: Reduced host fitness following infection; trade-off between parasite virulence and transmission under the influence of multiple infections.

Evolutionary stable strategy: A strategy maintained by natural selection which, once adopted, cannot be dethroned by any alternative strategy.

Exaptation, domestication: The emergence of new functions through co-optation, leading to adaptation.

Game theory and the evolution of cooperation: Evolutionary strategy illustrated by the prisoner's dilemma, where two players cooperate in the absence of communication.

Neutralist theory: Mostly neutral mutations with negligible influence on selective value. Importance of genetic drift.

Punctuated equilibria: Long periods of equilibrium punctuated by periods of rapid species evolution.

Box 9.3. *Examples of evolutionary models and theories of interest to genome and TE ecology. Adapted from Song and Schaack (2018)*

9.3.3. *Comments*

Recent developments in ecological approaches consider that the two theories of niche and stochasticity are not exclusive but can be complementary. Through simulations and analytical considerations, Fisher and Mehta (2014) conclude that the niche theory is favored in communities with large population sizes and relatively constant environments, while the neutralist theory is favored in small populations with fluctuating environments. Thus, we would move from one phase to the other according to the relative importance of ecological selection and drift. Working on experimental populations of *C. elegans* of different sizes Bergthorsson et al. (2020) have shown that natural selection is responsible for the low level of TE expression and that this expression increases in relation to the size of the ancestral population, the increase being strongest in small populations. This is reminiscent of the work of Lynch and Conery (2003), which concludes that genome complexity emerged in the course of evolution following the reduction in population size.

The distinction between the ecology of genomes and the ecology of TEs should make it easier to answer the issues raised by distinguishing between ecological effects and evolutionary processes. However, it should not be forgotten that species within their ecosystems and TEs within their genomes are subject to factors in the cellular and extra-cellular environments that can exert considerable effects on their behavior, modifying their interactions and disrupting population regulation (Figure 9.6). For example, genomic stresses can lead to TE transposition explosions that can generate new genetic diversity and promote colonization of new territories (Vieira et al. 1999; Stapley et al. 2015; Specchia and Pinna 2017), rapid adaptation and even speciation (Capy et al. 2000; Belyayev 2014; Kofler et al. 2015; Ricci et al. 2018; Chen et al. 2020).

9.4. Conclusion

The community of TEs differs from the ecological community in that selection acts directly on the individuals carrying the TEs, that is, on the genome-TE ecosystem itself. It is the set of TEs and other repeated sequences that, in addition to genes, forms the genome ecosystem. The presence of TEs in a genome enables the ecosystem's plasticity, that is, its evolution or extinction, in the knowledge that it is the ecosystem as a whole that is selected via the phenotypes induced by its components, notably by the mobility of its TEs. The distinction between the ecology of genomes and the ecology of TEs, which makes it possible to estimate the respective shares of the purely ecological effects of evolutionary processes, is necessary, but it is important not to forget that genomes can be subjected to drastic changes in the conditions of their external environment, which are likely to have a considerable impact on the interactions between their components.

An ecological approach to the genome will require the development of models that take these specificities into account (Boxes 9.2 and 9.3). It is expected that these studies will provide a new, integrative vision of how genomes function, and in particular of their part in TEs, for which recent studies show their capacity to create innovations important for adaptation (Quadrana et al. 2019), plant diversity (Alseekh et al. 2020), diversity in fish (Carleton et al. 2020), species evolution (Platt et al. 2018; Gilbert et al. 2021; Gozashti et al. 2023) and selected lines in agronomy (Ahmadi et al. 2020), but also for the appearance of deleterious mutations and diseases from which the human species is not exempt. Thus, Hallin et al. (2020), in seeking analogies between different levels from molecular to ecological, conclude that biological processes at different levels of organization are governed by common rules. It is to be hoped that taking into account the full range of possible reactions of genomes to their changing environment (Saylor et al. 2019) will lead to a more global and comprehensible vision of the functioning of our own genome, within which even a virus can disrupt the transcription of TEs (Roy et al. 2020). We know that in the face of environmental disturbances, ecosystems can suddenly degrade or even disappear. What then of genomic ecosystems, subjected to disturbances that may alter gene regulation via network systems (Feschotte 2008; Mita and Boeke 2016; Venner et al. 2017) involving the mobilization of TEs, the effects of which are even felt in our brains? New studies will require an integrative field approach, evolutionary and ecological genomic analyses, as well as physiological and functional approaches. Understanding how organisms interact with their environment to produce phenotypic variation is a recurring question in biology. The ecological genomics approach currently under development should open up new perspectives in this field.

9.5. References

Ahmadi, A., De Toma, I., Vilor-Tejedor, N., Eftekhariyan Ghamsari, M.R., Sadeghi, I. (2020). Transposable elements in brain health and disease. *Ageing Research Reviews*, 64, 101153.

Alseekh, S., Scossa, F., Fernie, A.R. (2020). Mobile transposable elements shape plant genome diversity. *Trends in Plant Science*, 25(11), 1062–1064.

Anxolabéhère, D., Nouaud, D., Periquet, G., Tchen, T. (1985). *P* element distribution in Eurasian populations of *Drosophila melanogaster*. *Proceedings of the National Academy of Sciences of America*, 82, 5418–5422.

Aravin, A.A., Hannon, G.J., Brennecke, J. (2007). The Piwi-piRNA pathway provides an adaptive defense in the transposon arms race. *Science*, 318, 761–764.

Arkhipova, I.R. (2018). Neutral theory, transposable elements, and eukaryotic genome evolution. *Molecular Biology and Evolution*, 35(6), 1332–1337.

Arnault, C., Loevenbruck, C., Biémont, C. (1997). Transposable element mobilization is not induced by heat shocks in *Drosophila melanogaster*. *Naturwissenschaften*, 84(9), 410–414.

Barron, M.G., Fiston-Lavier, A.S., Petrov, D.A., Gonzalez, J. (2014). Population genomics of transposable elements in *Drosophila*. *Annual Review of Genetics*, 48, 561–581.

Belyayev, A. (2014). Bursts of transposable elements as an evolutionary driving force. *Journal of Evolutionary Biology*, 27(12), 2573–2584.

Bergthorsson, U., Sheeba, C.J., Konrad, A., Belicard, T., Beltran, T., Katju, V., Sarkies, P. (2020). Long-term experimental evolution reveals purifying selection on piRNA-mediated control of transposable element expression. *BMC Biology*, 18(1), 162.

Biémont, C. (2010). A brief history of the status of transposable elements: From junk DNA to major players in evolution. *Genetics*, 186, 1085–1093.

Biémont, C. and Vieira, C. (2006). Junk DNA as an evolutionary force. *Nature*, 443, 521–524.

Bohme, U., Otto, T.D., Cotton, J.A., Steinbiss, S., Sanders, M., Oyola, S.O., Nicot, A., Gandon, S., Patra, K.P., Herd, C. et al. (2018). Complete avian malaria parasite genomes reveal features associated with lineage-specific evolution in birds and mammals. *Genome Research*, 28(4), 547–560.

Bonnivard, E., Bazin, C., Denis, B., Higuet, D. (2000). A scenario for the hobo transposable element invasion, deduced from the structure of natural populations of *Drosophila melanogaster* using tandem TPE repeats. *Genetical Research*, 75, 13–23.

Bourgeois, Y. and Boissinot, S. (2019). On the population dynamics of junk: A review on the population genomics of transposable elements. *Genes*, 10(6), 419.

Bourque, G., Burns, K.H., Gehring, M., Gorbunova, V., Seluanov, A., Hammell, M., Imbeault, M., Izsvak, Z., Levin, H.L., Macfarlan, T.S. et al. (2018). Ten things you should know about transposable elements. *Genome Biology*, 19, 199.

Bouyer, D., Kramdi, A., Kassam, M., Heese, M., Schnittger, A., Roudier, F., Colot, V. (2017). DNA methylation dynamics during early plant life. *Genome Biology*, 18, 179.

Brégliano, J.C. and Kidwell. M.G. (1983). Hybrid dysgenesis determinants. In *Mobile Genetic Elements*, Shapiro, J. (ed.). Academic Press, London.

Brennecke, J., Malone, C.D., Aravin, A.A., Sachidanandam, R., Stark, A., Hannon, G.J. (2008). An epigenetic role for maternally inherited piRNAs in transposon silencing. *Science*, 322, 1387–1392.

Brookfield, J.F.Y. (2005a). The ecology of the genome – Mobile DNA elements and their hosts. *Nature Review Genetics*, 6, 128–136.

Brookfield, J.F.Y. (2005b). Evolutionary forces generating sequence homogeneity and heterogeneity within retrotransposon families. *Cytogenetic and Genome Research*, 110, 383–391.

Bucheton, A. (1990). I transposable elements and I-R hybrid dysgenesis in *Drosophila*. *Trends Genetics*, 6, 16–21.

Bucheton, A., Vaury, C., Chaboissier, M.C., Abad, P., Pélisson, A., Simonelig, M. (1992). *I* elements and the *Drosophila* genome. *Genetica*, 86, 175–190.

Cappucci, U., Noro, F., Casale, A.M., Fanti, L., Berloco, M., Alagia, A.A., Grassi, L., Le Pera, L., Piacentini, L., Pimpinelli, S. (2019). The Hsp70 chaperone is a major player in stress-induced transposable element activation. *Proceedings of the National Academy of Sciences of America*, 116(36), 17943–17950.

Capy, P., Anxolabéhère, D., Langin, T. (1994). The strange phylogenies of transposable elements: Are horizontal transfers the only explantation? *Trends Genet.*, 10, 7–12.

Capy, P., Bazin, C., Higuet, D., Langin, T. (1997). *Evolution and Impact of Transposable Elements*. Kluwer Academic Publishers, Dordrecht.

Capy, P., Gasperi, G., Biémont, C., Bazin, C. (2000). Stress and transposable elements: Co-evolution or useful parasites? *Heredity*, 85, 101–106.

Carleton, K.L., Conte, M.A., Malinsky, M., Nandamuri, S.P., Sandkam, B.A., Meier, J.I., Mwaiko, S., Seehausen, O., Kocher, T.D. (2020). Movement of transposable elements contributes to cichlid diversity. *Molecular Ecology*, 29(24), 4956–4969.

Charlesworth, B. and Barton, N. (2004). Genome size: Does bigger mean worse? *Current Biology*, 14(6), 233–235.

Chase, J.M. and Leibold, M.A. (2003). *Ecological Niches: Linking Classical and Contemporary Approaches*. Chicago University Press, Chicago.

Chen, J., Lu, L., Robb, S.M.C., Collin, M., Okumoto, Y., Stajich, J.E., Wessler, S.R. (2020). Genomic diversity generated by a transposable element burst in a rice recombinant inbred population. *Proceedings of the National Academy of Sciences of the United States of America*, 117(42), 26288–26297.

Daborn, P.J., Yen, J.L., Bogwitz, M.R., Le Goff, G., Feil, E., Jeffers, S., Tijet, N., Perry, T., Heckel, D., Batterham, P. et al. (2002). A single *P450* allele associated with insecticide resistance in *Drosophila*. *Science*, 297, 2253–2256.

Daniels, S.B., Peterson, K.R., Strausbaugh, L.D., Kidwell, M.G., Chovnick, A. (1990). Evidence for horizontal transmission of the P transposable element between Drosophila species. *Genetics*, 124, 339–355.

Daubin, V. (2004). Comment on "The origins of genome complexity." *Science*, 306(5698), 978.

DeBarry, J.D., Ganko, E.W., McCarthy, E.M., McDonald, J.F. (2006). The contribution of LTR retrotransposon sequences to gene evolution in Mus musculus. *Molecular Biology and Evolution*, 23(3), 479–481.

Deceliere, G., Charles, S., Biémont, C. (2005). The dynamics of transposable elements in structured populations. *Genetics*, 169(1), 467–474.

Doolittle, W.F. and Sapienza, C. (1980). Selfish genes, the phenotype paradigm and genome evolution. *Nature*, 284, 601–603.

Doolittle, W.F., Kirkwood, T., Dempster, M. (1984). Selfish DNAs with self-restraint. *Nature*, 307(5951), 501–502.

Duc, C., Yoth, M., Jensen, S., Mouniée, N., Bergman, C.M., Vaury, C., Brasset, E. (2019). Trapping a somatic endogenous retrovirus into a germline piRNA cluster immunizes the germline against further invasion. *Genome Biology*, 20(1), 127.

Durand, P.M., Oelofse, A.J., Coetzer, T.L. (2006). An analysis of mobile genetic elements in three Plasmodium species and their potential impact on the nucleotide composition of the *P. falciparum* genome. *BMC Genomics*, 7, 282.

Engels, W.R. (1983). The P family of transposable elements in *Drosophila*. *Annual Review of Genetics*, 17, 315–344.

Esnault, C., Lee, M., Ham, C., Levin, H.L. (2019). Transposable element insertions in fission yeast drive adaptation to environmental stress. *Genome Research*, 29(1), 85–95.

Evans, T.A. and Erwin, J.A. (2021). Retroelement-derived RNA and its role in the brain. *Seminars in Cell Developmental Biology*, 114, 68–80.

Feschotte, C. (2008). Transposable elements and the evolution of regulatory networks. *Nature Reviews Genetics*, 9(5), 397–405.

Fisher, C.K. and Mehta, P. (2014). The transition between the niche and neutral regimes in ecology. *Proceedings of the National Academy of Sciences of the United States of America*, 111(36), 13111–13116.

Fouche, S., Badet, T., Oggenfuss, U., Plissonneau, C., Francisco, C.S., Croll, D. (2020). Stress-driven transposable element de-repression dynamics and virulence evolution in a fungal pathogen. *Molecular Biology and Evolution*, 37(1), 221–239.

García-Aljaro, C., Ballesté, E., Muniesa, M. (2017). Beyond the canonical strategies of horizontal gene transfer in prokaryotes. *Current Opinion in Microbiology*, 38, 95–105.

Garcia Guerreiro, M.P. (2012). What makes transposable elements move in the *Drosophila* genome? *Heredity*, 108(5), 461–468.

Gardner, M.J., Hall, N., Fung, E., White, O., Berriman, M., Hyman, R.W., Carlton, J.M., Pain, A., Nelson, K.E., Bowman, S. et al. (2002). Genome sequence of the human malaria parasite *Plasmodium falciparum*. *Nature*, 419(6906), 498–511.

Gilbert, C. and Feschotte, C. (2018). Horizontal acquisition of transposable elements and viral sequences: Patterns and consequences. *Current Opinion in Genetics and Development*, 49, 15–24.

Gilbert, C., Peccoud, J., Cordaux, R. (2021). Transposable elements and the evolution of insects. *Annual Review of Entomology*, 66, 355–372.

Gladyshev, E.A., Meselson, M., Arkhipova, I.R. (2008). Massive horizontal gene transfer in bdelloid rotifers. *Science*, 320, 1210–1213.

Gonzalez, J., Karasov, T.L., Messer, P.W., Petrov, D.A. (2010). Genome-wide patterns of adaptation to temperate environments associated with transposable elements in *Drosophila. PloS Genetics*, 6(4), e1000905.

Good, A.G., Meister, G.A., Brock, H.W., Grigliatti, T.A., Hickey, D.A. (1989). Rapid spread of transposable P elements in experimental populations of *Drosophila melanogaster. Genetics*, 122, 387–396.

Gozashti, L., Feschotte, C., Hoekstra, H.E. (2023) Transposable element interactions shape the ecology of the Deer Mouse genome. *Molecular Biology and Evolution*, 40(4), msad069.

Grandbastien, M.A. (1998). Activation of plant retrotransposons under stress conditions. *Trends in Plant Science*, 3(5), 181–187.

Grandbastien, M.A. (2015). LTR retrotransposons, handy hitchhikers of plant regulation and stress response. *Biochimica et Biophysica Acta – Gene Regulatory Mechanisms*, 1849(4), 403–416.

Grandbastien, M.A., Lucas, H., Morel, J.B., Mhiri, C., Vernhettes, S., Casacuberta, J.M. (1997). The expression of the tobacco Tnt1 retrotransposon is linked to plant defense responses. *Genetica*, 100(1–3), 241–252.

Guio, L., Barron, M.G., Gonzalez, J. (2014). The transposable element Bari-Jheh mediates oxidative stress response in Drosophila. *Molecular Ecology*, 23(8), 2020–2030.

Guio, L., Vieira, C., González, J. (2018). Stress affects the epigenetic marks added by natural transposable element insertions in *Drosophila melanogaster. Scientific Reports*, 8(1), 12197.

Haig, D. (2016). Transposable elements: Self-seekers of the germline, team-players of the soma. *Bioessays*, 38(11), 1158–1166.

Hallin, J., Cisneros, A.F., Henault, M., Fijarczyk, A., Dandage, R., Bautista, C., Landry, C.R. (2020). Similarities in biological processes can be used to bridge ecology and molecular biology. *Evolutionary Applications*, 13(6), 1335–1350.

Hastings, A., Byers, J.E., Crooks, J.A., Cuddington, K., Jones, C.G., Lambrinos, J.G., Talley, T.S., Wilson, W.G. (2007). Ecosystem engineering in space and time. *Ecology Letters*, 10(2), 153–164.

Horváth, V., Merenciano, M., González, J. (2017). Revisiting the relationship between transposable elements and the eukaryotic stress response. *Trends in Genetics*, 33(11), 832–841.

Hubbell, S.P. (2001). *The Unified Neutral Theory of Biodiversity and Biogeography*. Princeton University Press, Princeton.

Jones, C., Lawton, J., Shachak, M. (1994). Organisms as ecosystem engineers. *Oikos*, 69(3), 373–386.

Kaneko, H., Dridi, S., Tarallo, V., Gelfand, B.D., Fowler, B.J., Cho, W.G., Kleinman, M.E., Ponicsan, S.L., Hauswirth, W.W., Karikó, K. et al. (2011). *DICER*1 deficit induces Alu RNA toxicity in age-related macular degeneration. *Nature*, 471(7338), 325–330.

Kidwell, M.G. (1983). Evolution of hybrid dysgenesis determinants in *Drosophila melanogaster*. *Proceedings of the National Academy of Sciences of the United States of America*, 80, 1655–1659.

Kidwell, M.G. and Lisch, D. (1997). Transposable elements as sources of variation in animals and plants. *Proceedings of the National Academy of Sciences of the United States of America*, 94, 7704–7711.

Kimura, M. (1983). *The Neutral Theory of Molecular Evolution*. Cambridge University Press, New York.

Kimura, M. and Crow, J. (1963). Measurement of effective population number. *Evolution*, 17(3), 279–288.

Kimura, M. and Crow, J. (1964). Number of alleles that can be maintained in finite population. *Genetics*, 49(4), 725–738.

Kofler, R., Hill, T., Nolte, V., Betancourt, A.J., Schloetterer, C. (2015a). The recent invasion of natural *Drosophila simulans* populations by the P-element. *Proceedings of the National Academy of Sciences of the United States of America*, 112, 6659–6663.

Kofler, R., Nolte, V., Schloetterer, C. (2015b). Tempo and mode of transposable element activity in *Drosophila*. *PloS Genetics*, 11(7), e1005406.

Kremer, S.C., Linquist, S., Saylor, B., Elliott, T.A., Gregory, T.R., Cottenie, K. (2020). Transposable element persistence via potential genome-level ecosystem engineering. *BMC Genomics*, 21(1), 367.

Lanciano, S. and Mirouze, M. (2018). Transposable elements: All mobile, all different, some stress responsive, some adaptive? *Current Opinion in Genetics and Development*, 49, 106–114.

Le Rouzic, A. and Capy, P. (2005). The first steps of transposable elements invasion: Parasitic strategy vs. genetic drift. *Genetics*, 169, 1033–1043.

Le Rouzic, A. and Capy, P. (2006). Population genetics models of competition between transposable element subfamilies. *Genetics*, 174, 785–793.

Le Rouzic, A., Dupas, S., Capy, P. (2007). Genome ecosystem and transposable elements species. *Gene*, 390, 214–220.

Lefébure, T., Morvan, C., Malard, F., François, C., Konecny-Dupré, L., Guéguen, L., Weiss-Gayet, M., Seguin-Orlando, A., Ermini, L. Sarkissian, C.D. et al. (2017). Less effective selection leads to larger genomes. *Genome Research*, 27(6), 1016–1028.

Levins, R. (1969). Some demographic and genetic consequences of environmental heterogeneity for biological control. *American Entomologist*, 15, 237–240.

Li, W., Prazak, L., Chatterjee, N., Grüninger, S., Krug, L., Theodorou, D., Dubnau, J. (2013). Activation of transposable elements during aging and neuronal decline in *Drosophila. Nature Neuroscience*, 16(5), 529–531.

Linquist, S., Saylor, B., Cottenie, K., Elliott, T.A., Kremer, S.C., Gregory, T.R. (2013). Distinguishing ecological from evolutionary approaches to transposable elements. *Biological Reviews*, 88(3), 573–584.

Linquist, S., Cottenie, K., Elliott, T.A., Saylor, B., Kremer, S.C., Gregory, T.R. (2015). Applying ecological models to communities of genetic elements: The case of neutral theory. *Molecular Ecology*, 24(13), 3232–3242.

Lokta, A.J. (1925). *Elements of Physical Biology*. Williams Wilkins, Baltimore.

Loreto, E.L.S., Carareto, C.M.A., Capy, P. (2008). Revisiting horizontal transfer of transposable elements in *Drosophila. Heredity*, 100(6), 545–554.

Lynch, M. (2004). Response to comment on "The origins of genome complexity". *Science*, 306(5698), 978b.

Lynch, M. and Conery, J.S. (2003). The origins of genome complexity. *Science*, 302(5649), 1401–1404.

Makarevitch, I., Waters, A.J., West, P.T., Stitzer, M., Hirsch, C.N., Ross-Ibarra, J., Springer, N.M. (2015). Transposable elements contribute to activation of maize genes in response to abiotic stress. *PLoS Genetics*, 11(1).

Mauricio, R. (2005). Can ecology help genomics: The genome as ecosystem? *Genetica*, 123(1–2), 205–209.

McClintock, B. (1984). The significance of responses of the genome to challenge. *Science*, 226, 792–801.

McCue, A.D., Nuthikattu, S., Reeder, S.H., Slotkin, R.K. (2012). Gene expression and stress response mediated by the epigenetic regulation of a transposable element small RNA. *PLoS Genetics*, 8(2), 18.

McDonald, J.F. (1993). *Transposable Elements and Evolution*. Kluwer Academic Publishers, Dordrecht.

de Melo, E.S. and Wallau, G.L. (2020). Mosquito genomes are frequently invaded by transposable elements through horizontal transfer. *PLoS Genetics*, 16(11), e1008946.

Mérel, V., Boulesteix, M., Fablet, M., Vieira, C. (2020). Transposable elements in *Drosophila*. *Mobile DNA*, 11(1), 23.

Mérel, V., Gibert, P., Buch, I., Rodriguez Rada, V., Estoup, A., Gautier, M., Fablet, M., Boulesteix, M., Vieira, C. (2021). The worldwide invasion of *Drosophila suzukii* is accompanied by a large increase of transposable element load and a small number of putatively adaptive insertions. *Molecular Biology and Evolution*, 38(10), 4252–4267.

Mhiri, C., Morel, J.B., Vernhettes, S., Casacuberta, J.M., Lucas, H., Grandbastien, M.A. (1997). The promoter of the tobacco Tnt1 retrotransposon is induced by wounding and by abiotic stress. *Plant Molecular Biology*, 33(2), 257–266.

Mita, P. and Boeke, J.D. (2016). How retrotransposons shape genome regulation. *Current Opinion in Genetics and Development*, 37, 90–100.

Muller, S., Raman Pandey, R., Pillai, R.S. (2013). Les piARN forgent un système immunitaire pour le génome. *Médecine/Sciences*, 29(5), 487–494.

Ochoa Thomas, E., Zuniga, G., Sun, W., Frost, B. (2020). Awakening the dark side: Retrotransposon activation in neurodegenerative disorders. *Current Opinion in Neurobiology*, 61, 65–72.

Orgel, L.E. and Crick, F.H.C. (1980). Selfish DNA: The ultimate parasite. *Nature*, 284, 604–607.

Pasquesi, G.I.M., Perry, B.W., Vandewege, M.W., Ruggiero, R.P., Schield, D.R., Castoe, T.A. (2020). Vertebrate lineages exhibit diverse patterns of transposable element regulation and expression across tissues. *Genome Biology and Evolution*, 12(5), 506–521.

Pearce, T. (2011). Ecosystem engineering, experiment, and evolution. *Biology Philosophy*, 26(6), 793–812.

Perrat, P.N., DasGupta, S., Wang, J., Theurkauf, W., Weng, Z., Rosbash, M., Waddell, S. (2013). Transposition-driven genomic heterogeneity in the *Drosophila* brain. *Science*, 340(6128), 91–95.

Platt, R.N., Vandewege, M.W., Ray, D.A. (2018). Mammalian transposable elements and their impacts on genome evolution. *Chromosome Research*, 26(1–2), 25–43.

Quadrana, L., Etcheverry, M., Gilly, A., Caillieux, E., Madoui, M.A., Guy, J., Silveira, A.B., Engelen, S., Baillet, V., Wincker, P. et al. (2019). Transposition favors the generation of large effect mutations that may facilitate rapid adaption. *Nature Communications*, 10, 3421.

Rebollo, R., Romanish, M.T., Mager, D.L. (2012). Transposable elements: An abundant and natural source of regulatory sequences for host genes. *Annual Review of Genetics*, 46, 21–42.

Rech, G.E., Bogaerts-Marquez, M., Barron, M.G., Merenciano, M., Luis Villanueva-Canas, J., Horvath, V., Fiston-Lavier, A.S., Luyten, I., Venkataram, S., Quesneville, H. et al. (2019). Stress response, behavior, and development are shaped by transposable element-induced mutations in Drosophila. *PLoS Genetics*, 15(2), e1007900.

Ricci, M., Peona, V., Guichard, E., Taccioli, C., Boattini, A. (2018). Transposable elements activity is positively related to rate of speciation in mammals. *Journal of Molecular Evolution*, 86(5), 303–310.

Robillard, E., Le Rouzic, A., Zhang, Z., Capy, P., Hua-Van, A. (2016). Experimental evolution reveals hyperparasitic interactions among transposable elements. *Proceedings of the National Academy of Sciencesof the United States of America*, 113(51), 14763–14768.

Roy, M., Viginier, B., Saint-Michel, E., Arnaud, F., Ratinier, M., Fablet, M. (2020). Viral infection impacts transposable element transcript amounts in *Drosophila*. *Proceedings of the National Academy of Sciences of the United States of America*, 117(22), 12249–12257.

Saint-Leandre, B., Clavereau, I., Hua-Van, A., Capy, P. (2017). Transcriptional polymorphism of *pi* RNA regulatory genes underlies the *mariner* activity in *Drosophila simulans* testes. *Molecular Ecology*, 26(14), 3715–3731.

Saint-Leandre, B., Capy, P., Hua-Van, A., Filee, J. (2020). piRNA and transposon dynamics in *Drosophila*: A female story. *Genome Biology and Evolution*, 12(6), 931–947.

Saylor, B., Elliott, T.A., Linquist, S., Kremer, S.C., Gregory, T.R., Cottenie, K. (2013). A novel application of ecological analyses to assess transposable element distributions in the genome of the domestic cow. *Bos taurus*. *Genome*, 56(9), 521–533.

Saylor, B., Kremer, S.C., Gregory, T.R., Cottenie, K. (2019). Genomic environments and their influence on transposable element communities [Preprint]. *bioRxiv*. doi: 10.1101/667121.

Schwarz, F., Wierzbicki, F., Senti, K.A., Kofler, R. (2021). Tirant stealthily invaded natural *Drosophila melanogaster* populations during the last century. *Molecular Biology and Evolution*, 38(4), 1482–1497.

Serra, F., Becher, V., Dopazo, H. (2013). Neutral theory predicts the relative abundance and diversity of genetic elements in a broad array of eukaryotic genomes. *PLoS ONE*, 8(6), e63915.

Sessegolo, C., Burlet, N., Haudry, A. (2016). Strong phylogenetic inertia on genome size and transposable element content among 26 species of flies. *Biology Letters*, 12(8), 20160407.

Sfera, A., Fayard, L., Osorio, C., Price, A. (2018). Epigenetic interventions for brain rejuvenation: Anchoring age-related transposons. *Neural Regeneration Research*, 13(4), 635.

Song, M.J. and Schaack, S. (2018). Evolutionary conflict between mobile DNA and host genomes. *American Naturalist*, 192(2), 263–273.

Specchia, V. and Pinna, M. (2017). The potential link between mobile DNA and the invasiveness of the species. *Journal of RNAi and Gene Silencing*, 13(1), 557–561.

Stapley, J., Santure, A.W., Dennis, S.R. (2015). Transposable elements as agents of rapid adaptation may explain the genetic paradox of invasive species. *Molecular Ecology*, 24(9), 2241–2252.

Stitzer, M.C., Anderson, S.N., Springer, N.M., Ross-Ibarra, J. (2021). The genomic ecosystem of transposable elements in maize. *PLoS Genetics*, 17(10), e1009768.

Sukthaworn, S., Panyim, S., Udomkit, A. (2020). Homologues of Piwi control transposable elements and development of male germline in *Penaeus monodon*. *Comparative Biochemistry and Physiology A Molecular Integrative Physiology*, 250, 110807.

Szitenberg, A., Cha, S., Opperman, C.H., Bird, D.M., Blaxter, M.L., Lunt, D.H. (2016). Genetic drift, not life history or RNAi, determine long-term evolution of transposable elements. *Genome Biology and Evolution*, 8(9), 2964–2978.

Tan, H., Qurashi, A., Poidevin, M., Nelson, D.L., Li, H., Jin, P. (2012). Retrotransposon activation contributes to fragile X premutation rCGG-mediated neurodegeneration. *Human Molecular Genetics*, 21(1), 57–65.

Treiber, C.D. and Waddell, S. (2017). Resolving the prevalence of somatic transposition in *Drosophila*. *Elife*, 6, e28297.

Treiber, C.D. and Waddell, S. (2020). Transposon expression in the *Drosophila* brain is driven by neighboring genes and diversifies the neural transcriptome. *Genome Research*, 30(11), 1559–1569.

Venner, S., Feschotte, C., Biémont, C. (2009). Dynamics of transposable elements: Towards a community ecology of the genome. *Trends in Genetics*, 25, 317–323.

Venner, S., Miele, V., Terzian, C., Biémont, C., Daubin, V., Feschotte, C., Pontier, D. (2017). Ecological networks to unravel the routes to horizontal transposon transfers. *PLoS Biology*, 15(2), e2001536.

Vieira, C., Lepetit, D., Dumont, S., Biémont, C. (1999). Wake up of transposable elements following *Drosophila simulans* worldwide colonization. *Molecular Biology and Evolution*, 16, 1251–1255.

Villanueva-Cañas, J.L., Rech, G.E., Cara, M.A.R., González, J. (2017). Beyond SNPs: How to detect selection on transposable element insertions. *Methods in Ecology and Evolution*, 8(6), 728–737.

Villanueva-Cañas, J.L., Horvath, V., Aguilera, L., González, J. (2019). Diverse families of transposable elements affect the transcriptional regulation of stress-response genes in *Drosophila melanogaster*. *Nucleic Acids Research*, 47(13), 6842–6857.

Vinogradov, A.E. (2004a). Testing genome complexity. *Science*, 304(5669), 389–390.

Vinogradov, A.E. (2004b). Evolution of genome size: Multilevel selection, mutation bias or dynamical chaos? *Current Opinion in Genetics and Development*, 14(6), 620–626.

Volterra, V. (1926). *Variazioni e fluttuazioni del numero d'individui in species animali conviventi*. Thesis, Accademia Nazionale dei Lincei, Rome.

Wallau, G.L., Capy, P., Loreto, E., Le Rouzic, A., Hua-Van, A. (2016). VHICA, a new method to discriminate between vertical and horizontal transposon transfer: Application to the *Mariner* family within *Drosophila*. *Molecular Biology and Evolution*, 33(4), 1094–1109.

Wang, D., Qu, Z., Yang, L., Zhang, Q., Liu, Z.H., Do, T., Adelson, D.L., Wang, Z.Y., Searle, I., Zhu, J.K. (2017). Transposable elements (TEs) contribute to stress-related long intergenic noncoding RNAs in plants. *The Plant Journal*, 90(1), 133–146.

Wicker, T., Gundlach, H., Spannagl, M., Uauy, C., Borrill, P., Ramirez-Gonzalez, R.H., De Oliveira, R., Mayer, K.F.X., Paux, E., Choulet, F. (2018). Impact of transposable elements on genome structure and evolution in bread wheat. *Genome Biology*, 19, 103.

Wright, S. (1931). Evolution in Mendelian populations. *Genetics*, 16(2), 97–159.

Yannopoulos, G., Stamatis, N., Monastirioti, M., Louis C. (1987). *Hobo* is responsible for the induction of hybrid dysgenesis by strains of *Drosophila melanogaster* bearing the male recombination factor 23.5 MRF. *Cell*, 49, 487–495.

10

Transposable Elements as Tools

Chengyi SONG[1] and Zoltán IVICS[2]

[1] *College of Animal Science & Technology, Yangzhou University, China*

[2] *Division of Medical Biotechnology, Paul-Ehrlich-Institute, Langen, Germany*

Transposable elements, particularly cut-and-paste DNA transposons, can be considered as natural, non-viral gene delivery vehicles and are valuable, widely used tools for an array of genetic engineering applications. This chapter illustrates the general considerations and experimental approaches that have been successfully used to identify active DNA transposons and to develop them into tools for genetic manipulations. We describe the use of DNA transposons as an efficient gene transfer tool applied for germline transgenesis and insertional mutagenesis (such as gene trapping and enhancer trapping) in important model organisms. We also review highly specialized applications that are uniquely enabled by transposon systems, for example, elimination of selectable marker genes in transgenesis or reprogramming factors from the genome of induced pluripotent stem cells (iPSCs). Finally, a general overview of transposon system applications in gene and cell therapy is provided, with an update of major developments, current clinical tests and future perspectives.

10.1. Introduction

DNA transposons, belonging to the Class II transposable elements and often referred to as “jumping genes”, can move and integrate to different locations within the genome by mechanisms involving a DNA intermediate. DNA transposons have been found in both prokaryotic and eukaryotic organisms, and display great diversity at superfamily and family levels, with diverse distinct transposition mechanisms, such as cut-and-paste, rolling-circle and self-synthesis, from which the cut-and-paste

Transposable Elements and Genome Evolution,
coordinated by Aurélie HUA-VAN and Pierre CAPY. © ISTE Ltd 2024.

mechanism is the most widespread (Feschotte and Pritham 2007). The evolutionary profiles of the cut-and-paste DNA transposons have been well defined, and represent high family diversity, distinct taxonomic distribution and structural organization (Yuan and Wessler 2011; Hickman and Dyda 2016). The typical structure of the cut-and-paste DNA transposons includes a gene encoding a transposase that catalyzes the movement of the transposon and terminal inverted repeats (TIRs) at each end of the transposon (Figure 10.1(A)). Generally, cut-and-paste transposition involves at least two transposase molecules that bind to the TIRs of the transposon, mediate synapsis of the transposon ends and deliberate the transposon from its DNA sequence environment by cleavage at the ends of the transposon. Then, the excised transposon–transposase complex becomes integrated in the target DNA. In this manner, the transposon is cut out from one position and pasted into another position by a mechanism mediated by the transposase. Host cell DNA repair mechanisms contribute to the generation of transposon footprints at the excision sites of most cut-and-paste transposons; whereas, the gaps that are introduced on either side of the inserted transposon at the integration site are resolved to result in target site duplications (TSDs). Both the footprints and the TSDs are hallmarks that are characteristic to every transposon family (Figure 10.1(B)) (Yuan and Wessler 2011; Hickman and Dyda 2016).

The structural organization and mobility mechanism of cut-and-paste DNA transposons facilitate vector engineering for gene transfer. Namely, transposition can be experimentally controlled by separating the two functional components of the transposon and by providing a donor plasmid harboring a transgene of interest to be mobilized, flanked by TIRs, and a helper plasmid that expresses the transposase to drive the transposition reaction. Alternatively, the transposase can be provided by mRNA or protein synthesized in vitro. Thus, a DNA molecule of interest (e.g. a mutagenic unit or a therapeutic gene construct) cloned between the TIRs in a transposon-based vector can be stably integrated into the genome in a regulated and highly efficient manner. The major advantage of using cut-and-paste transposons as genetic tools are that their high activities and simple structural organization permit easy vector engineering for transgenesis and mutagenesis and vector manufacturing for gene and cell therapy, and also facilitate the downstream processing of transposon-based cell therapy and the subsequent analysis of gene function and activity (such as in enhancer and gene trapping) in an easy, controlled and scalable manner (Kebriaei et al. 2017; Tipanee et al. 2017b; Friedrich et al. 2019; Weber et al. 2019). Particularly, the significant improvement of transposon-based techniques by optimizing the design of transposase and transposon vectors promotes the development and application of DNA transposons as efficient gene delivery tools in life science (Mátés et al. 2009; Yusa et al. 2011; Hudecek et al. 2017) as an alternative to viral-based delivery systems.

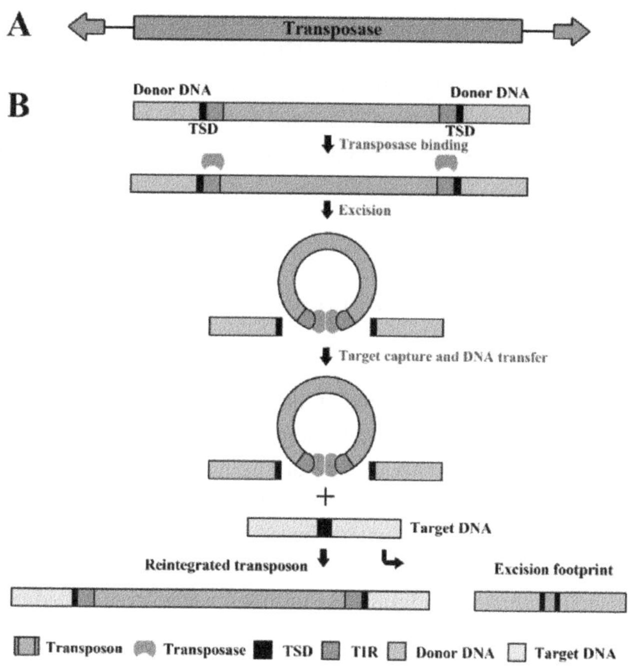

Figure 10.1. *Structure of cut-and-paste transposons and their transposition mechanism. (A) The transposon is a mobile genetic element containing a transposase coding sequence (green box), flanked by terminal inverted repeats (TIRs; orange arrows on the left and right). (B) The transposase (green spheres) binds to its sites within the transposon TIRs (orange boxes). Excision takes place in a synaptic complex and separates the transposon from the donor DNA (gray box). The excised element integrates into a target site in the target DNA (yellow box). This process generates target site duplications (TSDs, black boxes), which flank the newly integrated transposon. Most cut-and-paste transposons generate a transposon excision footprint in the donor DNA. For a color version of this figure, see www.iste.co.uk/huavan/transposable.zip*

On the one hand, DNA transposons, as powerful genetic tools, are currently still under development to satisfy multiple application purposes in life science. On the other hand, they have been widely utilized for efficient gene transfer in cultured cells and in primary cell types, including stem cells, and have been exploited for transgenesis and mutagenesis for a wide range of organisms. Several well-characterized transposons, such as *Sleeping Beauty* (*SB*), *piggyBac* (*PB*) and *Tol2*, have been widely used for transgenesis and mutagenesis in important animal models (Sandoval Villegas et al. 2021), while *SB* and *PB* have been successfully applied for preclinical or clinical trials in human gene therapy, including cell engineering (Amberger and Ivics 2020). Furthermore, transposon-based technologies toward specific targeting by fusing transposase with other nucleases are also exploited, and mutant transposase as excision tools has been established, and can be used to remove undesired transgene materials after genetic manipulation.

10.2. Development of DNA transposons as genetic tools

10.2.1. *Discovery and de novo engineering of genetic tools based on active DNA transposons*

The protocol for genetic tool development based on active DNA transposons mainly includes four steps: rebuilding or identification of the putatively active DNA transposons by bioinformatic analysis, vector engineering, transposition activity test in cells and application evaluation (Figure 10.2). The activity of DNA transposons in a specific genome can be evaluated based on the copy number, structural organization and estimated age using bioinformatic analysis. Young transposons display an intact structure, including TIRs flanking a transposase coding region, low pairwise sequence divergence between individual copies and low copy number in the genome. To find an active member for a specific family of DNA transposons in nature, first the taxonomic distribution can be investigated by TBlastN or Blast to identify homologous sequences in the NCBI genome databases by using a known transposase or transposon sequence as a query. Then, the classification of these transposases can be confirmed using a phylogenetic tree; the copy number and structural organization of these transposons in each genome can be defined by alignments, and the activity of the DNA transposon can be assessed by using divergence levels, or the formula for insertion age estimation: $T = k/2r$, where T represents the insertion time in millions of years, k represents the number of nucleotide substitutions at each site and r represents the neutral mutation rate in species lineages (Kimura 1980). Once young DNA transposons (better <2 million years old, or <2% divergence) with intact structure (TIR and transposase) are obtained, the TIRs and transposase encoded by intact copies in the specific genome can be cloned or synthesized for the following vector engineering, which includes constructing a dual plasmid system (donor and helper plasmids) that has been well established and verified for transposition activity testing (Ivics et al. 1997; Shen et al. 2021). The donor plasmid containing a promoter and drug (antibiotic) resistance gene cassette flanked by TIRs can confer drug resistance in cells upon transposition into chromosomes. The helper plasmid harbors an expression cassette of transposase driven by a promoter. Then, the engineered vectors are submitted to a cell-based test by using a dual plasmid-based transposition assay. The donor plasmid is mixed with the helper plasmid and co-transfected into cultured cells with subsequent drug selection. The active transposase displaying substantial transposition activity in cells will generate significantly higher numbers of resistant colonies, compared with the control (cells lacking a transposase source). On the other hand, active DNA transposons can also be obtained, optimized or rebuilt by using transposase and TIR consensus sequences derived from multiple species or multiple, homologous families of transposons, which has been described for *SB* and *PB* in detail (Ivics et al. 1997; Mátés et al. 2009; Yusa et al. 2011). In addition, it is also necessary to evaluate the safety and cargo capacity of the identified active transposon, which are also crucial biological features of DNA transposons and can significantly influence their application potential. Finally, new DNA transposons may be evaluated for their potential in versatile gene delivery and gene discovery applications, ranging from transgenesis to functional genomics (such as enhancer and gene annotation based on insertion mutagenesis mediated with transposons) in animal models and gene therapy in humans (Kawakami et al. 2017; Bourque et al. 2018).

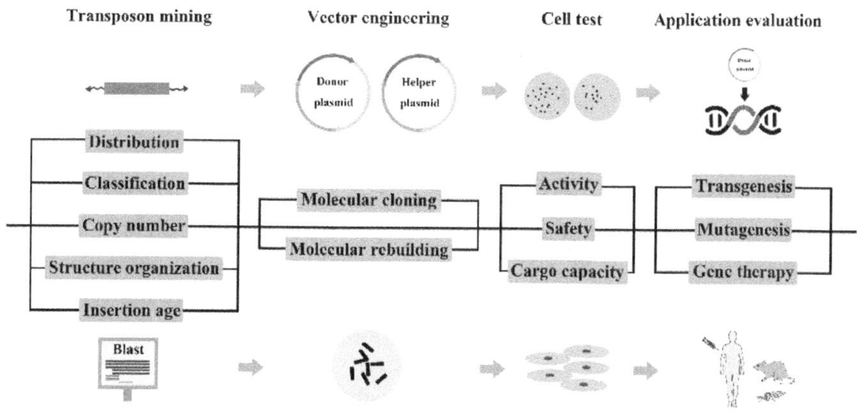

Figure 10.2. *Experimental pipeline for the development of genetic tools based on active DNA transposons. For a color version of this figure, see www.iste.co.uk/huavan/transposable.zip*

10.2.2. *DNA transposons popularly applied as genetic tools*

ISY100, *Tn5*, *Tn10* and *Mu* represent a common choice for transgenesis and mutagenesis in prokaryotes (Choi and Kim 2009; Conte et al. 2019; Ruiz and van Sinderen 2019). All of these transposons belong to the cut-and-paste DNA transposons, which, in eukaryotes, can be classified into 17 superfamilies, that is, *Tc1/mariner*, *PB*, *hAT*, *Zator*, *Merlin*, *PIF/Harbinger*, *MULE*, *P* element, *Kolobok*, *Novosib*, *Sola1*, *Sola2*, *Sola3*, *CMC* (*CACTA*, *Mirage* and *Chapaev*), *Transib*,

Academ and *Ginger*, based on the alignment of the catalytic domain of transposase (Yuan and Wessler 2011), which contains a triad of three acidic amino acids (DDE or DDD) that catalyzes strand cleavage and transfer during cut-and-paste transposition. At least 26 transposons, including *Transib* (Hencken et al. 2012), *Mutator* (Lisch 2015) and *En/Spm* (Ram et al. 2019), have been exploited for gene transfer applications (Mátés et al. 2007; Ni et al. 2008; Ivics et al. 2009; Skipper et al. 2013; Atkinson 2015; Handler and Schetelig 2020); most of these belong to *Tc1/mariner* (*Tc1*, *Tc3*, *Mos1*, *Minos*, *Passport*, *Frog Prince*, *Hsmar1*, *Himar1* and *SB*) and *hAT* (*hobo*, *Ac*/Ds, *Tam3*, *Tol1*, *Tol2*, *SPIN*, *TcBuster*, *Hermes*, *Herves* and *hopper*) transposons; few of these are from other groups, such as the *PB* and *P* elements. In addition, *Helraiser*, a reconstructed *Helitron* transposon from the bat genome has been shown to enable highly efficient copy-and-paste transposition in mammalian cells (Grabundzija et al. 2016). The *Tc1/mariner* superfamily, which may represent the most extensively reported DNA transposons, was named based on the first *Tc1* (Transposon *Caenorhabditis elegans* number 1) element identified in *Caenorhabditis elegans* and the first *mariner* element detected in *Drosophila mauritiana*. It may represent the highest diversity and the most widespread DNA transposons in nature. Ten families of this group, that is, *DD34D/mariner*, *DD34E/Tc1*, *DD35E/TR*, *DD36E/IC*, *DDE7E/TRT*, *DD38E/IT*, *DD37D/maT*, *DD39D/GT*, *DD41D/VS* and *DD×D/pogo*, have been recognized (Tellier et al. 2015; Sang et al. 2019; Zong et al. 2020). Evolution profiles of *DD35E/TR*, *DD36E/IC*, *DDE7E/TRT*, *DD38E/IT*, *DD37D/maT*, *DD39D/GT*, *DD41D/VS* and *DD×D/pogo* have recently been well defined (Zhang et al. 2016; Sang et al. 2019; Gao et al. 2020b; Shen et al. 2020; Zong et al. 2020; Wang et al. 2021) and they display differential taxonomic distribution (Figure 10.3). One domesticated gene (*SETMAR*) derived from a *DD34D/mariner* transposon has been identified in the human genome (Tellier et al. 2015). However, the evolutionary dynamics and complexities of the family and intrafamily structures of *Tc1/mariner*, such as *DD34D/mariner* and *DD34E/Tc1*, remain poorly understood.

Over 10 elements (including *Tc1*, *Tc3*, *Famar1*, *Minos*, *Mos1*, *ISY100*, *Fot1*, *Impala* and *Thm3*) of this group are known to be active in their natural form, and five, revived and reconstructed from inactive elements based on phylogenetic analyses, that is, *SB*, *Passport*, *Frog Prince*, *Himar1* and *Hsmar1*, are from the *DD34E/Tc1* family (Mátés et al. 2007; Ni et al. 2008; Ivics et al. 2009; Skipper et al. 2013). However, only the synthetic *SB* transposon exhibits high activity and displays great potential as a genetic tool, is popularly applied in vertebrates for transgenesis and mutagenesis (Ivics et al. 2009; Kawakami et al. 2017; Bourque et al. 2018), and in gene therapy for humans (Kebriaei et al. 2017). Recently, a highly active *Tc1/mariner* transposon in native form in the zebrafish genome, thus named *ZB*, was shown capable of robust transposition in vertebrate cells, which can be translated to highly efficient mutagenesis and enhancer trap (ET) screening in zebrafish and mice (Shen et al. 2021). *ZB* therefore represents an alternative, powerful tool for transgenesis and mutagenesis in vertebrates and possibly a new gene therapy delivery system for humans.

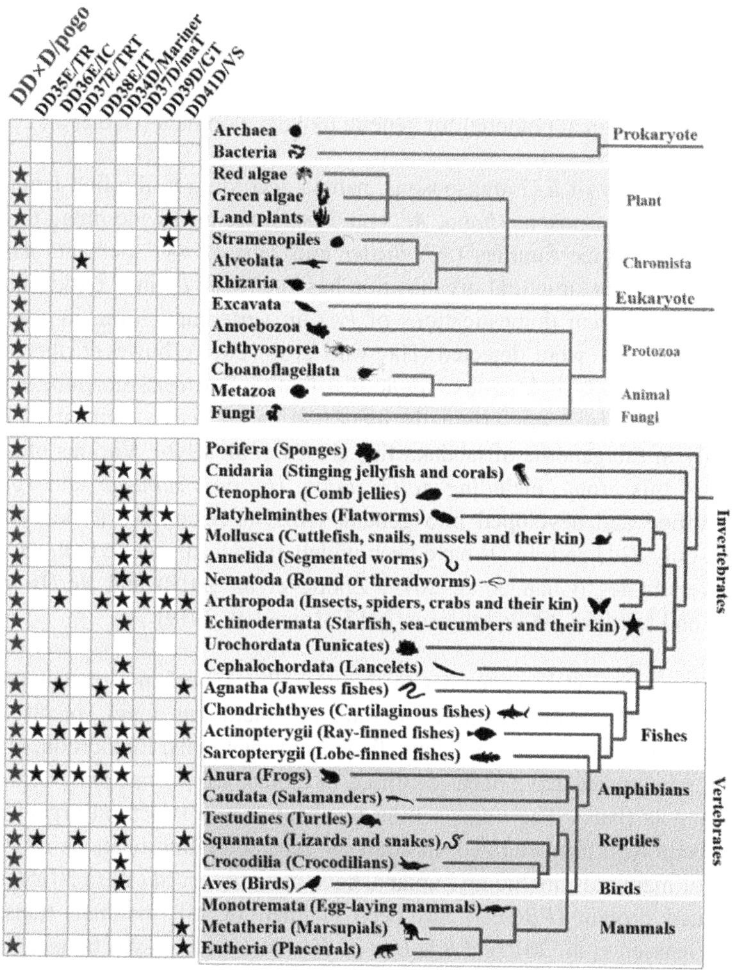

Figure 10.3. *Taxonomic distribution of Tc1/mariner and DD×D/pogo transposons. DD35E/TR, DD36E/IC, DDE7E/TRT, DD38E/IT, DD37D/maT, DD39D/GT and DD41D/VS are well-defined families, belonging to the superfamily of Tc1/mariner; whereas, DD×D/pogo was recently defined as a separate superfamily. For a color version of this figure, see www.iste.co.uk/huavan/transposable.zip*

In addition, phylogenetic analysis demonstrated that the *DD×D/pogo* transposons, long believed to belong to the *Tc1/mariner* superfamily (Mátés et al. 2007; Atkinson 2015), together with *Tc1/mariner*, *DD34E/Gambol* and *Zator* transposases, form four distinct monophyletic clades with high bootstrap supports, suggesting that they are separate superfamilies of the *IS630-Tc1-mariner* (*ITm*) group. The *pogo* transposons represent high diversity and wide distribution in eukaryotes with intact copies detected in the genomes of multiple lineages (Figure 10.3), indicating that these transposons may also be highly active in some genomes and have great potential for genetic tool development (Gao et al. 2020a).

The superfamily of *hAT* transposons, named according to the first letters of the three "founder" transposons (*hobo*, *Ac*, and *Tam3*), are also widespread throughout the tree of life. Three families (*Ac*, *Buster* and *Tip*) of *hAT* elements have been recognized and their structural organization has been well defined (Atkinson 2015). Furthermore, recurrent domestications of *hAT* in vertebrates, even in mammals – including humans – were detected (Hayward et al. 2013); however, the evolution profile of this group, such as taxonomic distribution and classification, is still not fully understood. *Tol2* was originally identified as an active element of the *hAT* superfamily in the genome of medaka fish (Koga et al. 1996). Various other active members of this group, including *Ac/Ds*, *hobo*, *Hermes*, *TcBuster* and *Tgf2*, have been identified and developed into genetic tools in eukaryotes (Atkinson 2015). However, only *Tol2* and *Ac/Ds* have been widely used as a genetic tool, particularly *Tol2* in vertebrates (Chan et al. 2019; Zhong et al. 2019) and *Ac/Ds* in plants (Lazarow and Lütticke 2009; Atkinson 2015; Ram et al. 2019).

The *PB* transposon is a member of the *piggyBac* superfamily that is present in the genomes of a wide range of organisms, including fungi, plants, insects, crustaceans, urochordates, amphibians, fishes and mammals. In addition, putatively active *PB* transposons have been identified in moths, silkworms, ants, xenopus and even mammals (bats). The transposition mechanism and structural organization of *PB* have been well characterized. In addition, repeated domestications of *PB* in the genomes of mammals, including humans, have been observed (Yusa 2015), and one domesticated protein (*PBGD5*) has been claimed to still maintain transposition activity (Henssen et al. 2015). *PB* is one of the most widely used transposons for a variety of applications, such as transgenesis and mutagenesis, and has potential as a gene therapy vehicle (Yusa 2015).

10.2.3. *Insertion preference and cargo capacity of DNA transposons*

The genome-wide integration preference and cargo capacity of the element, as important biological properties of DNA transposons, can greatly affect the utility of transposon vectors for different applications (Kawakami et al. 2017). On a genomic scale, *PB*, *Tol2*, *TcBuster*, *SPIN* and *ZB* transposons exhibit biased integration near transcription start sites and transcriptional regulatory regions in human cells (Li et al. 2013; Gogol-Döring et al. 2016; Kawakami et al. 2017; Shen et al. 2021), while *SB* transposons exhibit a close-to-random integration profile (Table 10.1) (Gogol-Döring et al. 2016). Consistently, for functional genomic analysis (such as gene and enhancer trapping), *TcBuster*, *SPIN*, *PB*, *ZB* and *Tol2* are more suited than *SB*; whereas, for gene therapy, *SB* may be safer than other transposon systems.

A decrease in transposition efficiency in cultured cells was observed for some DNA transposons with the increase in cargo size (Izsvák et al. 2000; Ivics et al. 2009). To address this constraint in the *SB* system, a sandwich *SB* vector, consisting of two full *SB* transposons flanking the gene to be mobilized, was designed to boost the amount of *SB* binding sites and thus enhance transposition efficiency of large fragments (Zayed et al. 2004). However, *SB*, *PB* and *Tol2* display a surprisingly large cargo capacity, *SB* and *PB* have been shown to transpose bacterial artificial chromosomes (BACs) over 100 kb in length at reasonable efficiencies in human embryonic stem cells (Rostovskaya et al. 2012), while both *PB* and *Tol2* transposons can mediate BAC (more than 70 kb) transgenesis in zebrafish and mice (Suster et al. 2009; Rostovskaya et al. 2013).

Tn	Integration preference	Largest cargo capacity	Footprint	Hyperactive transposase	Gene therapy application
SB	Almost random	Over 100 kb	C(A/T)GTA	SB100X	Clinical I/II
ZB	Enriched at transcriptional start sites	ND	C(A/T)GTA	None	ND
PB	Enriched at transcriptional start sites	Over 100 kb	None	hyPBase	Clinical I/II
TcBuster	Transcription units	ND	8 bp random sequence	$TcBuster_{co}V_{596}A$	Preclinical
Tol2	Transcription units	Over 70 kb	8 bp random sequence	None	Preclinical

Tn: transposon; ND: not determined.

Table 10.1. *DNA transposons and their main characteristics*

10.3. DNA transposons as efficient gene transfer tools applied in important model organisms

Efficient and long-term transposon-mediated transgene integration and expression have been shown in cultured cells from fish, frogs, chickens and mammals, and efficient transposition has been observed for well-established DNA transposons in genetically modified cell lines, embryonic stem cells (ESCs), hematopoietic stem cells (HSCs), iPSCs and cultured human deciduous tooth dental pulp cells and hepatocytes in vivo (Izsvák et al. 2000; Yusa 2015; Hickman and Dyda 2016; Kumar 2020; Sato et al. 2020).

In addition, transposon systems offer a powerful tool for the generation of transgenic animals and plants. *Ac/Ds*, *Mutator* and *En/Spm* represent a common choice for transgenesis and mutagenesis of plants (Ram et al. 2019). Over 20 DNA transposons, such as *P* element, *Mos1*, *Minos*, *SB*, *PB* and *Tol2*, have been exploited for gene transfer applications in animals, while *SB*, *PB* and *Tol2* are the most successfully applied DNA transposons in the important models of vertebrates (such as frogs, zebrafish, medaka, chickens and mice (Sandoval Villegas et al. 2021)). Efficient germline transgenesis is achieved by co-injection of a transposon carrying a transgene cassette of interest together with an mRNA synthesized in vitro as a transposase source, popularly applied for vertebrate transgenesis. This technique offers elevated germline transmission rates of the transgene due to the fast availability of the transposase in the injected oocytes or eggs, and reduces transgene mosaicism in the embryo. Transposon systems have also been reported as useful for transgenesis of domestic animals, and particularly promising for livestock, such as pigs, cattle and sheep. Transgenesis in livestock achieved by traditional pronuclear DNA injection is an inefficient and expensive process hampered by poor predictability of the patterns and levels of transgene expression and low survival rate of embryos after injection. One of the hurdles is that the lipid droplet layer of livestock zygotes hinders the visualization of pronuclei. Thus, generally, zygotes must be centrifuged before microinjection into pronuclei to produce transgenic animals (Gogol-Döring et al. 2016), which is laborious and may cause embryo damage, resulting in a decrease both of the survival rate of embryos and transgenesis efficiency. This can be overcome with cytoplasmic injection by using transposon delivery systems. With the aid of transposase, the plasmid-based transposon containing the transgene can be efficiently delivered from the cytoplasm of zygotes into nuclei and integrated into the genome. Commonly, transposase mRNA (or protein) synthesized in vitro is co-injected with a plasmid-based transposon, and in some cases, both components are delivered as DNA.

Transposon (particularly *SB*, *PB* and *Tol2*) delivery systems have been successfully employed for germline transgenesis in fish, frogs, birds and mice, and domestic animals, and significant increases in the ratio of transgenic animals per microinjected zygotes mediated with transposons have been consistently reported by using pronuclear injection or cytoplasmic injection (Suster et al. 2009; Ivics et al. 2014; Tang et al. 2015; Doran et al. 2016; Garrels et al. 2016b; Kalds et al. 2019). Both *SB* and *PB* have been popularly applied to mammalian transgenesis and used to generate transgenic livestock (including pigs, cattle and sheep) successfully by using cytoplasmic injection (Ivics et al. 2014; Li et al. 2014; Garrels et al. 2016a; Yum et al. 2016; Bevacqua et al. 2017). *SB*, *PB* and *Tol2* have been used to efficiently create transgenic mice by using cytoplasmic injection (Sumiyama et al. 2010; Suzuki et al. 2015; Garrels et al. 2016b). *Tol2* has been extensively used to make transgenic fish, including cichlids, tilapia, tetra and killifish (Fujimura and Kocher 2011; Valenzano et al. 2011; Juntti et al. 2013; Stahl et al. 2019). However, as an alternative method for DNA microinjection, *Tol2* has not yet been widely used in mouse transgenesis. Moreover, it has been found that the gene transfer efficiency of *Tol2* in medaka is low due to an unknown suppressing mechanism.

Transgenic procedures for avian species have lagged far behind livestock due to their oviparity and the physiological properties of the ovum. Thus, germline transmission in avian species largely depends on primordial germ cell (PGC) transplantation methods (Han 2009). Transgenic technology has been successfully adapted to the production of transgenic chickens with PGC modification, with both the *PB* and *Tol2* transposon systems and transplantation (Doran et al. 2016). Direct transfection of PGCs in vivo mediated by the *Tol2* transposon has been successfully applied to produce transgenic chickens (Doran et al. 2016), suggesting this method may be promising for transgenesis for other avian species for which PGC culture has not been established.

10.4. Insertional mutagenesis based on engineered transposons

The unique ability of DNA transposons to change their position in the genome also provides them with great application potential for functional genomic analyses, such as insertion mutagenesis for the functional annotation of regulatory elements and genes, which aims to obtain mutant phenotypes by introducing loss-of-function or gain-of-function mutations in genomes of model organisms in a random and genome-wide manner (Ivics et al. 2009; Kawakami et al. 2017; Kumar 2020). Mutating genes and regulatory elements by insertion of mutagenic cassettes mediated by transposons has numerous advantages over other strategies of mutagenesis, including high throughput, easy identification of mutated alleles and providing insight into genetic networks and pathways based on phenotypes. Insertional mutagenesis using engineered transposons may be one of the most productive and versatile approaches towards disrupting and manipulating genes and regulatory elements on a genome-wide scale (Kumar 2020). Transposons have long been used as screening tools in both prokaryotes (e.g. *Escherichia coli*) and eukaryotes (e.g. fungi, animals and plants) (Jorgensen and Mango 2002; Page and Grossniklaus 2002; St Johnston 2002; Shuman and Silhavy 2003; Kile and Hilton 2005; Korzh 2007; Candela and Hake 2008).

The engineered transposons contain the mutagenic cassettes, generally including reporter genes, promoter trap and polyadenylation (polyA) trap elements, or ET elements or enhancer block (EB) elements, flanked by TIRs of transposons. These engineered vectors can be used for ET, EB and gene trap (GT) in model organisms (Korzh 2007; Ivics et al. 2009; Bessa et al. 2014; Kawakami et al. 2017). This approach is particularly potent when applied to transparent embryos, such as those of worms, insects, zebrafish, medaka and chickens, in which developmental events can be followed in vivo.

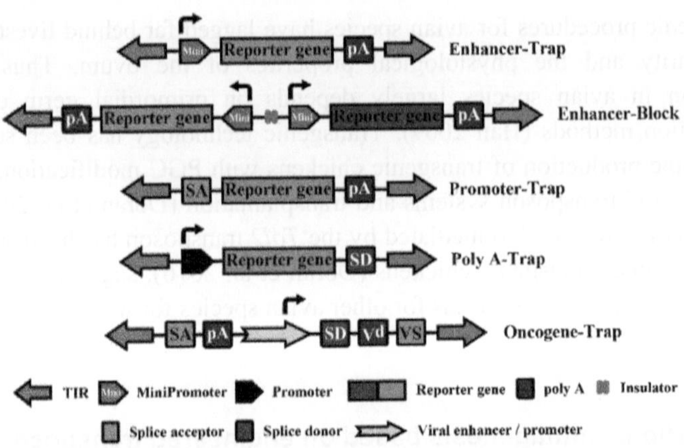

Figure 10.4. *Transposon-based vectors for functional genomics. Catalogue of transposon designs for different genetic screens. They all have transposon TIRs (orange arrows) to facilitate genetic screens. SA, splice acceptor; pA, polyadenylation sequence; SD, splice donor; Mini, Minimal promoter. For a color version of this figure, see www.iste.co.uk/huavan/transposable.zip*

Generally, a transposon-based ET vector, which is an insertional construct designed for enhancer screening, contains an ET box harboring a minimal promoter, a reporter gene and a polyA, and it is flanked by transposon TIRs (Figure 10.4). The minimal promoter is unable to drive the reporter gene expression alone; however, it can be activated by neighboring *cis*-acting chromosomal enhancer elements. The ET box function can be extended to enhancer blocking by introducing an insulator in the ET box, separating two fluorescent reporter genes (Figure 10.4), which cannot only trap enhancers, but can also block the regulatory activities by simultaneously disrupting the contacts between the upstream enhancers and the downstream genes they target or vice versa (Bessa et al. 2014). ET studies in zebrafish suggested that ET efficiency across different transposons is substantially different; *Tol2* and *ZB* represented relatively high ET efficiency (about 55%), while ET mediated by *SB* (about 40%) and *PB* (about 30%) is relatively low, indicating that ET efficiency in different species across different transposons may vary significantly (Shen et al. 2018, 2021). *SB* and *ZB* have been tested for ET in mice and display great potential as a simple and efficient way to annotate the regulatory landscape of mammalian chromosomes in vivo (Ruf et al. 2011; Shen et al. 2021). In addition, strong basic expression noise, which may interfere with the annotation of the targeted enhancer, was observed for some minimal promoter in zebrafish, and the gata minimal promoter, displaying low basic expression noise, was suggested as a good choice for ET in zebrafish (Chan et al. 2019). Therefore, the optimization of a minimal promoter for ET in mice is highly recommended.

Several types of GT vectors, such as promoter, polyadenylation (polyA) and oncogene trap vectors, have been developed for insertional mutagenesis mediated by transposon delivery systems. Enhancing the mutagenic effect of an insertion is the main consideration for vector design (Kawakami et al. 2017). Generally, a promoter trap vector harbors a splice acceptor (SA) site, a reporter gene (usually, a fluorescent protein, β-galactosidase or antibiotic resistance) and a polyA signal, which can cause partial or full loss of function of the targeted gene by disrupting splicing (Figure 10.4). A polyA trap box contains an internal promoter, a reporter gene and a splice donor (SD) site, but lacks a polyA signal (Figure 10.4). These vectors allow transcription initiation from inside the vector and polyadenylation of the transcript by a polyA signal donated by a targeted, endogenous gene – in case the insertion is in the right orientation. To enhance GT efficiency, an internal ribosome entry site (IRES) sequence can be inserted upstream or downstream of the reporter gene, which allows for the expression of the reporter cassette, irrespective of the reading frame of the disrupted gene in the promoter trap or represses non-sense-mediated RNA decay (NMD) of chimeric transcripts in a polyA trap. NMD can result in a bias of preferential detection of the last intron insertion of the trapped genes (Shigeoka et al. 2005). The oncogene trap, which was developed to identify proto-oncogenes and novel tumor suppressor genes by combining GT and polyA trap features, contains both SA and polyA signals in both orientations to disrupt transcription of the targeted gene efficiently (Figure 10.4), which is useful for knocking down or out tumor suppressor genes. Additionally, a strong enhancer or promoter and splice donor sequence may also be included in the oncogene trap vector to initiate the overexpression of a full-length (or truncated) protein product of the trapped gene (Bermejo Rodríguez and Pérez Mancera 2015).

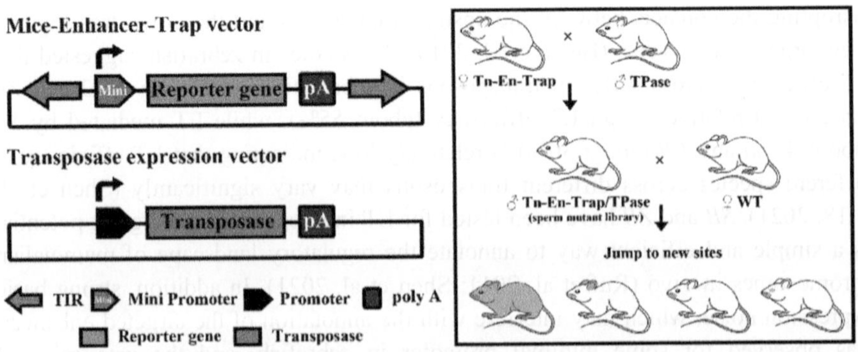

Figure 10.5. *Sperm mutagenesis. Left: Mice enhancer trap (ET) vector and transposase expression vector used for mouse sperm mutagenesis. The ET vector containing TIRs (orange arrows), Minimal promoter (Mini, gray arrow), reporter gene (green box) and polyA (pA, dark blue box). The transposase expression vector contains a germline cell–specific promoter (black arrow), transposase coding sequence (blue box) and polyA (pA, blue box). Right: The procedure used for the generation of a sperm mutant library, which is a double-transgenic mouse, generated by crossing ET transgene mice with transposase transgene mice, containing both a transposon cassette and a transposase expression cassette. Mutant libraries can produce pups with new integration sites after mating with female wild-type (WT) mice. For a color version of this figure, see www.iste.co.uk/huavan/transposable.zip*

These transposon-based GTs have been employed to mutagenize ESCs (Leeb et al. 2014) and the germlines of important animal models, such as flies, zebrafish and mice (Wilson et al. 1989; Sun et al. 2007; Kawakami et al. 2010; Trinh and Fraser 2013). Particularly, an efficient transposon-based sperm mutagenesis system has been developed and tested for GT in frogs and rats (Furushima et al. 2012; Lane et al. 2013) and verified for ET in mice (Ruf et al. 2011; Shen et al. 2021), which greatly facilitates the ET and GT analysis. The sperm mutagenesis system, which was evaluated for ET in mice recently (Shen et al. 2021), was developed by using two transgenic lines (transposase and transposon). The transposase transgenic mice (TPase) host a transposase expression cassette, which can drive transposase expression specifically in the testis by using a germline cell specific promoter. The transposon transgenic mouse carries a transposon-based ET transgene (Tn-En-Trap).

The transposon-based ET transgene harbored an ET box flanked by transposon TIRs (Figure 10.5). The transposon and transposase transgenic lines were crossed to generate the double-transgenic males, which contain both transposase and transposon transgenes, and were referred to as a "sperm mutant library" because they were the source of sperm carrying new mobilization events caused by ongoing transposition in the developing germ cells of these mice. The double-transgenic males are bred to wild-type (WT) females to generate offspring for ET screening by PCR and GFP detection. The ET box can be replaced with a GT box for promoter, polyA, and the oncogene trap. The general strategy used to carry out this procedure is shown in Figure 10.5. This strategy has been successfully applied for gene function and enhancer annotation (Furushima et al. 2012; Ruf et al. 2011; Shen et al. 2021). The options of the transposon systems may have a great impact on the trap efficiency and depend on the model organisms (Shen et al. 2018, 2021).

In addition, in vivo cancer screening with *SB* and *PB* transposons is the most widespread application of transposons as mutagenic tools in vertebrates. *SB* and *PB* are complementary tools with the deployment of both being promising to lead to saturating mutagenesis and maximal gene discovery in cancer, and have led to the discovery of new cancer candidate genes and insights into the molecular mechanism of tumorigenesis (Beckmann and Largaespada 2020; Noorani et al. 2020).

10.5. Application of transposons in human gene therapy

Gene therapy, which was successfully demonstrated in the early 1990s, is treating human diseases by the introduction of therapeutic genes into target cells or tissues. The major gene delivery systems used in ongoing clinical trials of gene therapy are based on viral vectors. However, the limited cargo size of viral vectors hampers the transfer of larger therapeutic genes, and viral vector manufacturing is cumbersome, costly and complicated because it requires living cells for their production with high-quality control and downstream processing. In addition, the immunogenicity, cellular toxicity and insertional oncogenesis, which have been observed for viral vector-based gene therapy, are the main safety concerns and remain major challenges regarding viral vector platforms in gene therapy (Bulcha et al. 2021).

DNA transposons, as a nonviral vector delivery system, have been considered to circumvent some of these limitations. For gene transfer applications with DNA transposons, a binary system based on two distinct plasmids has been well developed; whereby, the transposase is physically separated from the transposon DNA containing the gene of interest flanked by the TIRs. This feature makes transposon systems easily controllable DNA delivery vehicles and significantly improves their safety for applications in gene therapy (Woodard and Wilson 2015; Kebriaei et al. 2017). Codelivery of the transposon and transposase into the target cells or tissues enables efficient transgene integration in the genome via a conventional cut-and-paste transposition mechanism. Ideally, the transposase should only be transiently available to enable transposition. Sustained expression should be avoided, because it may impose risks of continuous mobilization, which may result in insertion mutagenesis and adverse genomic impacts. As a safer alternative, the codelivery of the transposase in the form of mRNA (or protein), by replacing the transposase plasmid, has been popularly applied for gene therapy. This cannot only minimize the risk of excision and re-insertions of transposons, but also reduce cellular toxicity significantly (Hudecek et al. 2017; Kebriaei et al. 2017; Tipanee et al. 2017a). Due to the stable genomic integration of the transgene, transposition-mediated gene delivery can lead to sustained and efficient transgene expression in target cells or tissues. In addition, the manufacturing of the transposon vector (in the form of plasmid DNA) is much easier and inexpensive, compared with viral vectors. These advantages are important concerns for implementation and scale-up in the practice of gene therapy. Furthermore, transposon-mediated gene transfer systems have the advantages of lower immunogenicity, larger cargo capacity for the transgene and a superior biosafety profile to that of viral-based vectors (Woodard and Wilson 2015; Hudecek et al. 2017; Kebriaei et al. 2017; Tipanee et al. 2017a). A significant improvement for the transposon-based delivery system is the use of minicircle vectors as carriers of the transposon components. This technology allows for a significant reduction of transposon vector size by removing most of the backbone sequences from parental plasmids – an approach that has been evaluated for *SB* transposon system in human T cells with very promising results in the reduction of cellular toxicity and a dramatic enhancement of gene delivery efficiency (Kebriaei et al. 2017; Holstein et al. 2018; Hudecek and Ivics 2018).

Well-characterized DNA transposons, such as *SB* and *PB*, have been considered for gene therapy ex vivo and in vivo, and reported extensively (VandenDriessche et al. 2009; Di Matteo et al. 2012). Prime examples for the use of *SB* and *PB* transposons in gene therapy include the treatment of hematologic disorders, lysosomal storage diseases, pulmonary disorders, dermatologic diseases, a variety of metabolic disorders, neurologic disorders, muscle disorders and cancer (Hudecek et al. 2017; Tipanee et al. 2017a).

10.5.1. *Gene therapy in vivo*

Gene therapy in vivo could be achieved by using a transposon-based delivery system. However, efficient delivery of transposon vectors in vivo is a challenge because, unlike viruses, naked nucleic acids (DNA and mRNA) lack the capacity to pass through the cell membrane through infection. Several methods (such as polyethylenimine-based transfection, electroporation, hydrodynamic injection, nanoparticle packaging and transposon viral hybrid technology) aiming to improve the transposon delivery in vivo have been validated (Tipanee et al. 2017a, 2017b). Hydrodynamic injection is considered one of the most common methods for in vivo transposon delivery through high-speed injection of a large volume of DNA solution. Hydrodynamic injection transfection of transposons has been utilized to direct long-term therapeutic gene expression in the liver, lung, brain, pulmonary endothelium and kidney for various disease models (Tipanee et al. 2017a, 2017b). In addition, transposon viral hybrid technologies have been developed and evaluated, which can overcome cellular barriers and enhance the gene transfer rate. These technologies represent a promising approach for transposon-based delivery in vivo. Viral transposon hybrids have been established for adenovirus, adeno-associated virus, herpes simplex virus, integration-deficient lentivirus vectors and baculovirus, where both *SB* and *PB* transposons provide stable gene integration (Woodard and Wilson 2015; Vargas et al. 2016; Kebriaei et al. 2017; Narayanavari and Izsvák 2017). Moreover, inducible gene expression systems that are stable in vivo can be established by transposon technology, allowing temporal control of transgene expression (Saridey et al. 2009).

Transposon-based gene therapy in vivo has been successfully exploited for at least 15 disease models, that is, sickle cell disease, hemophilia A, hemophilia B, mucopolysaccharidosis, tyrosinemia type I, familial hypercholesterolemia, diabetes type I, Crigler–Najjar syndrome type 1, pulmonary fibrosis, pulmonary hypertension, thrombotic thrombocytopenic purpura, colorectal cancer, glioblastoma, von Willebrand disease, unilateral ureteral obstruction and cervical and ovarian solid tumors (Tipanee et al. 2017a, 2017b).

10.5.2. *Ex vivo cell engineering with transposons*

Ex vivo cell and gene therapy, which is defined as the treatment or prevention of disease by administration of cells that have been selected, manipulated or altered outside the body, is emerging for a variety of human diseases. Several approaches have been used to deliver transposons into the desired target cells to achieve sustained transgene expression, and most approaches can yield high transposition efficiency in easy-to-transfect cell lines. However, most clinically relevant primary target cells, such as $CD34^+$ human HSCs, immune cells and muscle cell progenitors are relatively refractory to these types of transfection methods. Minicircle-based transposon vectors, owing to the depletion of the bacterial plasmid backbone, can decrease levels of cytotoxicity and improve gene delivery efficiency in vitro significantly by combining with nucleofection. Minicircle-based transposon platforms may accelerate both preclinical and clinical implementation of transposon-based delivery systems (Hudecek et al. 2017; Monjezi et al. 2017; Prommersberger et al. 2021).

Transposons have been used to modify clinically relevant cells, including iPSCs, human ESCs (hESCs), HSCs and human T lymphocytes (Woodard and Wilson 2015; Narayanavari and Izsvák 2017; Tipanee et al. 2017b; Magnani et al. 2020). Immunotherapy based on redirected human T cells expressing chimeric antigen receptors (CARs) holds great promise for cancer therapy. *SB*, *PB* and *Tol2* transposons were explored to generate CAR-T cells redirected against different antigens and cancer types, including acute lymphocytic leukemia, myelomonocytic leukemia, breast cancer and cholangiocarcinoma (Woodard and Wilson 2015; Tipanee et al. 2017b; Magnani et al. 2020). Clinical experience suggests that *SB*-based CAR-T engineering is feasible, safe and allows stable engraftment of modified T cells with efficiencies comparable to those of viral vectors, and particularly by using advanced *SB* transposon technology, where the transposon cassette is vectorized as minicircle DNA and the hyperactive *SB100X* transposase is provided in the form of mRNA (Magnani et al. 2020; Prommersberger et al. 2021).

SB and *PB* transposons have been applied for gene therapy research ex vivo for over 10 disease models, including sickle cell disease, Fanconi anemia type C, mucopolysaccharidosis, Alzheimer disease, Huntington disease, age-related macular degeneration, junctional epidermolysis bullosa, epidermolysis bullosa, muscular dystrophy, hepatocellular carcinoma, B-cell malignancies, sarcoma, β-thalassemia, α1-antitrypsin deficiency, myeloid and lymphoid malignancies, and HER2-positive solid tumors (Tipanee et al. 2017a, 2017b).

10.5.3. *Induced pluripotent stem cell reprogramming with transposons*

iPSCs (induced pluripotent stem cell), which are generated by reprogramming differentiated somatic cells, can be used to model diseases in vitro, and for drug screening and cell therapy. However, the most important and promising application of iPSCs is in cell-based therapy. The iPSC technology promises to provide an unlimited source of cells for innovative therapies, and to treat so far incurable diseases. Furthermore, application of iPSCs can solve the immunological rejection problem of cell transplantation process and avoid the social and ethical controversies associated with hESCs (Rowe and Daley 2019; Saito Diaz and Zeltner 2019; Kumar et al. 2020). In addition, iPSCs from domesticated animals also hold enormous scientific, economic and societal value (Su et al. 2020).

iPSC reprogramming technology is still relatively new, and diverse methods have been developed to generate iPSCs. Viral-based reprogramming technologies of iPSCs were the first and most common approach used, but they have the disadvantage of permanently incorporating their genetic material in the genome and contributing to teratoma formation (Liu et al. 2020). Transposons are uniquely associated with a cut (excision) reaction, which can be harnessed for transgene removal following reprogramming (see next section for further details). Compared with viral-based reprogramming technologies, transposons – as a non-viral vector delivery system – have numerous additional advantages, such as a larger cargo size of up to 100 kb, a close-to-random integration profile, no or very limited footprint after removal of the transposon, less innate immunogenicity, no requirement for a specialized biosafety facility and cost-effective production of the basic plasmids. Such approaches provide significant advances in the safety and efficacy of iPSCs, which may then be applied for downstream scientific and clinical applications. Thus, transposon systems are becoming developed and more popular for iPSC reprogramming (Tipanee et al. 2017a; Kumar et al. 2020). Both *SB* and *PB* transposon systems have been used successfully to obtain reprogrammed iPSCs from human somatic cells, but also somatic cells from murine models and cells from larger model species, such as pigs, horses, bats, monkeys, rats, cattle and buffalos (Tipanee et al. 2017a; Kumar et al. 2020).

10.6. Transposase as an excision tool

The use of a selectable marker gene (positive or/and negative selection), allowing the enrichment of cells harboring genetically modified target genes, is essential for the generation of transgenic organisms, but might also lead to potential environmental and food safety problems. Therefore, the production of marker-free transgenic animals and plants is becoming an extremely attractive alternative and could contribute to the public acceptance of genetically modified organisms (GMOs) (Yau and Stewart 2013). In addition, the generation of iPSCs via ectopic expression of reprogramming factors is simple, and is advanced by the use of numerous distinct systems for factor delivery. However, once reprogramming is complete, overexpression of the transcription factors is no longer a requirement, and residual expression and reactivation of reprogramming factors, uncontrolled silencing of transgenes, apoptosis, cell senescence and strong immunogenicity from the reprogramming source are putative safety risks (Hu 2014). Thus, eliminating the selectable marker gene or the reprogramming source to reduce potential risks for GMOs or regenerative medicine is necessary in practice. An effective method for elimination is to use the site-specific recombination that includes *Cre/loxP* or *FLP/FRT*. However, the recombinase-mediated toxicity and pseudo-recognition sequences of recombinase in the human genome pose a threat to clinical applications (Baer and Bode 2001). This limitation has been addressed by using an engineered transposon system, where the transposase mutants, which are competent in transposon excision (Exc$^+$) but have the feature of deficiency in transposon integration (Int$^-$), were generated. Both *PB* and *SB* transposase mutants (Exc$^+$/Int$^-$) have been developed and can be used to generate reprogramming factor-free iPSCs (Wang et al. 2017; Kesselring et al. 2020). *PB* transposase mutants (Exc$^+$/Int$^-$) have been applied to transgene excision in algae (Kasai et al. 2017). The excision site of *PB* can be precisely repaired without leaving a footprint, while the excision of *SB* will leave canonical footprints of a few nucleotides (Wang et al. 2017; Kesselring et al. 2020).

10.7. Toward specific gene targeting by fusing transposases with other nucleases

Transposon systems are non-viral gene delivery tools capable of integrating defined DNA segments into host chromosomes, which is simpler, safer and cheaper to produce, handle and store than retroviral vectors, but with a main drawback of unspecific or random integration (Kawakami et al. 2017; Kumar 2020; Sato et al. 2020). Different strategies have been devised to achieve enhanced specificity of transposon-mediated gene targeting. These include the fusion of transposase with zinc finger proteins (ZFP), transcription activator-like effector proteins (TALE) and CRISPR-associated (Cas) protein (Luo et al. 2017).

In practice, the coevolution of transposons and the CRISPR system has been observed, and at least five classes of mobile genetic elements show multifaceted evolutionary relationships with the CRISPR system (Koonin and Makarova 2017). Particularly, casposons, sharing several features with self-synthesizing eukaryotic DNA transposons of the *Polinton*/*Maverick* class, including TIRs and genes for B family DNA polymerases, can encode Cas1 protein as a recombinase and could have contributed to the origin of the CRISPR-Cas adaptive immunity systems in archaea and bacteria. Comparative genomic analysis has suggested that casposons display clear signs of recent mobility and some copies may still be active (Koonin and Makarova 2017). Additionally, individual Cas proteins, such as the Cas4 nuclease, were recruited by bacteriophages and transposons. ISC (insertion of Cas9-like sequences), a novel group of bacterial and archaeal DNA transposons that encode Cas9 homologs, has been identified. Recruitment of CRISPR-Cas systems by *Tn7*-like transposons has been reported (Peters et al. 2017). The existence of RNA-guided transposition with promising targeting frequencies has been further confirmed (Klompe et al. 2019; Strecker et al. 2019), which highlights the potential of using RNA-guided *Tn7*-like transposons for specific genome editing.

The targeting specificity mediated with transposons was significantly improved by utilizing the CRISPR-associated, RNA-guided *Tn7*-like transposons in prokaryotic genomes. However, adopting this technology for a eukaryotic cellular environment has been challenging. CRISPR/Cas protein and transposase fusions have been tested; the human *Hsmar1* transposase was fused with catalytically inactive Cas9 (dCas9), and this fusion protein enabled a 15-fold enrichment of transposon insertions into a 600-bp target region in an in vitro plasmid-to-plasmid assay; however, no targeted transposition was detected with this system in bacterial cells (Bhatt and Chalmers 2019). A previous study failed to target the *PB* transposon into the *HPRT* gene with CRISPR/Cas9 components in human cells, even though some targeting was observed with other DNA binding domains (Luo et al. 2017). However, in a recent study, some integrations were successfully biased to the *CCR5* locus using a dCas9-PB fusion (Hew et al. 2019). The *SB* transposase has been retargeted by directly fusing it to Gal4, TetR, AAV Rep and different ZFPs (Ivics et al. 2007; Yant et al. 2007; Ammar et al. 2012; Voigt et al. 2012). In addition, a hyperactive version of the *SB* transposase (SB100X) was fused with dCas9 and shown to enable guide RNA (gRNA)-dependent enrichment of integration near sites specified by the gRNAs in human cells (Kovač et al. 2020). Nonetheless, the available evidence suggests that, while retargeting of transposon integration by heterologous DNA binding domains fused to the transposase is possible, there is a major limitation to these approaches: the low ratio of targeted to non-targeted insertions (off-target insertions).

10.8. Conclusion

Substantial progress has recently been achieved towards the development of improved transposon-based systems, and particularly those based on highly active members, such as *SB*, *PB*, *Tol2* and *TcBuster*, as tools for genetic manipulation, including transgenesis, mutagenesis, cell engineering and gene therapy. The transposon-based technology platform not only provides an easy and efficient method for germline transgenesis in laboratory animals and in large animal species for biotechnology, but also facilitates the sophisticated annotation of functional genomics, such as enhancer and cancer gene discovery by insertional mutagenesis.

Furthermore, transposon-based technology is now both accessible and practical for gene therapy, and the DNA transposons are versatile and have been validated for gene therapy ex vivo and in vivo for many disease targets. Particularly, a significant improvement of gene delivery efficiency, as well as the safety profile for the *SB* transposon system by the use of minicircle vectors as carriers of the transposon components has been confirmed, and this vector engineering technology can also be extended to other transposon systems (e.g. *ZB* and *PB*). In addition, the transposon and viral technologies have been successfully combined, and the hybrid technologies may significantly improve the efficiency and reduce the immune complications for gene therapy by merging the excellent nucleic acid delivery properties of a non-integrating viral vector and the integrative properties of transposons, especially for approaches in vivo. Compared with other DNA transposons, *SB*-based gene therapy constitutes one of the most promising clinical applications to-date. *SB* is currently being tested in 14 clinical trials worldwide and has led to the foundation of multiple companies (Kebriaei et al. 2017). Transposon-based CAR-T cell engineering technology has met with clear success and holds great promise for the future treatment of cancer.

10.9. References

Amberger, M. and Ivics, Z. (2020). Latest advances for the Sleeping Beauty transposon system: 23 years of insomnia but prettier than ever: Refinement and recent innovations of the Sleeping Beauty transposon system enabling novel, nonviral genetic engineering applications. *Bioessays*, 42(11), 2000136.

Ammar, I., Gogol-Döring, A., Miskey, C., Chen, W., Cathomen, T., Izsvak, Z., Ivics, Z. (2012). Retargeting transposon insertions by the adeno-associated virus Rep protein. *Nucleic Acids Res.*, 40(14), 6693–6712.

Atkinson, P.W. (2015). hAT Transposable In *Mobile DNA III*, Craig, N.L., Chandler, M., Gellert, M., Lambowitz, A.M., Rice, P.A., Sandmeyer, S.B. (eds). ASM Press, Washington, DC.

Baer, A. and Bode, J. (2001). Coping with kinetic and thermodynamic barriers: RMCE, an efficient strategy for the targeted integration of transgenes. *Curr. Opin. Biotechnol.*, 12(5), 473–480.

Beckmann, P.J. and Largaespada, D.A. (2020). Transposon insertion mutagenesis in mice for modeling human cancers: Critical insights gained and new opportunities. *Int. J. Mol. Sci.*, 21(3), 1172.

Bermejo Rodríguez, C. and Pérez Mancera, P.A. (2015). Use of DNA transposons for functional genetic screens in mouse models of cancer. *Curr. Opin. Biotechnol.*, 35103–35110.

Bessa, J., Luengo, M., Rivero Gil, S., Ariza Cosano, A., Maia, A.H., Ruiz Ruano, F.J., Caballero, P., Naranjo, S., Carvajal, J.J., Gómez Skarmeta, J.L. (2014). A mobile insulator system to detect and disrupt cis-regulatory landscapes in vertebrates. *Genome Res.*, 24(3), 487–495.

Bevacqua, R.J., Fernandez Martin, R., Canel, N.G., Gibbons, A., Texeira, D., Lange, F., Vans Landschoot, G., Savy, V., Briski, O., Hiriart, M.I. (2017). Assessing Tn5 and Sleeping Beauty for transpositional transgenesis by cytoplasmic injection into bovine and ovine zygotes. *PLoS One*, 12(3), e0174025.

Bhatt, S. and Chalmers, R. (2019). Targeted DNA transposition in vitro using a dCas9-transposase fusion protein. *Nucleic Acids Res.*, 47(15), 8126–8135.

Bourque, G., Burns, K.H., Gehring, M., Gorbunova, V., Seluanov, A., Hammell, M., Imbeault, M., Izsvák, Z., Levin, H.L., Macfarlan, T.S. (2018). Ten things you should know about transposable elements. *Genome Biol.*, 19(1), 1–12.

Bulcha, J.T., Wang, Y., Ma, H., Tai, P.W., Gao, G. (2021). Viral vector platforms within the gene therapy landscape. *Signal Transduction Targeted Ther.*, 6(1), 1–24.

Candela, H. and Hake, S. (2008). The art and design of genetic screens: Maize. *Nat. Rev. Genet.*, 9(3), 192–203.

Chan, S., Shen, D., Sang, Y., Wang, S., Wang, Y., Chen, C., Gao, B., Song, C. (2019). Development of enhancer-trapping and-detection vectors mediated by the Tol2 transposon in zebrafish. *Peer J*, 7, e6862.

Choi, K.H. and Kim, K.J. (2009). Applications of transposon-based gene delivery system in bacteria. *J. Microbiol. Biotechnol.*, 19(3), 217–228.

Conte, E., Mende, L., Grainge, I., Colloms, S.D. (2019). A Mini-ISY100 transposon delivery system effective in γ proteobacteria. *Front. Microbiol.*, 10, 280.

Di Matteo, M., Belay, E., Chuah, M.K., VandenDriessche, T. (2012). Recent developments in transposon-mediated gene therapy. *Expert Opin. Biol. Ther.*, 12(7), 841–858.

Doran, T.J., Cooper, C.A., Jenkins, K.A., Tizard, M.L. (2016). Advances in genetic engineering of the avian genome: "Realising the promise". *Transgenic Res.*, 25(3), 307–319.

Feschotte, C. and Pritham, E.J. (2007). DNA transposons and the evolution of eukaryotic genomes. *Annu Rev Genet*, 41, 331–368.

Friedrich, M.J., Bronner, I.F., Liu, P., Bradley, A., Rad, R. (2019). piggyBac transposon-based insertional mutagenesis in mice. In *Cancer Driver Genes*, Starr, T.K. (ed.). Humana New York, New York.

Fujimura, K. and Kocher, T.D. (2011). Tol2-mediated transgenesis in tilapia (*Oreochromis niloticus*). *Aquaculture*, 319(3–4), 342–346.

Furushima, K., Jang, C.W., Chen, D.W., Xiao, N., Overbeek, P.A., Behringer, R.R. (2012). Insertional mutagenesis by a hybrid piggyBac and Sleeping Beauty transposon in the rat. *Genetics*, 192(4), 1235–1248.

Gao, B., Wang, Y., Diaby, M., Zong, W., Shen, D., Wang, S., Chen, C., Wang, X., Song, C. (2020a). Evolution of pogo, a separate superfamily of IS630-Tc1-mariner transposons, revealing recurrent domestication events in vertebrates. *Mobile DNA*, 11(1), 1–15.

Gao, B., Zong, W., Miskey, C., Ullah, N., Diaby, M., Chen, C., Wang, X., Ivics, Z., Song, C. (2020b). Intruder (DD38E), a recently evolved sibling family of DD34E/Tc1 transposons in animals. *Mobile DNA*, 11(1), 1–15.

Garrels, W., Talluri, T.R., Apfelbaum, R., Carratalá, Y.P., Bosch, P., Pötzsch, K., Grueso, E., Ivics, Z., Kues, W.A. (2016a). One-step multiplex transgenesis via Sleeping Beauty transposition in cattle. *Sci. Rep.*, 6(1), 1–6.

Garrels, W., Talluri, T.R., Ziegler, M., Most, I., Forcato, D.O., Schmeer, M., Schleef, M., Ivics, Z., Kues, W.A. (2016b). Cytoplasmic injection of murine zygotes with Sleeping Beauty transposon plasmids and minicircles results in the efficient generation of germline transgenic mice. *Biotechnol. J.*, 11(1), 178–184.

Gogol-Döring, A., Ammar, I., Gupta, S., Bunse, M., Miskey, C., Chen, W., Uckert, W., Schulz, T.F., Izsvák, Z., Ivics, Z. (2016). Genome-wide profiling reveals remarkable parallels between insertion site selection properties of the MLV retrovirus and the piggyBac transposon in primary human CD4+ T cells. *Mol. Ther.*, 24(3), 592–606.

Grabundzija, I., Messing, S.A., Thomas, J., Cosby, R.L., Bilic, I., Miskey, C., Gogol-Döring, A., Kapitonov, V., Diem, T., Dalda, A. (2016). A Helitron transposon reconstructed from bats reveals a novel mechanism of genome shuffling in eukaryotes. *Nat. Commun.*, 7(1), 1–12.

Han, J.Y. (2009). Germ cells and transgenesis in chickens. *Comp. Immunol. Microbiol. Infect. Dis.*, 32(2), 61–80.

Handler, A.M. and Schetelig, M.F. (2020). The hAT-family transposable element, *hopper*, from *Bactrocera dorsalis* is a functional vector for insect germline transformation. *BMC Genomics*, 21(2), 1–9.

Hayward, A., Ghazal, A., Andersson, G., Andersson, L., Jern, P. (2013). ZBED evolution: Repeated utilization of DNA transposons as regulators of diverse host functions. *PLoS One*, 8(3), e59940.

Hencken, C.G., Li, X., Craig, N.L. (2012). Functional characterization of an active Rag-like transposase. *Nat. Struct. Mol. Biol.*, 19(8), 834–836.

Henssen, A.G., Henaff, E., Jiang, E., Eisenberg, A.R., Carson, J.R., Villasante, C.M., Ray, M., Still, E., Burns, M., Gandara, J. (2015). Genomic DNA transposition induced by human PGBD5. *eLife*, 4e10565.

Hew, B.E., Sato, R., Mauro, D., Stoytchev, I., Owens, J.B. (2019). RNA-guided piggyBac transposition in human cells. *Synth. Biol.*, 4(1), ysz018.

Hickman, A.B. and Dyda, F. (2016). DNA transposition at work. *Chem. Rev.*, 116(20), 12758–12784.

Holstein, M., Mesa Nuñez, C., Miskey, C., Almarza, E., Poletti, V., Schmeer, M., Grueso, E., Flores, J.C.O., Kobelt, D., Walther, W. (2018). Efficient non-viral gene delivery into human hematopoietic stem cells by Minicircle Sleeping Beauty transposon vectors. *Mol. Ther.*, 26(4), 1137–1153.

Hu, K. (2014). All roads lead to induced pluripotent stem cells: The technologies of iPSC generation. *Stem Cells Dev.*, 23(12), 1285–1300.

Hudecek, M. and Ivics, Z. (2018). Non-viral therapeutic cell engineering with the Sleeping Beauty transposon system. *Current Opinion in Genetics & Development*, 52, 100–108.

Hudecek, M., Izsvák, Z., Johnen, S., Renner, M., Thumann, G., Ivics, Z. (2017). Going non-viral: The Sleeping Beauty transposon system breaks on through to the clinical side. *Crit. Rev. Biochem. Mol. Biol.*, 52(4), 355–380.

Ivics, Z., Hackett, P.B., Plasterk, R.H., Izsvák, Z. (1997). Molecular reconstruction of Sleeping Beauty, a *Tc1*-like transposon from fish, and its transposition in human cells. *Cell*, 91(4), 501–510.

Ivics, Z., Katzer, A., Stüwe, E.E., Fiedler, D., Knespel, S., Izsvák, Z. (2007). Targeted Sleeping Beauty transposition in human cells. *Mol. Ther.*, 15(6), 1137–1144.

Ivics, Z., Li, M.A., Mátés, L., Boeke, J.D., Nagy, A., Bradley, A., Izsvák, Z. (2009). Transposon-mediated genome manipulation in vertebrates. *Nat. Methods*, 6(6), 415–422.

Ivics, Z., Garrels, W., Mátés, L., Yau, T.Y., Bashir, S., Zidek, V., Landa, V., Geurts, A., Pravenec, M., Rülicke, T. (2014). Germline transgenesis in pigs by cytoplasmic microinjection of Sleeping Beauty transposons. *Nat. Protoc.*, 9(4), 810–827.

Izsvák, Z., Ivics, Z., Plasterk, R.H. (2000). Sleeping Beauty, a wide host-range transposon vector for genetic transformation in vertebrates. *J. Mol. Biol*, 302(1), 93–102.

Jorgensen, E.M. and Mango, S.E. (2002). The art and design of genetic screens: *Caenorhabditis elegans*. *Nat. Rev. Genet.*, 3(5), 356–369.

Juntti, S.A., Hu, C.K., Fernald, R.D. (2013). *Tol2*-mediated generation of a transgenic haplochromine cichlid, *Astatotilapia burtoni*. *PLoS One*, 8(10), e77647.

Kalds, P., Zhou, S., Cai, B., Liu, J., Wang, Y., Petersen, B., Sonstegard, T., Wang, X., Chen, Y. (2019). Sheep and goat genome engineering: From random transgenesis to the CRISPR era. *Frontiers in genetics*, 10, 750.

Kasai, Y., Matsuzaki, K., Ikeda, F., Yoshimitsu, Y., Harayama, S. (2017). Precise excision of a selectable marker gene in transgenic *Coccomyxa* strains by the *piggyBac* transposase. *Algal Res.*, 27, 152–161.

Kawakami, K., Abe, G., Asada, T., Asakawa, K., Fukuda, R., Ito, A., Lal, P., Mouri, N., Muto, A., Suster, M.L. (2010). zTrap: Zebrafish gene trap and enhancer trap database. *BMC Dev. Biol.*, 10(1), 1–10.

Kawakami, K., Largaespada, D.A., Ivics, Z. (2017). Transposons as tools for functional genomics in vertebrate models. *Trends Genet.*, 33(11), 784–801.

Kebriaei, P., Izsvák, Z., Narayanavari, S.A., Singh, H., Ivics, Z. (2017). Gene therapy with the Sleeping Beauty transposon system. *Trends Genet.*, 33(11), 852–870.

Kesselring, L., Miskey, C., Zuliani, C., Querques, I., Kapitonov, V., Laukó, A., Fehér, A., Palazzo, A., Diem, T., Lustig, J. (2020). A single amino acid switch converts the Sleeping Beauty transposase into an efficient unidirectional excisionase with utility in stem cell reprogramming. *Nucleic Acids Res.*, 48(1), 316–331.

Kile, B.T. and Hilton, D.J. (2005). The art and design of genetic screens: Mouse. *Nat. Rev. Genet.*, 6(7), 557–567.

Kimura, M. (1980). A simple method for estimating evolutionary rates of base substitutions through comparative studies of nucleotide sequences. *J. Mol. Evol.*, 16(2), 111–120.

Klompe, S.E., Vo, P.L., Halpin Healy, T.S., Sternberg, S.H. (2019). Transposon-encoded CRISPR–Cas systems direct RNA-guided DNA integration. *Nature*, 571(7764), 219–225.

Koga, A., Suzuki, M., Inagaki, H., Bessho, Y., Hori, H. (1996). Transposable element in fish. *Nature*, 383(6595), 30.

Koonin, E.V. and Makarova, K.S. (2017). Mobile genetic elements and evolution of CRISPR-Cas systems: All the way there and back. *Genome Biol. Evol.*, 9(10), 2812–2825.

Korzh, V. (2007). Transposons as tools for enhancer trap screens in vertebrates. *Genome Biol.*, 8(1), 1–6.

Kovač, A., Miskey, C., Menzel, M., Grueso, E., Gogol-Döring, A., Ivics, Z. (2020). RNA-guided retargeting of Sleeping Beauty transposition in human cells. *eLife*, 9, e53868.

Kumar, A. (2020). Jump around: Transposons in and out of the laboratory. *F1000Research*, 9, 135, doi: 10.12688/f1000research.21018.1.

Kumar, D., Anand, T., Talluri, T.R., Kues, W.A. (2020). Potential of transposon-mediated cellular reprogramming towards cell-based therapies. *World J. Stem Cells*, 12(7), 527.

Lane, M.A., Kimber, M., Khokha, M.K. (2013). Breeding based remobilization of Tol2 transposon in *Xenopus tropicalis*. *PLoS One*, 8(10), e76807.

Lazarow, K. and Lütticke, S. (2009). An Ac/Ds-mediated gene trap system for functional genomics in barley. *BMC Genomics*, 10(1), 1–13.

Leeb, M., Dietmann, S., Paramor, M., Niwa, H., Smith, A. (2014). Genetic exploration of the exit from self-renewal using haploid embryonic stem cells. *Cell Stem Cell*, 14(3), 385–393.

Li, X., Ewis, H., Hice, R.H., Malani, N., Parker, N., Zhou, L., Feschotte, C., Bushman, F.D., Atkinson, P.W., Craig, N.L. (2013). A resurrected mammalian hAT transposable element and a closely related insect element are highly active in human cell culture. *Proc. Natl. Acad. Sci. USA*, 110(6), E478–E487.

Li, Z., Zeng, F., Meng, F., Xu, Z., Zhang, X., Huang, X., Tang, F., Gao, W., Shi, J., He, X. (2014). Generation of transgenic pigs by cytoplasmic injection of piggyBac transposase-based pm GENIE-3 plasmids. *Biol. Reprod.*, 90(5), 93, 1-10.

Lisch, D. (2015). Mutator and MULE transposons, In *Mobile DNA III*, Craig, N.L., Chandler, M., Gellert, M., Lambowitz, A.M., Rice, P.A., Sandmeyer, S.B. (eds). ASM Press, Washington, DC.

Liu, G., David, B.T., Trawczynski, M., Fessler, R.G. (2020). Advances in pluripotent stem cells: History, mechanisms, technologies, and applications. *Stem Cell Rev. Rep.*, 16(1), 3–32.

Luo, W., Galvan, D.L., Woodard, L.E., Dorset, D., Levy, S., Wilson, M.H. (2017). Comparative analysis of chimeric ZFP-, TALE and Cas9-piggyBac transposases for integration into a single locus in human cells. *Nucleic Acids Res.*, 45(14), 8411–8422.

Magnani, C.F., Tettamanti, S., Alberti, G., Pisani, I., Biondi, A., Serafini, M., Gaipa, G. (2020). Transposon-based CAR T cells in acute leukemias: Where are we going? *Cells*, 9(6), 1337.

Mátés, L., Izsvák, Z., Ivics, Z. (2007). Technology transfer from worms and flies to vertebrates: Transposition-based genome manipulations and their future perspectives. *Genome Biol.*, 8(1), 1–19.

Mátés, L., Chuah, M.K., Belay, E., Jerchow, B., Manoj, N., Acosta-Sanchez, A., Grzela, D.P., Schmitt, A., Becker, K., Matrai, J. (2009). Molecular evolution of a novel hyperactive Sleeping Beauty transposase enables robust stable gene transfer in vertebrates. *Nat. Genet.*, 41(6), 753–761.

Monjezi, R., Miskey, C., Gogishvili, T., Schleef, M., Schmeer, M., Einsele, H., Ivics, Z., Hudecek, M. (2017). Enhanced CAR T-cell engineering using non-viral Sleeping Beauty transposition from minicircle vectors. *Leukemia*, 31(1), 186–194.

Narayanavari, S.A. and Izsvák, Z. (2017). Sleeping Beauty transposon vectors for therapeutic applications: Advances and challenges. *Cell Gene Ther. Insights*, 3(2), 131–158.

Ni, J., Clark, K.J., Fahrenkrug, S.C., Ekker, S.C. (2008). Transposon tools hopping in vertebrates. *Briefings Funct. Genomics Proteomics*, 7(6), 444–453.

Noorani, I., Bradley, A., de la Rosa, J. (2020). CRISPR and transposon in vivo screens for cancer drivers and therapeutic targets. *Genome Biol.*, 21(1), 1–22.

Page, D.R. and Grossniklaus, U. (2002). The art and design of genetic screens: *Arabidopsis thaliana*. *Nat. Rev. Genet.*, 3(2), 124–136.

Peters, J.E., Makarova, K.S., Shmakov, S., Koonin, E.V. (2017). Recruitment of CRISPR-Cas systems by Tn7-like transposons. *Proc. Natl. Acad. Sci. USA*, 114(35), E7358–E7366.

Prommersberger, S., Reiser, M., Beckmann, J., Danhof, S., Amberger, M., Quade Lyssy, P., Einsele, H., Hudecek, M., Bonig, H., Ivics, Z. (2021). CARAMBA: A first-in-human clinical trial with SLAMF7 CAR-T cells prepared by virus-free Sleeping Beauty gene transfer to treat multiple myeloma. *Gene Ther.*, 28(9), 560–571.

Ram, H., Soni, P., Salvi, P., Gandass, N., Sharma, A., Kaur, A., Sharma, T.R. (2019). Insertional mutagenesis approaches and their use in rice for functional genomics. *Plants*, 8(9), 310.

Rostovskaya, M., Fu, J., Obst, M., Baer, I., Weidlich, S., Wang, H., Smith, A.J., Anastassiadis, K., Stewart, A.F. (2012). Transposon-mediated BAC transgenesis in human ES cells. *Nucleic Acids Res.*, 40(19), e150.

Rostovskaya, M., Naumann, R., Fu, J., Obst, M., Mueller, D., Stewart, A.F., Anastassiadis, K. (2013). Transposon mediated BAC transgenesis via pronuclear injection of mouse zygotes. *Genesis*, 51(2), 135–141.

Rowe, R.G. and Daley, G.Q. (2019). Induced pluripotent stem cells in disease modelling and drug discovery. *Nat. Rev. Genet.*, 20(7), 377–388.

Ruf, S., Symmons, O., Uslu, V.V., Dolle, D., Hot, C., Ettwiller, L., Spitz, F. (2011). Large-scale analysis of the regulatory architecture of the mouse genome with a transposon-associated sensor. *Nat. Genet.*, 43(4), 379.

Ruiz, L. and van Sinderen, D. (2019). Implementation of transposon mutagenesis in Bifidobacterium. In *Microbial Transposon Mutagenesis*, Ricke, S.C., Hong Park, S., Davis, M.L. (eds). Humana New York, New York.

Saito Diaz, K. and Zeltner, N. (2019). Induced pluripotent stem cells for disease modeling, cell therapy and drug discovery in genetic autonomic disorders: A review. *Clin. Auton. Res.*, 29(4), 367–384.

Sandoval Villegas, N., Nurieva, W., Amberger, M., Ivics, Z. (2021). Contemporary transposon tools: A review and guide through mechanisms and applications of Sleeping Beauty, piggyBac and Tol2 for Genome Engineering. *Int. J. Mol. Sci.*, 22(10), 5084.

Sang, Y., Gao, B., Diaby, M., Zong, W., Chen, C., Shen, D., Wang, S., Wang, Y., Ivics, Z., Song, C. (2019). Incomer, a DD36E family of Tc1/mariner transposons newly discovered in animals. *Mobile DNA*, 10(1), 1–12.

Saridey, S.K., Liu, L., Doherty, J.E., Kaja, A., Galvan, D.L., Fletcher, B.S., Wilson, M.H. (2009). PiggyBac transposon-based inducible gene expression in vivo after somatic cell gene transfer. *Mol. Ther.*, 17(12), 2115–2120.

Sato, M., Inada, E., Saitoh, I., Watanabe, S., Nakamura, S. (2020). piggyBac-based non-viral in vivo gene delivery useful for production of genetically modified animals and organs. *Pharmaceutics*, 12(3), 277.

Shen, D., Xue, S., Chan, S., Sang, Y., Wang, S., Wang, Y., Chen, C., Gao, B., Mueller, F., Song, C. (2018). Enhancer trapping and annotation in Zebrafish mediated with Sleeping Beauty, piggyBac and Tol2 transposons. *Genes*, 9(12), 630.

Shen, D., Gao, B., Miskey, C., Chen, C., Sang, Y., Zong, W., Wang, S., Wang, Y., Wang, X., Ivics, Z. (2020). Multiple invasions of visitor, a DD41D family of Tc1/mariner transposons, throughout the evolution of vertebrates. *Genome Biol. Evol.*, 12(7), 1060–1073.

Shen, D., Song, C., Miskey, C., Chan, S., Guan, Z., Sang, Y., Wang, Y., Chen, C., Wang, X., Müller, F. (2021). A native, highly active Tc1/mariner transposon from zebrafish (ZB) offers an efficient genetic manipulation tool for vertebrates. *Nucleic Acids Res.*, 49(4), 2126–2140.

Shigeoka, T., Kawaichi, M., Ishida, Y. (2005). Suppression of nonsense-mediated mRNA decay permits unbiased gene trapping in mouse embryonic stem cells. *Nucleic Acids Res.*, 33(2), e20.

Shuman, H.A. and Silhavy, T.J. (2003). The art and design of genetic screens: *Escherichia coli*. *Nat. Rev. Genet.*, 4(6), 419–431.

Skipper, K.A., Andersen, P.R., Sharma, N., Mikkelsen, J.G. (2013). DNA transposon-based gene vehicles-scenes from an evolutionary drive. *J. Biomed. Sci.*, 20(1), 1–23.

St Johnston, D. (2002). The art and design of genetic screens: *Drosophila melanogaster. Nat. Rev. Genet.*, 3(3), 176–188.

Stahl, B.A., Peuß, R., McDole, B., Kenzior, A., Jaggard, J.B., Gaudenz, K., Krishnan, J., McGaugh, S.E., Duboue, E.R., Keene, A.C. (2019). Stable transgenesis in *Astyanax mexicanus* using the Tol2 transposase system. *Dev. Dyn.*, 248(8), 679–687.

Strecker, J., Ladha, A., Gardner, Z., Schmid Burgk, J.L., Makarova, K.S., Koonin, E.V., Zhang, F. (2019). RNA-guided DNA insertion with CRISPR-associated transposases. *Science*, 365(6448), 48–53.

Su, Y., Zhu, J., Salman, S., Tang, Y. (2020). Induced pluripotent stem cells from farm animals. *J. Anim. Sci.*, 98(11), skaa343.

Sumiyama, K., Kawakami, K., Yagita, K. (2010). A simple and highly efficient transgenesis method in mice with the Tol2 transposon system and cytoplasmic microinjection. *Genomics*, 95(5), 306–311.

Sun, L.V., Jin, K., Liu, Y., Yang, W., Xie, X., Ye, L., Wang, L., Zhu, L., Ding, S., Su, Y. (2007). PBmice: An integrated database system of piggyBac (PB) insertional mutations and their characterizations in mice. *Nucleic Acids Res.*, 36(suppl_1), D729–D734.

Suster, M.L., Sumiyama, K., Kawakami, K. (2009). Transposon-mediated BAC transgenesis in zebrafish and mice. *BMC Genomics*, 10(1), 1–7.

Suzuki, S., Tsukiyama, T., Kaneko, T., Imai, H., Minami, N. (2015). A hyperactive piggyBac transposon system is an easy-to-implement method for introducing foreign genes into mouse preimplantation embryos. *J. Reprod. Dev.*, 2014–2157.

Tang, L., González, R., Dobrinski, I. (2015). Germline modification of domestic animals. *Anim Reprod.*, 12(1), 93.

Tellier, M., Bouuaert, C.C., Chalmers, R. (2015). Mariner and the ITm superfamily of transposons. *Mobile DNA III*, 753–772.

Tipanee, J., Chai, Y.C., VandenDriessche, T., Chuah, M.K. (2017a). Preclinical and clinical advances in transposon-based gene therapy. *Biosci Rep*, 37(6), BSR20160614.

Tipanee, J., VandenDriessche, T., Chuah, M.K. (2017b). Transposons: Moving forward from preclinical studies to clinical trials. *Hum. Gene Ther.*, 28(11), 1087–1104.

Trinh, L.A. and Fraser, S.E. (2013). Enhancer and gene traps for molecular imaging and genetic analysis in zebrafish. *Dev. Growth Differ.*, 55(4), 434–445.

Valenzano, D.R., Sharp, S., Brunet, A. (2011). Transposon-mediated transgenesis in the short-lived African killifish *Nothobranchius furzeri*, a vertebrate model for aging. *G3: Genes, Genomes, Genet.*, 1(7), 531–538.

VandenDriessche, T., Ivics, Z., Izsvák, Z., Chuah, M.K. (2009). Emerging potential of transposons for gene therapy and generation of induced pluripotent stem cells. *Blood*, 114(8), 1461–1468.

Vargas, J.E., Chicaybam, L., Stein, R.T., Tanuri, A., Delgado Cañedo, A., Bonamino, M.H. (2016). Retroviral vectors and transposons for stable gene therapy: Advances, current challenges and perspectives. *J. Transl. Med.*, 14(1), 1–15.

Voigt, K., Gogol Döring, A., Miskey, C., Chen, W., Cathomen, T., Izsvák, Z., Ivics, Z. (2012). Retargeting sleeping beauty transposon insertions by engineered zinc finger DNA-binding domains. *Mol. Ther.*, 20(10), 1852–1862.

Wang, G., Yang, L., Grishin, D., Rios, X., Lillian, Y.Y., Hu, Y., Li, K., Zhang, D., Church, G.M., Pu, W.T. (2017). Efficient, footprint-free human iPSC genome editing by consolidation of Cas9/CRISPR and piggyBac technologies. *Nat. Protoc.*, 12(1), 88.

Wang, S., Diaby, M., Puzakov, M., Ullah, N., Wang, Y., Danley, P., Chen, C., Wang, X., Gao, B., Song, C. (2021). Divergent evolution profiles of DD37D and DD39D families of *Tc1/mariner* transposons in eukaryotes. *Mol. Phylogenet. Evol.*, 161, 107143.

Weber, J., de la Rosa, J., Grove, C.S., Schick, M., Rad, L., Baranov, O., Strong, A., Pfaus, A., Friedrich, M.J., Engleitner, T. (2019). PiggyBac transposon tools for recessive screening identify B-cell lymphoma drivers in mice. *Nat. Commun.*, 10(1), 1–16.

Wilson, C., Pearson, R.K., Bellen, H.J., O'Kane, C.J., Grossniklaus, U., Gehring, W.J. (1989). P-element-mediated enhancer detection: An efficient method for isolating and characterizing developmentally regulated genes in Drosophila. *Genes Dev.*, 3(9), 1301–1313.

Woodard, L.E. and Wilson, M.H. (2015). piggyBac-ing models and new therapeutic strategies. *Trends Biotechnol.*, 33(9), 525–533.

Yant, S.R., Huang, Y., Akache, B., Kay, M.A. (2007). Site-directed transposon integration in human cells. *Nucleic Acids Res.*, 35(7), e50.

Yau, Y.Y. and Stewart, C.N. (2013). Less is more: Strategies to remove marker genes from transgenic plants. *BMC Biotechnol.*, 13(1), 1–23.

Yuan, Y.W. and Wessler, S.R. (2011). The catalytic domain of all eukaryotic cut-and-paste transposase superfamilies. *Proc. Natl. Acad. Sci. USA*, 108(19), 7884–7889.

Yum, S.Y., Lee, S.J., Kim, H.M., Choi, W.J., Park, J.H., Lee, W.W., Kim, H.S., Kim, H.J., Bae, S.H., Lee, J.H. (2016). Efficient generation of transgenic cattle using the DNA transposon and their analysis by next-generation sequencing. *Sci. Rep.*, 6(1), 1–12.

Yusa, K. (2015). piggyBac transposon. In *Mobile DNA III*, Craig, N.L., Chandler, M., Gellert, M., Lambowitz, A.M., Rice, P.A., Sandmeyer, S.B. (eds). ASM Press, Washington, DC.

Yusa, K., Zhou, L., Li, M.A., Bradley, A., Craig, N.L. (2011). A hyperactive piggyBac transposase for mammalian applications. *Proc. Natl. Acad. Sci. USA*, 108(4), 1531–1536.

Zayed, H., Izsvák, Z., Walisko, O., Ivics, Z. (2004). Development of hyperactive sleeping beauty transposon vectors by mutational analysis. *Mol. Ther.*, 9(2), 292–304.

Zhang, H.H., Li, G.Y., Xiong, X.M., Han, M.J., Zhang, X.G., Dai, F.Y. (2016). *TRT*, a vertebrate and protozoan *Tc1*-like transposon: Current activity and horizontal transfer. *Genome Biol. Evol.*, 8(9), 2994–3005.

Zhong, Y., Huang, W., Du, J., Wang, Z., He, J., Luo, L. (2019). Improved *Tol2*-mediated enhancer trap identifies weakly expressed genes during liver and β cell development and regeneration in zebrafish. *J. Biol. Chem.*, 294(3), 932–940.

Zong, W., Gao, B., Diaby, M., Shen, D., Wang, S., Wang, Y., Sang, Y., Chen, C., Wang, X., Song, C. (2020). *Traveler*, a new DD35E family of *Tc1/mariner* transposons, invaded vertebrates very recently. *Genome Biol. Evol.*, 12(3), 66–76.

11

Genomic Characterization of Transposable Elements: Databases and Software

Gabriel DA LUZ WALLAU

Department of Entomology and Bioinformatic Core, Instituto Aggeu Magalhães (IAM), Fiocruz - Pernambuco, Recife, Brazil

11.1. Introduction

The massive scaling up of nucleic acid sequencing has revolutionized genetic studies of all fields, including those on mobile genetic elements (MGEs, which include plasmids, viruses/phages and transposable elements [TEs][1]). In a short period of time, many genomic projects were launched making a large amount of genetic information available through databases such as EMBL (Kanz et al. 2005) and Genbank (Clark et al. 2016). However, TEs characterization was usually not considered or sidelined in genomic analysis and the existing general databases provide limited support to integrate detailed information about these repetitive and complex elements.

1 Prokaryotic mobile integrative sequences are often referred to as MGEs, but here we will use the term TEs, whatever the origin of the sequence.

Transposable Elements and Genome Evolution,
coordinated by Aurélie HUA-VAN and Pierre CAPY. © ISTE Ltd 2024.

MGE and TE databases emerged during the 1990s as a response to store and share the ever-growing amount of sequences and associated information from a number of organisms' genomes. Among the first TE databases available were Repbase Update (UP), which was first developed to store repetitive sequences found in the human genome and later expanded to other eukaryotic species (Jurka et al. 1992), as well as ACLAME (Leplae et al. 2004) for storing prokaryotic MGE (including TE) sequences and associated information. The identification and characterization of TEs from genomic sequences can be performed by library/similarity/homology-based approaches, ab initio approaches or a combination of both. Homology-based approaches mostly rely on pairwise alignment of known TEs with new genomic sequences to find homologous elements based on arbitrary similarity threshold. Ab initio approaches rely on TE features such as its repetitiveness across the genome and inverted and repetitive sequence patterns from TE copies (see section 11.3).

The characterization and databasing of TEs has gone through major achievements in the last decades, but failed to reach the high level of organization of more popular genome-centered databases, mainly due to the absence of consensus among researchers and the lack of common practices adoptions. As a result, there is a plethora of databases and software solutions for TEs characterization available, which are highly heterogeneous regarding several aspects such as input data, methods for TE detection and information available. Moreover, there are limited benchmark studies providing key information that could help devise best practices in this field. In this chapter, we will approach this wide subject, focusing on the most common and representative databases and software to highlight key features and assumptions for TE characterization and illustrate best practices.

11.2. Databases

TEs and MGEs databases have an inherent complexity based on the need to store information about these paraphyletic elements and their associated host genome. Heterogeneous genome information is available for both TEs and hosts due to the different outputs of molecular methods used to characterize these elements or the host level in which a phenomenon is being investigated, such as the species or population level. Available TE/MGE databases reflect this complexity encompassing a range of more general databases storing sequences, either single copies or TE family/subfamily consensus sequences, TE family alignment and/or annotation information (Figure 11.1 and Table 11.1).

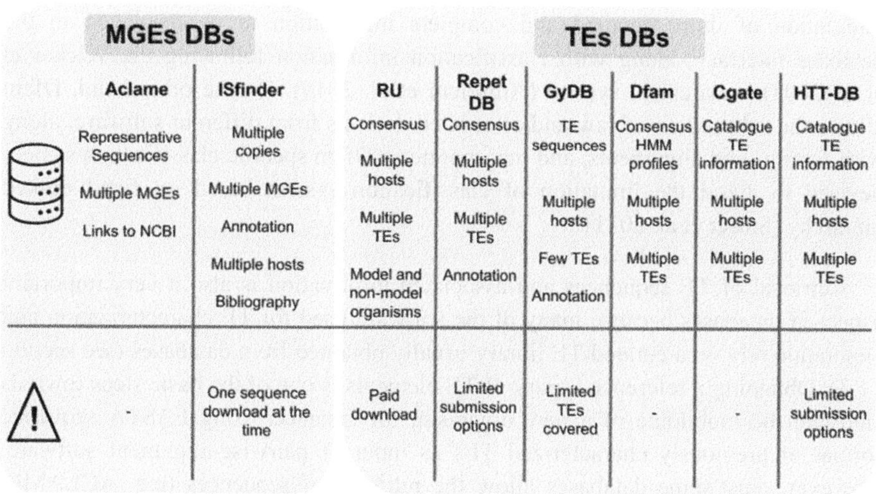

Figure 11.1. *Databases for mobile genetic elements of prokaryotic origin (MGEs DBs) and transposable elements from eukaryotes (TEs DBs) highlighting the main data stored and limitations*

One of the key features of any TE database is a sequence submission/storing system that allows the community to contribute submitting new characterized elements, assigning sequences unique IDs, storing sequences, appending and storing classification information and data download. Most of the databases depicted in Figure 11.1 allow the submission of new sequences either to compare against the sequences already stored or to store them as a new sequence entry. The main difference in the sequence submission processes is the requested submission format that varies widely from any TE nucleotide sequence, generated by different sequencing methodologies and representing a TE fragment derived from a single TE copy, a complete single TE copy or TE family/subfamily consensus, or the entire output from a TE characterization software. Most databases require sequence submission to be accompanied with a brief description of the methodology used to obtain them and associated host taxa information. For instance, Repbase allows the submission of fragmentary single copy TE sequences as well as consensus and provides an identification code associated with a TE classification (Kapitonov and Jurka 2008). RepetDB allows submission of highly structured data from the REPET package, which are more complex to obtain but that allow data validation and curation and the integration of more detailed and complete information to be displayed in the database interface, along with classification information following the Wicker et al.'s (2007) hierarchical system (Amselem et al. 2019). On the other hand, Dfam allows the submission of raw and/or curated datasets from different software, along with family seed alignments, and incorporates a Dfam specific classification system devised to avoid the limitation of classification system based on fixed ranked hierarchy (Storer et al. 2021).

Retrieval of TE sequences and associated information is also a very important aspect of databases because many of the software used for TE characterization and annotation rely on a curated TE library usually obtained from databases (see section 11.3). Obtaining a reference library of TE elements is one of the basic steps towards studying the mobilome of a new organism, for instance, using FASTA sequence format of previously characterized TEs as input of pairwise-alignment software. However, only some databases allow the retrieval of sequences (e.g. ACLAME, ISfinder, Dfam and RepetDB), while others only store links that lead to the corresponding sequences stored in general databases (e.g. GyDB, Cgate and HTT-DB). Although database linking is a feature that improves the connection between available complementary information, it is not practicable when accessing and downloading a large amount of data.

Lastly, several of these databases store both primary (originally deposited sequences) and secondary (TE consensus sequences from available genome and associated information) data, along with links to the original publication that describe in more detail how the TE was characterized and the biological phenomenon in which they are involved. All secondary information is crucial for the deep understanding of TE/MGE natural history.

From an ecological point of view, there are few databases that store and integrate information. The ACLAME database of prokaryotic MGEs is unique in this respect, since it stores information about past interaction between host species that allowed the transfer of MGE modules or families between genomes showing a common phenomenon of reticulate evolution of these elements. In addition, HTT-DB stores information about horizontal transfer of TEs between eukaryotic species and the ecological relationships that allow these events to occur more frequently.

MGE vs. TE database	Database name	General vs. specific – MGEs	General vs. specific – host genomes	Access link	Reference
Database for prokaryotic TEs	ACLAME	General for all MGEs (viruses, plasmids and transposons)	General for all prokaryotes	http://aclame.ulb.ac.be/	(Leplae et al. 2004, 2009)
	ISfinder	General database for Insertion sequences	General database for bacterial ISs	https://www-is.biotoul.fr/index.php	(Siguier et al. 2006)
Database for eukaryotic TEs					
	Repeat Update (RU)	General database for all TEs	General database for eukaryotes	https://www.girinst.org/repbase/	(Bao et al. 2015)
	RepetDB	General database for all TEs	General database for eukaryotes	http://urgi.versailles.inra.fr/repetdb/begin.do	(Amselem et al. 2019)
	GyDB	Specific TE database focused on some TE groups	General database for eukaryotes and prokaryotes	https://gydb.org/index.php?title=Main_Page	(Llorens et al. 2011)
	Dfam	General database for all TEs	General database for eukaryotes and bacterial	https://dfam.org/	(Wheeler et al. 2013; Hubley et al. 2015; Storer et al. 2021)
TE Databases focused on ecological phenomena					
	Cgate	General TE database for all TEs	Specific information on model organisms	https://ritarebollo.wixsite.com/cgate	(Rebollo et al. 2012)
	HTT-DB	General TE database for all eukaryotic TEs	General database for eukaryotes	http://pluriserver.servehttp.com/httdatabase/ http://lpa.saogabriel.unipampa.edu.br:8080/httdatabase	(Dotto et al. 2015, 2018)

Table 11.1. *List of representative TE databases with broad (non-restricted to taxa and phenomenon) and limited scope (limited to taxa and/or phenomenon)*

11.3. Search strategies for transposable element characterization

TE characterization, the process of detection and annotation of TEs in a given genome, follows several steps similar to host genes characterization, such as detection of coding and non-coding regions and conserved protein domains. However, the inherent higher complexity of TEs makes their characterization much more challenging. TEs have a number of specific features (i.e. repetitiveness, variable number of highly similar and/or degraded copies, nested TE copies with different evolutionary histories, repetitive nucleotide stretches into the TE structure and/or generated as a result of TE activity) that are taken into account by software particularly tailored for TE characterization. The detection of TEs encompasses the identification of a given sequence/locus that corresponds to a potential TE sequence, while TE annotation/classification is the process of information transfer to a new TE from homologous previously known TEs.

There are two major TE detection approaches based on different premises: (i) ab initio approach that is based on the average higher repetitiveness of TEs compared to other regions of the genome (especially in eukaryotes). Two main strategies to recover TEs based on their repetitiveness were developed, depending on the input sequences: raw sequencing reads or genome assembly. Repetitive sequences are overrepresented in raw sequencing reads when compared to single or low copy gene families, which encompasses most of the eukaryotic organism core genes; hence, several algorithms and pipelines (dnaPipeTE, RepeatExplorer and TEdna – Figure 11.2 and Table 11.2) explore the k-mer frequency or clustering pattern of sequenced reads to tease apart high and low repetitive elements from whole-genome sequencing reads. On the other hand, several software (RepeatScout, REPET package – Figure 11.2 and Table 11.2) receive genome assemblies as input data and break the genome by user-defined smaller k-mers that are further self-aligned to detect repetitive sequences (that align with several other k-mers from distinct parts of the genome and that reach a minimum number threshold of aligned sequences). (ii) Library-based approach that relies on a previously curated TE sequence library that is employed for similarity search of new related TEs.

TE annotation/classification approaches rely on information that can be obtained from the TE detection step, such as similarity values to previously characterized TEs and presence/absence of structural features such as terminal repeats. Once the similarity between a potential new TE meets a previously established threshold, it can be associated with complementary structural information to transfer or add annotation to newly detected TEs. Annotation and classification of TEs can be performed on potential TE sequences that are detected by an ab initio approach but that have no taxonomic information associated. Annotation/classification information can be readily transferred from the homologous TEs detected by a sequence similarity-based approach or through specific software that employ the detection of structural features, along with TE similarity search (REPCLASS and PASTEC – Figure 11.2).

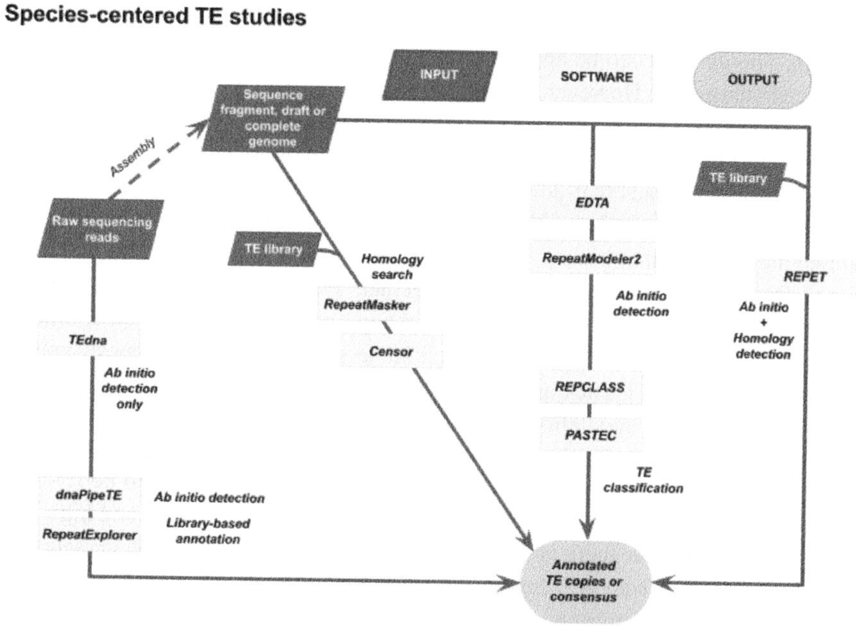

Figure 11.2. *Main strategies for TE/MGE characterization for species-centered studies depicting input data, software and outputs of software for TE detection, annotation and classification. Parallelograms indicate inputs; rectangles represent a process/software; ovals depict an end point, and arrows show the information flow. Annotated TE copies or consensus are related to both genome annotation (start and end of each TE loci across the genome) and/or taxonomic annotation of TEs, which depends on the chosen software output. For a color version of this figure, see www.iste.co.uk/huavan/transposable.zip*

Both ab initio and library-based approaches have known strengths and weaknesses; ab initio strategy can, in theory, characterize a broad and more diverse set of TEs, since it is based only on TEs repetitiveness and associated TE structural features, that is, every genomic element that shows a higher repetitiveness (multiple copies) than the average repetitive profile of the genome (single or low copy genes) should be detected. This strategy can only be reliably applied on well-assembled genomic sequences (restricted to model organisms) because a highly fragmented genome would disturb the repetitive (TEs) and non-repetitive (host genes) balance leading to a high number of false positives. However, there are several developed software that receive as input raw sequencing reads, avoiding the genome assembly problem. Therefore, it can be applied to virtually any organism with available whole genome shotgun data. However, ab initio approaches based on raw reads or genome assemblies suffer from a common limitation: the detection of old and low copy number TE families. Ancient TE families, even in high copy numbers, would generate divergent reads that do not cluster and/or do not bear the same k-mers and are expected to be missing from these software's outputs. Moreover, low copy number TE families that have a comparable copy number with host genes would be missed from ab initio-based software altogether. Finally, other repetitive genomic elements such as satellite DNAs and multiple copy gene families are also commonly found in these software's output, although some software such as RepeatExplorer, dnaPipeTE and REPET package implemented different modules to tag or remove those sequences at the annotation and classification step.

Software that implement a homology-based approach are known to be robust to detect ancient, degraded and low copy number, as well as more recent and high copy number TEs, as long as there is enough similarity between a reference TE sequence and a potential TE sequence; however, it has a number of caveats. Library-based approaches rely on arbitrary similarity thresholds and are highly dependent on how comprehensive the previously known TE database is. For instance, detection and annotation of TEs can be compromised on genomes from highly divergent taxa, since these elements may have diverged substantially from previously characterized elements. It is particularly true for non-model organisms that bear distantly related TEs compared to model organisms curated libraries. Moreover, new TE families/superfamilies may not be captured by a library-based approach alone. One complementary strategy to detect and annotate highly divergent TE families is based on the search of TE families/superfamilies hidden Markov models profiles as the ones available in the Dfam database (Wheeler et al. 2013).

Currently, there is a consensus in the scientific community that a combination of ab initio- and homology-based approach is the best strategy for a comprehensive characterization of TEs in any given genome (Hoen et al. 2015; Arkhipova 2017).

Several groups developed pipelines integrating different pieces of software to perform TE characterization under a more comprehensive framework (Table 11.2) such as REPET package and RepeatModeler 2.

11.4. Nature of input sequences

Nucleotide and amino acid sequences can be used for the different steps of TE characterization, but new developments have been made to cope with increasing nucleotide data available from high-throughput sequencing machines. Most software developed for TE detection receive as input nucleotide sequences from draft or complete genome sequences (Figure 11.2 and Table 11.2); however, sequencing projects that aim to obtain a high-quality complete genome assembly are rare and require a substantial amount of resources and years of gap closing and annotation refinement. Therefore, different software has been implemented to characterize TEs directly from sequencing reads, which allows the investigation of the mobilome of several non-model species (Figure 11.2 and Table 11.2).

11.4.1. *Raw reads*

High-quality sequencing reads can be directly provided as input for different software: TEdna uses the paired-end information to assemble and solve highly repetitive elements using De Bruijn graphs, RepeatExplorer characterizes repetitive sequences based on the clustering of high similar and abundant reads, explores the paired-end information to extend and sort individual TE families and annotates the resulting clusters based on similarity search and TE conserved domains. The dnaPipeTE assembles highly repetitive reads with Trinity transcriptome assembler (Grabherr et al. 2011), classifies assembled TEs based on a reference database (e.g. Repbase or user provided library) and estimates TE family abundance and age. Even though these algorithms were developed based on the same principle (high TE representation in the sequenced reads), there is no benchmark established so far, and recent results have shown that each software is able to assemble and annotate different sets of TEs (Amorim et al. 2020).

11.4.2. *Draft or complete genome assembly*

Both draft and complete genome assemblies can be used as input on TE characterization software, but a reliable TE characterization can only be performed on more complete genomes. It is mainly because draft genome assemblies are highly fragmented, which can inflate the true TE copy number (several TE copies may be fragmented, which makes the identification of structural features such as LTRs and TIRs more difficult) and directly affect any ab initio software that estimates the difference between the background genome repetitiveness and repetitive elements.

11.5. Population genomics of transposable elements

The population dynamics of TEs has long been a field of heated discussion mainly based on theoretical models (see Chapter 8). Active TEs are usually polymorphic, that is, the insertion sites will vary between individuals and populations. The insertion frequency at a particular site can also vary between populations as a result of selection, genetic drift or other demographic parameters. It is also expected that a new insertion will segregate at low frequency in the population (few individuals carry the insertion), while old insertions are sometimes able to reach a higher insertion frequency (up to fixation). However, characterizing polymorphism and diversity in populations cannot only rely on sequencing and assembling new genomes; it is quite straightforward to estimate from short-read sequencing data. Nowadays, with the cementing of reference genome projects with high-quality annotation of genes and TEs and the deluge of short-read sequencing data at the population level for many model and non-model organisms, several pieces of software were developed to detect and genotype TE insertions in natural and experimental settings (Figure 11.3 and Table 11.2). These software take as input a reference genome, the annotation coordinates of each TE copy and quality filtered sequencing reads. Those reads can come from one or several individuals of one population and from several populations. Two main software solutions exist, Tlex3 and PoPoolation TE2, that were developed and tested on Drosophila data. Although T-lex 3 can detect TSD (target site duplication), PoPoolation TE2 can identify new insertions sites not present in the reference genome. Another difference is that T-lex 3 detection is based on split-read detection (reads that map, both on the genome (masked for the TE) and on the TE). PoPoolation TE2 is based on paired-end sequencing and uses a discordant-read approach (one read of a pair map on the genome, and the other read on the TE). After reads mapping, several filters are applied to evaluate the presence/absence of each of TE copy or footprint of insertion/excision of these elements to estimate their frequency at the population level, as well as between populations (Figure 11.3).

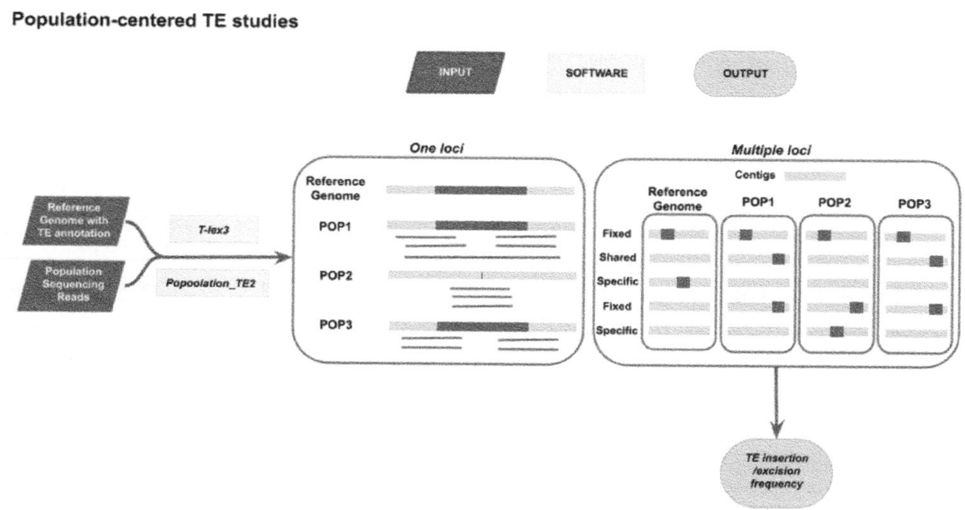

Figure 11.3. *Software for population-centered studies to detect and estimate the presence/absence and insertion frequency based on the reads covering the TE boundary and host flanking sequences. Darker rectangles are TE copies, light gray rectangles are host genome sequences flanking TE copies and smaller gray rectangles are sequenced reads. For a color version of this figure, see www.iste.co.uk/huavan/transposable.zip*

	Search approach	Software	Key features	Access link	References
Genome centered	Library, Homology	RepeatMasker	Mask repeats based on reference library, allows annotation and extraction of each TE copy using with "One code to find them all" script*	https://www.repeatmasker.org/	(Smit et al. 2013; *Bailly-Bechet et al. 2014)
		Censor	Similarity search of DNA/RNA sequences computationally translated or not.	https://www.girinst.org/censor/index.php	(Kohany et al. 2006)
	Ab initio or de novo structural features	EDTA	TE detection pipeline that runs several ab initio software that detect TEs based on structural features	https://github.com/oushujun/EDTA	(Ou et al. 2019)
		dnaPipeTE	General repeats finder Based on assembling TEs from high frequency K-mers	https://github.com/clemgoub/dnaPipeTE	(Goubert et al. 2015)
		TEdna	Based on assembling TEs from high frequency K-mers – do not provide TE quantification	http://urgi.versailles.inra.fr/Tools/Tedna https://github.com/mzytnicki/tedna	(Zytnicki et al. 2014)
		RepeatExplorer	General repeats finder based on high similar and interconnected group or reads forming a graph	https://galaxy-elixir.cerit-sc.cz/	(Novák et al. 2013)
	Ab initio and homology/ signature-based	REPET package	TE detection pipeline and multiple information TE annotation. Produces TE family consensus suitable for submitting to RepetDB	https://urgi.versailles.inra.fr/Tools/REPET	(Flutre et al. 2011; Quesneville et al. 2005)
		RepeatModeler2	Automates running of different ab initio algorithms producing TE libraries suitable for use in RepatMasker	https://github.com/Dfam-consortium/RepeatModeler	(Flynn et al. 2020; Smit and Hubley 2008)
Population centered		T-lex 3	Genotype a TE insertion in population genomic data. Runs with pooled data and detect TSDs	https://github.com/GonzalezLab/T-lex3	(Bogaerts-Márquez et al. 2020, p. 3)
		Popoolation TE2	Genotype a TE insertion in population genomic data. Runs with pooled data.	https://sourceforge.net/p/popoolation-te2/wiki/Home/	(Kofler et al. 2016)
Classification of TEs		REPCLASS	Classify potential TE sequences based on homology, TE structural features and sequence patterns from remains of TE activity	https://github.com/feschottelab/REPCLASS	(Feschotte et al. 2009)
		PASTEC	REPET package module of multiple information TE annotation and classification including similarity, HMM profile search and structural TE features detection	https://urgi.versailles.inra.fr/Tools/PASTEClassifier	(Hoede et al. 2014)

Table 11.2. *List of representative software used for transposable elements search and characterization*

11.6. How to evaluate the most suitable database and TE/MGEs search strategy for your study?

Finding the best database and tool to detect, annotate and classify TEs is not an easy task considering the "dense forest" of possibilities much more thoroughly mapped by Lerat (2009), and briefly depicted in Tables 11.1 and 11.2. The first and foremost question that should be clearly answered before running software to characterize TEs is how comprehensive our characterization should be to test our hypothesis. A sub-question would be as follows: Is it necessary to perform a comprehensive characterization of the mobilome or would a more focused strategy characterizing specific TEs families be enough to test it? A study that intends to have a comprehensive estimate of a species' TEs should first evaluate which resources stored in databases are available from the target or closely related species. For instance, model organisms such as species from the *Drosophila* and *Caenorhabditis* genus are currently well represented in TE databases and only a library-based characterization of the mobilome using curated TE sequences would, most of the time, be enough to perform a genome-wide characterization of the mobilome of a closely related species. On the other hand, as more divergent host species genomes are analyzed, it is more likely that some TEs will not reach sequence similarity threshold to reference elements and will be missed, as was the case for several *Drosophila buzzatti* TEs (Rius et al. 2016). Therefore, genomic datasets of organisms that are sub-represented in TE databases should be thoroughly accessed with a combination of library and ab initio-based software to ensure a comprehensive mobilome characterization. It is important to note that using this combined approach, the TE characterization could be enhanced, even in model organisms.

Considering TE population studies, a previous mobilome characterization is mandatory and most studies published so far focus on model organisms that have a well-assembled genome available. Here, again a library-based TE detection and annotation should be enough to rapidly characterize the mobilome in a reference strain, allowing further populational studies, while species and populations where genomic studies are scarce would demand a more thorough TE characterization using a combination of strategies to obtain a well-curated TE reference dataset.

As a rule of thumb, running several software that search TE based on different databases and TE features and combining their output to generate a curated reference mobilome is primordial to evaluate how comprehensive your TE characterization may be and to reveal how extensively these repetitive and dynamic genomic components are shaping the host genome. More focused analysis employing only one or several databases and software is advisable only for well-known model organisms with an available precise TE characterization. Anyhow, there is no perfect solution for TE characterization, once it is common to detect false positive TE sequences. The complexity of such elements underscores the continued need for further software and database developments and curation to improve TE characterization.

11.7. Acknowledgments

I would like to thank Dr. Pierre Capy and Dr. Aurélie Hua-Van for the invitation to collaborate with this book chapter and my colleagues Alexandre Freitas da Silva, Filipe Zimmer Dezordi and Elgion Lucio da Silva Loreto for several suggestions to improve the original version of the text. Moreover, it is important to stress that by no means do the list of databases and tools included in this chapter represent the full spectrum available. The choice was mainly due to my personal experience with these tools and to provide a broad view of the tools available for undergraduate students. Anyhow, I apologize for not quoting other relevant tools because of the page limits to explore such a vast research area. An interesting resource that continuously catalogs TE data and software and include a more comprehensive list of it is the TE Hub (https://tehub.org/). I suggest that any newcomer to the TE research field navigates and get new updates from this important initiative.

11.8. References

Amorim, I.C., Melo, E.S., Moura, R.C., Wallau, G.L. (2020). Diverse mobilome of Dichotomius (Luederwaldtinia) schiffleri (Coleoptera: Scarabaeidae) reveals long-range horizontal transfer events of DNA transposons. *Molecular Genetics and Genomics*, 295(6), 1339–1353.

Amselem, J., Cornut, G., Choisne, N., Alaux, M., Alfama-Depauw, F., Jamilloux, V., Maumus, F., Letellier, T., Luyten, I., Pommier, C. et al. (2019). RepetDB: A unified resource for transposable element references. *Mobile DNA*, 10(1), 4–11.

Arkhipova, I.R. (2017). Using bioinformatic and phylogenetic approaches to classify transposable elements and understand their complex evolutionary histories. *Mobile DNA*, 8(1), 19.

Bailly-Bechet, M., Haudry, A., Lerat, E. (2014). "One code to find them all": A perl tool to conveniently parse RepeatMasker output files. *Mobile DNA*, 5(1), 13.

Bao, W., Kojima, K.K., Kohany, O. (2015). Repbase update, a database of repetitive elements in eukaryotic genomes. *Mobile DNA*, 6(1), 11.

Bogaerts-Márquez, M., Barrón, M.G., Fiston-Lavier, A.-S., Vendrell-Mir, P., Castanera, R., Casacuberta, J.M., González, J. (2020). *T-lex3*: An accurate tool to genotype and estimate population frequencies of transposable elements using the latest short-read whole genome sequencing data. *Bioinformatics*, 36(4), 1191–1197.

Clark, K., Karsch-Mizrachi, I., Lipman, D.J., Ostell, J., Sayers, E.W. (2016). GenBank. *Nucleic Acids Research*, 44(Database issue), D67–D72.

Dotto, B.R., Carvalho, E.L., Freitas, A., Duarte, L.F., Pinto, P.M., Ortiz, M.F., Wallau, G.L. (2015). HTT-DB: Horizontally transferred transposable elements database. *Bioinformatics*, 31(17), 2915-2917.

Dotto, B.R., Carvalho, E.L., da Silva, A.F., Dezordi, F.Z., Pinto, P.M., Campos, T.L., Rezende, A.M., Wallau, G. da L. (2018). HTT-DB: New features and updates. *Database: The Journal of Biological Databases and Curation*, 102.

Feschotte, C., Keswani, U., Ranganathan, N., Guibotsy, M.L., Levine, D. (2009). Exploring repetitive DNA landscapes using REPCLASS, a tool that automates the classification of transposable elements in eukaryotic genomes. *Genome Biology and Evolution*, 205–220.

Flutre, T., Duprat, E., Feuillet, C., Quesneville, H. (2011). Considering transposable element diversification in de novo annotation approaches. *PLoS ONE*, 6(1), e16526.

Flynn, J.M., Hubley, R., Goubert, C., Rosen, J., Clark, A.G., Feschotte, C., Smit, A.F. (2020). RepeatModeler2 for automated genomic discovery of transposable element families. *Proceedings of the National Academy of Sciences*, 117(17), 9451–9457.

Goubert, C., Modolo, L., Vieira, C., ValienteMoro, C., Mavingui, P., Boulesteix, M. (2015). De novo assembly and annotation of the Asian tiger mosquito (*Aedes albopictus*) repeatome with dnaPipeTE from raw genomic reads and comparative analysis with the yellow fever mosquito (*Aedes aegypti*). *Genome Biology and Evolution*, 7(4), 1192–1205.

Grabherr, M.G., Haas, B.J., Yassour, M., Levin, J.Z., Thompson, D.A., Amit, I., Adiconis, X., Fan, L., Raychowdhury, R., Zeng, Q. et al. (2011). Trinity: Reconstructing a full-length transcriptome without a genome from RNA-Seq data. *Nature Biotechnology*, 29(7), 644–652.

Hoede, C., Arnoux, S., Moisset, M., Chaumier, T., Inizan, O., Jamilloux, V., Quesneville, H. (2014). PASTEC: An automatic transposable element classification tool. *PLoS One*, 9(5), e91929.

Hoen, D.R., Hickey, G., Bourque, G., Casacuberta, J., Cordaux, R., Feschotte, C., Fiston-Lavier, A.-S., Hua-Van, A., Hubley, R., Kapusta, A. et al. (2015). A call for benchmarking transposable element annotation methods. *Mobile DNA*, 6(1), 13.

Hubley, R., Finn, R.D., Clements, J., Eddy, S.R., Jones, T.A., Bao, W., Smit, A.F. A., Wheeler, T.J. (2015). The Dfam database of repetitive DNA families. *Nucleic Acids Research*, 44(D1), D81–89.

Jurka, J., Walichiewicz, J., Milosavljevic, A. (1992). Prototypic sequences for human repetitive DNA. *Journal of Molecular Evolution*, 35(4), 286–291.

Kanz, C., Aldebert, P., Althorpe, N., Baker, W., Baldwin, A., Bates, K., Browne, P., van den Broek, A., Castro, M., Cochrane, G. et al. (2005). The EMBL nucleotide sequence database. *Nucleic Acids Research*, 33(Database issue), D29–D33.

Kapitonov, V. V. and Jurka, J. (2008). A universal classification of eukaryotic transposable elements implemented in Repbase. *Nature Reviews Genetics*, 9(5), 411–412.

Kofler, R., Gómez-Sánchez, D., Schlötterer, C. (2016). PoPoolationTE2: Comparative population genomics of transposable elements using Pool-Seq. *Molecular Biology and Evolution*, 33(10), 2759–2764.

Kohany, O., Gentles, A.J., Hankus, L., Jurka, J. (2006). Annotation, submission and screening of repetitive elements in Repbase: RepbaseSubmitter and Censor. *BMC Bioinformatics*, 7(1), 474.

Leplae, R., Hebrant, A., Wodak, S.J., Toussaint, A. (2004). ACLAME: A CLAssification of Mobile genetic Elements. *Nucleic Acids Research*, 32(Database issue), D45–D49.

Leplae, R., Lima-Mendez, G., Toussaint, A. (2009). ACLAME: A CLAssification of Mobile genetic Elements, update 2010. *Nucleic Acids Research*, 38(Suppl. 1), 57–61.

Lerat, E. (2009). Identifying repeats and transposable elements in sequenced genomes: How to find your way through the dense forest of programs. *Heredity*, 104(6), 520–533.

Llorens, C., Futami, R., Covelli, L., Domínguez-Escribá, L., Viu, J.M., Tamarit, D., Aguilar-Rodríguez, J., Vicente-Ripolles, M., Fuster, G., Bernet, G.P. et al. (2011). The Gypsy Database (GyDB) of mobile genetic elements: Release 2.0. *Nucleic Acids Research*, 39(Database issue), D70–D74.

Novák, P., Neumann, P., Pech, J., Steinhaisl, J., Macas, J. (2013). RepeatExplorer: A Galaxy-based web server for genome-wide characterization of eukaryotic repetitive elements from next-generation sequence reads. *Bioinformatics*, 29(6), 792–793.

Ou, S., Su, W., Liao, Y., Chougule, K., Agda, J.R.A., Hellinga, A.J., Lugo, C.S.B., Elliott, T.A., Ware, D., Peterson, T. et al. (2019). Benchmarking transposable elements annotation methods for creation of a streamlined comprehensive pipeline. *Genome Biology*, 20, 275.

Quesneville, H., Bergman, C.M., Andrieu, O., Autard, D., Nouaud, D., Ashburner, M., Anxolabehere, D. (2005). Combined evidence annotation of transposable elements in genome sequences. *PLoS Computational Biology*, 1(2), 166–175.

Rebollo, R., Farivar, S., Mager, D.L. (2012). C-GATE – Catalogue of genes affected by transposable elements. *Mobile DNA*, 3(1), 9.

Rius, N., Guillén, Y., Delprat, A., Kapusta, A., Feschotte, C., Ruiz, A. (2016). Exploration of the *Drosophila buzzatii* transposable element content suggests underestimation of repeats in Drosophila genomes. *BMC Genomics*, 17(1), 344.

Siguier, P., Perochon, J., Lestrade, L., Mahillon, J., Chandler, M. (2006). ISfinder: The reference centre for bacterial insertion sequences. *Nucleic Acids Research*, 34(Database issue), 32–36.

Smit, A.F.A. and Hubley, R. (2008). RepeatModeler Open-1.0 [Online]. Available at: http://www.repeatmasker.org.

Smit, A.F.A., Hubley, R., Green, P. (2013). RepeatMasker Open-4.0 [Online]. Available at: http://www.repeatmasker.org.

Storer, J., Hubley, R., Rosen, J., Wheeler, T. J., Smit, A.F. (2021). The Dfam community resource of transposable element families, sequence models, and genome annotations. *Mobile DNA*, 12(1), 2.

Wheeler, T.J., Clements, J., Eddy, S.R., Hubley, R., Jones, T.A., Jurka, J., Smit, A.F.A., Finn, R.D. (2013). Dfam: A database of repetitive DNA based on profile hidden Markov models. *Nucleic Acids Research*, 41(Database issue), D70–D82.

Wicker, T., Sabot, F., Hua-van, A.A., Bennetzen, J.L., Capy, P., Chalhoub, B., Flavell, A., Leroy, P., Morgante, M., Panaud, O. et al. (2007). A unified classification system for eukaryotic transposable elements. *Nature Reviews. Genetics*, 8(12), 973–982.

Zytnicki, M., Akhunov, E., Quesneville, H. (2014). Tedna: A transposable element de novo assembler. *Bioinformatics*, 30(18), 2656–2658.

List of Authors

Christian BIÉMONT
Laboratoire Biométrie et Biologie Évolutive (LBBE)
CNRS
Université Claude Bernard Lyon 1
Villeurbanne
France

Pierre CAPY
Évolution, Génomes, Comportement et Écologie (EGCE)
CNRS
IRD
Université Paris-Saclay
Gif-sur-Yvette
France

Mick CHANDLER
Department of Biochemistry and Molecular & Cellular Biology
Georgetown University Medical Center
Washington, DC
USA

Benoît CHÉNAIS
BiOSSE (Biologie des Organismes : Stress, Santé, Environnement)
Université du Mans
France

Marta CORONADO-ZAMORA
Institute of Evolutionary Biology
CSIC UPF
Barcelona
Spain

Christopher ELLISON
Department of Genetics
Human Genetics Institute of New Jersey
Rutgers University
Piscataway
USA

Josefa GONZÁLEZ
Institute of Evolutionary Biology
CSIC UPF
Barcelona
Spain

Aurélie HUA-VAN
Évolution, Génomes, Comportement et Écologie (EGCE)
CNRS
IRD
Université Paris-Saclay
Gif-sur-Yvette
France

Zoltán IVICS
Division of Medical Biotechnology
Paul-Ehrlich-Institute
Langen
Germany

Arnaud LE ROUZIC
Évolution, Génomes, Comportement et Écologie (EGCE)
CNRS
IRD
Université Paris-Saclay
Gif-sur-Yvette
France

Emmanuelle LERAT
Laboratoire Biométrie et Biologie Évolutive (LBBE)
CNRS
Université Claude Bernard Lyon 1
Villeurbanne
France

Gabriel DA LUZ WALLAU
Department of Entomology and Bioinformatic Core
Instituto Aggeu Magalhães (IAM)
Fiocruz – Pernambuco
Recife
Brazil

Karen E. ROSS
Protein Information Resource
Department of Biochemistry and Molecular & Cellular Biology
Georgetown University Medical Center
Washington, DC
USA

Chengyi SONG
College of Animal Science & Technology
Yangzhou University
China

Alessandro M. VARANI
Department of Agricultural and Environmental Biotechnology
School of Agricultural and Veterinary Sciences
UNESP - São Paulo State University
Jaboticabal
Brazil

Chantal VAURY
Institute for Genetics, Reproduction and Development (iGReD)
CNRS
Université Clermont Auvergne
Clermont-Ferrand
France

Cristina VIEIRA
Laboratoire Biométrie et Biologie Évolutive (LBBE)
CNRS
Université Claude Bernard Lyon 1
Villeurbanne
France

Index

E

G, H

I, J, K

L, M, N

Printed and bound by CPI Group (UK) Ltd, Croydon, CR0 4YY
13/09/2026
14961070-0005